South Asia in a Globalising World

DEVELOPING AREAS RESEARCH GROUP
THE ROYAL GEOGRAPHICAL SOCIETY
(WITH THE INSTITUTE OF BRITISH GEOGRAPHERS)

DARG Regional Development Series No. 3
Series Editor: David Simon

South Asia in a Globalising World

A reconstructed regional geography

edited by Robert W. Bradnock and Glyn Williams

An imprint of **Pearson Education**

Harlow, England · London · New York · Reading, Massachusetts · San Francisco · Toronto · Don Mills, Ontario · Sydney
Tokyo · Singapore · Hong Kong · Seoul · Taipei · Cape Town · Madrid · Mexico City · Amsterdam · Munich · Paris · Milan

Pearson Education Limited
Edinburgh Gate
Harlow
Essex CM20 2JE
and Associated Companies throughout the world

Visit us on the World Wide Web at:
www.pearsoneduc.com

ISBN 0130-25947-0

British Library Cataloguing-in-Publication Data
A catalogue record for this book is available from the British Library

10   9   8   7   6   5   4   3   2   1
11   10   09   08   07   06   05   04   03   02

Typeset by 68 in 10/11 pt Palatino
Printed in Malaysia, KVP

Contents

Contents

List of Figures and Tables

Figures

Tables

List of Figures and Tables

Contributors

Robert W. Bradnock is Senior Lecturer in Geography with Reference to South Asia at King's College London and has published extensively on the political, historical and environmental geography of South Asia. His research is currently focused on issues of geopolitics and environmental change in South Asia, especially in the Bengal delta. These issues are set in the wider context of inter-state relationships and resource development in South Asia as a whole. He is series editor of the Kings' SOAS series *Studies in Development Geography* (London, Ashgate), and his recent publications include *Subsidence, Submergence and Sea-level Rise: The political ecology of environmental change in the Bengal Delta* (with Patricia Saunders), Chapter 4 in Philip A. Stott and Sian Sullivan, *Political Ecology: Science, myth and power* (London, Edward Arnold, 1999) 'Regional geopolitics in a globalising world: Kashmir in geopolitical perspective', *Geopolitics*, 3(2); 'South Asia: how can regional co-operation be achieved?', in Nicholas Hopkinson (ed.), *Current Issues in Diplomacy and Foreign Policy* Vol. III (2002), pp. 7–35 (London, The Stationery Office), and *Urban Growth and Development in Asia*, 2 vols, ed. with G.P. Chapman and A.K. Dutt (Aldershot, Ashgate, 2000).

Graham Chapman is Professor of Geography, Department of Geography, Lancaster University, UK and is researching into water and development in South Asia, the geopolitics of South Asia, and environmentalism in developing countries. His recent publications include: *Water: and the quest for sustainable development in the Ganges Valley* (Mansell Publishing, London (edited with M. Thompson, 1995); *Timescales and Environmental Change* (co-ed with T. Driver, Routledge, London, 1996); *Environmentalism and the Mass Media and the North-South Divide* (with Keval Kumar, Caroline Fraser and Ivor Gaber, Routledge, London, 1997); *Urban Growth and Development in Asia*, 2 vols (ed. with A.K. Dutt and R.W. Bradnock, London, Ashgate, 2000); '"Other" cultures, "other" environments, and the mass media', in J. Smith (ed.), *The Daily Globe: Environmental change, the public and the media* (London, Earthscan, 2000); *The Geopolitics of South Asia: From early empires to India, Pakistan and Bangladesh* (Aldershot, Ashgate, 2000).

Stuart Corbridge is Professor of Geography at the London School of Economics and Professor of International Studies at the University of

Miami. He is the author or co-author of four books, including *Reinventing India: Liberalization, Hindu nationalism and popular democracy* (with John Harriss, Polity Press, 2000). His main area of interest is eastern India, where he has worked recently with Sanjay Kumar on a study of the DFID-funded Eastern India Rainfed Farming Project, and with Manoj Srivastava, René Véron and Glyn Williams on studies of state performance, empowerment and pro-poor governance in Bihar, Jharkhand and West Bengal.

Vandana Desai is Lecturer in Development Geography at Royal Holloway, University of London. Her research interests are NGOs, empowerment, gender, community participation, slum housing, displacement and development. Her recent publications include *Community Participation and Slum Housing* (Sage, 1995) and the *Arnold Companion to Development Studies* (forthcoming), coedited with Rob Potter.

Yolanda Foster is a Research Fellow at Canterbury Christ Church University, and is currently completing her PhD, 'Desires, imaginaries and discourses of the nation in Sri Lanka' at the School of Oriental and African Studies, University of London. She was previously a researcher at the Social Scientists' Association (SSA) Colombo, Sri Lanka. Her forthcoming publications include book chapters on 'Hybridity in the work of the No Order Artists' (SSA, Colombo) and 'Politics of development in a complex political emergency: the Jaffna Rehabilitation Plan' (India, Macmillan).

John Harriss is Professor of Development Studies at the London School of Economics, and Director of the LSE's Development Studies Institute. A long-time student of the politics and political economy of India, he has done extensive research on agrarian change and rural politics, on the working class, and on big business mostly in the southern State of Tamil Nadu. His more recent books include *Depoliticizing Development: the World Bank and social capital* (LeftWord, 2001), and *Reinventing India: Liberalization, Hindu nationalism and popular democracy* (with Stuart Corbridge, Polity Press, 2000).

Arif Hasan is an architect-planner working in Karachi with two community research, development, extension and advocacy projects, the Orangi Pilot Project Research and Training Institute and the Urban Resource Centre. These projects have won considerable international acclaim and are been replicated in many Pakistani and other Asian cities. He also teaches at various Karachi architecture and planning institutes and is a member of the governing boards of local government planning and development agencies. He is the author of a number of books and is a recipient of prestigious national and international awards.

Shahnaz Huq-Hussain is a Professor of Geography at the University of Dhaka, Bangladesh. She obtained her PhD from SOAS and MSc from LSE, University of London. Her published work includes over 25 research papers and two books on gender. Her current interest is on gender, population and environmental issues. Prof. Huq-Hussain has been selected for a Fulbright research award to study 'The Analysis of the Feminist Environmental

Conservation/Protection Movement in USA and its Relevance to Bangladesh' for the year 2001–2002 at the University of California, Davis, USA.

Craig Jeffrey is a Lecturer in the Department of Geography, University of Edinburgh. His research interests include the sociology of Indian development and the politics of education in South Asia. His doctoral research focused on the social and political strategies of a rural elite in western Uttar Pradesh. In collaboration with Professor Roger Jeffery and Professor Patricia Jeffery at the Department of Sociology (University of Edinburgh), he is currently undertaking research in UP on how the rise of secondary schooling is changing patterns and processes of social exclusion and social inequality.

Sarah Jewitt, Department of Geography, University of Nottingham, teaches on environment and development-related issues. Previously she lectured at SOAS, having undertaken doctoral research at the University of Cambridge on forest use and forest management in the Jharkhand region of India. Her primary research interests include community (particularly gender)-based environmental management (with particular reference to forests), the use and management of non-timber forest produce and the displacement (or otherwise) of 'traditional' environmental knowledge and management systems. Recent publications include: 'Autonomous and joint forest management in India's Jharkhand: lessons for the future?' in R. Jeffrey (ed.), *The Social Construction of Indian Forests* (New Delhi; Manohar, 1998); 'Mothering earth? Gender and environmental protection in the Jharkhand, India', *The Journal of Peasant Studies*, **27**(2) (2002), pp. 94–131; 'Unequal knowledges: de-romanticizing eco-feminist and neo-populist interpretations of local agro-ecological knowledge systems', *Development and Change*, **31**, 961–85: 'Political ecology of forest management: gender and indigenous silvicultural knowledges in the Jharkhand, India (with S. Kumar, 2000), in P. Stott and S. Sullivan, *Political Ecology: Science power and myth* and *Environment, Knowledge and Gender: Local development in India's Jharkhand* (Aldershot, Ashgate) (SOAS Development Geography Series).

Emma Mawdsley is Lecturer in Geography at Durham University. She completed her PhD on the separate State movement of Uttaranchal, spending some 16 months in the Himalayas exploring a wide variety of issues connected to it, including questions of environment, gender, politics and social movements. Her research interest in this region continues, but more recently she has worked on a DFID-funded project exploring knowledge, power and development agendas in India, Ghana, Mexico and Europe. A Carnegie Fellowship has allowed her to start researching 'middle class environmental values' in India.

Patricia L. Saunders is a freelance researcher in development, environment and peace issues in the governmental, intergovernmental NGO and academic sectors. Recent publications include 'Environmental refugees: the origins of a construct', and 'Sea level rise, subsidence and submergence:

the political ecology of environmental change in the Bengal delta' (with R.W. Bradnock) in P.A. Stott and S. Sullivan (eds) *Political Ecology: Science, myth and power* (London, Edward Arnold, 1999).

Glyn Williams is Lecturer in Human Geography at Keele University. His research centres on issues of rural poverty and the politics of development, and he has conducted extensive ethnographic fieldwork in eastern India. He has recently undertaken (with Manoj Srivastava, René Véron and Stuart Corbridge) a DFID-funded project, 'Enhancing Pro-Poor Governance in Bihar and West Bengal', which used the group's earlier academic research to promote government reform and social change in both States.

Lakshman Yapa, originally from Sri Lanka, is an associate professor of Geography at Pennsylvania State University. He is the Director of the Philadelphia Field Project, which is a Pennsylvania State University course in service learning designed to create a new university and community partnership in addressing urban poverty.

Series Preface

In the late 1980s, the Developing Areas Research Group (DARG) of the then Institute of British Geographers produced a series of three edited student texts under the general editorship of Prof. Denis Dwyer (University of Keele). These volumes, focusing on Latin America, Asia and Tropical Africa respectively, were published by Longman and achieved wide circulation, thereby also contributing to the financial security of DARG and enabling it to expand its range of activities. These are now out of print. The Latin American volume was revised and published in a second edition in 1996.

However, there have been dramatic changes in the global political economy, in the nature of development challenges facing individual developing countries and regions, and in debates on development theory over the last decade or so. In a review of the situation in 1997/8, the DARG Committee and Matthew Smith, the former Geography Editor at Addison Wesley Longman (now Pearson Education) therefore felt that mere updating of the existing texts would not do these circumstances justice or catch the imagination of a new generation of students. Accordingly, we have launched an entirely new series.

This is both different in conception and larger, enabling us to address smaller, more coherent continental or subcontinental regions in greater depth. The organising principles of the current series are that the volumes should be thematic and issue-based rather than having a traditional sectoral focus, and that each volume should integrate perspectives on development theory and practice. The objective is to ensure topicality and clear coherence of the series, while permitting sufficient flexibility for the editors and contributors to each volume to highlight regional specificities and their own interests. Another important innovation is that the series was launched in January 1999 by a book devoted entirely to provocative contemporary analyses of *Development as Theory and Practice; current perspectives on development and development co-operation.* Edited by David Simon and Anders Närman, this provides a unifying foundation for the regionally focused texts and is designed for use in conjunction with one or more of the regional volumes.

The complete series is expected to include titles on Central America and the Caribbean, Southern and East Africa, West Africa, the Middle East and North Africa, South Asia, Pacific Asia, the transitional economies of central

Asia, and Latin America. While the editors and many contributors are DARG members, other expertise – not least from within the respective regions – is being specifically included to provide more diverse perspectives and representativeness. Once again, DARG is benefiting substantially from the royalties. In addition, a generous number of copies of each volume will be supplied to impoverished higher education institutions in developing countries in exchange for their departmental publications, thereby contributing in a small way to overcoming one pernicious effect of the debt crisis, namely the dearth of new imported literature available to staff and students in those countries.

David Simon
Royal Holloway, University of London

Series Editor
(Chair of DARG 1996–8)

Publisher's Acknowledgements

We are grateful to the following for permission to reproduce copyright material:

Table 1.1 adapted from *World Development Indicators Database* (July 2000) and UNDP (2000) *Human Development Report*, Oxford University Press, New York by permission of publisher; Figure 3.2 from Houghton, J. (1997) *Global Warming*, 2nd edition, Cambridge University Press by permission of the publisher and the author; Figure 3.4 from Bradnock, R.W. (1989) *Agricultural Change in South Asia*, John Murray, London by permission of John Murray (Publishers) Ltd; Table 4.2 from 'Nepal's water resources: the potential for exploitation in the upper Ganges catchment' in Chapman, G.P. and Thompson, M. (eds.), Continuum International Publishing Group, reprinted by permission of the publisher and editors; Table 12.5 from Huq-Hussain, S. (1998) 'Health conditions of factory workers: a case study of a sweater factory in Dhaka' in Islam, N. (ed.), *Recent Urban Studies*, Bangladesh, pp. 163–76 and Huq-Hussain, S. (2000) 'From villages to Dhaka: female migration as a strategy for poverty alleviation', *Aisan Migrant*, **13**, (2), April-June, Philippines, pp. 37–43 by permission of Sclabrini Migration Center, Quezon City, Philippines.

We are grateful to Blackwell Publishing Ltd for permission to reproduce a chaper adapted from 'A fist is stronger than five fingers: Caste and dominance in rural north India' by Jeffrey Craig published in *Transactions of The Institute of British Geographers* V26 (2001)© Royal Geographical Society (with The Institute of British Geographers).

In some instances we have been unable to trace the owners of copyright material, and we would appreciate any information that would enable us to do so.

List of Abbreviations

ADB	Asian Development Bank
AERC	Applied Economic Research Centre (University of Karachi)
CBR	Central Board of Revenue
FDA	Faisalabad Development Authority
HBFC	House Building Finance Corporation
ISAL	illegal subdivision on agricultural land
ISD	illegal subdivision
KAIRP	Katchi Abadi Improvement and Regularisation Programme
KBCA	Karachi Building Control Authority
KDA	Karachi Development Authority
KESC	Karachi Electric Supply Corporation
KHASDA	Karachi Health and Social Development Association
KMC	Karachi Metropolitan Corporation
OCT	Orangi Charitable Trust
OPP-RTI	Orangi Pilot Project-Research and Training Institute
SKAA	Sindh Katchi Abadis Authority
WAPDA	Water and Power Development Authority

CHAPTER 1

South Asia in a globalising world
A reconstructed regional geography

Glyn Williams and Robert W. Bradnock

In the years that have elapsed since the previous DARG series on regional development, writing an introduction to a volume on South Asia has become a more problematic task. Since the late 1980s, changes in the global economy have been dramatic. The collapse of the Soviet Union has heralded a period of unbridled neo-liberalism, which has survived both the South-East Asian crash of 1997/8 and the collapse of the Seattle Round of the World Trade Organization (WTO) talks in December 1999 relatively unscathed. We are repeatedly told that we live in an era of 'globalisation', where technology is eroding the importance of distance and the distinctiveness of place, producing new opportunities and risks.[1] The resultant realities of 'development' over the last decade or so have been very mixed for people living in the global South: while some have seen standards of living improve rapidly, others have experienced growing economic and ecological instability and even dramatic collapses in life expectancy.[2] The academic changes within development geography and development studies through this period have been no less dramatic. Although 'development' has always been a contested term, the rise of 'anti-development' and (somewhat belatedly) postmodern thinking has raised new criticisms of the objects and practices of development that are important for practitioners and academics alike. Together, these academic and geopolitical shifts have destroyed many of the certainties that underpinned geographical imaginations of development in the mid-twentieth century, from a faith in modernisation to the idea of the 'Third World' itself.

South Asia has experienced these changes in a variety of ways. At the most superficial level, urban dwellers in South Asia are seeing the rapid 'globalisation' of their consumption patterns: even in India, where import substitution and tariff barriers once held multinationals at bay, Coca-Cola, GAP and McDonald's are in great demand.[3] At the same time, South Asian economies now produce a range of their own thoroughly 'modern' products, from IT services to nuclear warheads. That these changes exist alongside the reproduction of extreme poverty – South Asia has over 500 million people living on less than a dollar a day, 43.5 per cent of the global total (World Bank 2000: 23) – draws our attention to the fact that old models of development need rethinking. The differences are not simply the result of a modern, globalised sector of South Asia 'developing' away from its traditional, localised

1

Table 1.1 South Asia: population and development indices

Country	Land area (1,000 km²)	Population			Human development					Economy	
		Total (millions) 1999	Growth p.a. 1999 (%)	Urban (% of total) 1999	Life expectancy (yrs) 1999	HDI[a] 1998	GDI[b] 1998	Adult literacy[c] 1998	Female literacy[c] 1998	GNP pc (US 1995$) 1998	Real GDP pc (PPP$)[d] 1998
Bangladesh	144	127.7	1.6	24.0	60.7	0.461	0.441	40%	29%	348	1,361
Bhutan	47	0.8	2.9	6.9	61.5	0.483	(0.330)[e]	42%	(28%)[e]	493	1,536
India	3,300	997.5	1.8	28.1	63.2	0.563	0.545	56%	44%	444	2,077
Maldives	0.3	0.3	2.5	26.0	67.9	0.728	0.720	96%	96%	1,247	4,083
Nepal	147	23.4	2.3	11.6	58.2	0.474	0.449	39%	22%	217	1,157
Pakistan	796	134.8	2.4	36.5	62.5	0.522	0.489	44%	29%	511	1,715
Sri Lanka	66	19.0	1.1	23.3	73.5	0.733	0.727	91%	88%	802	2,979

Notes: [a] The Human Development Index (HDI) has three components: life expectancy at birth, income and educational attainment. For details of the calculation of HDI (and GDI) see UNDP (2000).
[b] The Gender-related Development Index (GDI) adjusts the HDI downwards according to the degree of gender inequality in each of its three components.
[c] Literacy rates are for the 15+ population.
[d] Real GDP per capita is priced in purchasing power parity dollars.
[e] Figures for 1998 not available – these are figures for 1995.
Sources: Land area, population figures and life expectancy taken from World Bank, *World Development Indicators Database, July 2000.* All other figures taken from UNDP (2000).

'undeveloped' counterpart. If two South Asias appear to exist today, both are the result of common economic and geopolitical changes emanating from within the region and beyond (see Table 1.1 for national development indicators).

South Asian intellectuals have also played a very important role in the critical reappraisal of 'mainstream development'. The list of achievements here is long and varied, ranging from Amartya Sen's work on poverty (Sen 1981, 1989, 1992), the Subaltern Studies series of alternative Indian histories that have had a global influence on post-colonial debates since the 1980s (Guha 1997), to Vandana Shiva's eco-feminist critiques of development (Shiva 1988, 1991). The emergence of so many leading academics from, and within, the region challenges many Northern preconceptions about patterns of intellectual change. In contrast to the crippling patterns of asymmetry described by Reginald Cline-Cole in his review of Anglo-American/African professional interaction (Cline-Cole 1999), Northern development geography arguably still has much to do to 'catch up' with some of the intellectual centres of South Asia.

Writing 'a regional development text' today therefore requires some critical self-reflection if we are not to fall into the trap of writing redundant regional geographies that merely reproduce outdated stereotypes of South Asia. In this introduction, we attempt to avoid this pitfall by addressing three questions: how should one approach the study of a 'developing region' in these changing times, what issues should a volume such as this focus on, and how do we, the editors, hope that our book will be used?

Studying 'development' and 'the region'

Our first question has already been addressed in some depth by the contributors to *Development as Theory and Practice* (Simon and Närman 1999), the volume that introduced this DARG series. We would particularly direct readers to the chapters by Simon, Parnwell and Cline-Cole, and will confine ourselves here to brief comments. Over the 1990s, post-colonial, anti-development and postmodern commentaries have raised a number of important criticisms of 'mainstream development', which, as noted above, have fundamentally questioned the discipline's practices and objects of study.[4] These debates have in turn been inspired by sources as varied as the resurgence of social movements in the South to the intellectual deconstruction of development, where the work of Edward Said and Michel Foucault has provided important foundations. One subject that has been given greater academic attention as a result of the latter is the history of ideas of development (see, *inter alia*, Cowen and Shenton 1996; Sachs 1992; Rist 1997): here, the linkages between development and colonialism are traced, and development's role as a system of power and knowledge is critically assessed. We consider that such deconstruction is a worthwhile exercise, and provides a history of the discipline that students should be aware of. Although we have not directly addressed the evolution of concepts of development in South Asia in this volume, the chapters by Lakshman Yapa and Sarah Jewitt provide useful pointers for readers wishing to trace this history in relation to ideas of poverty and environmental conservation.

The same theoretical imperatives have also led to an increase in research that reflects critically on current development practice and the ways in which knowledge about developing areas is constructed. At its best, this work has produced subtle and context-specific analyses,[5] but at its worst, it represents an undifferentiated 'development' as scientific and economic reductionism, hiding an agenda of Western imperialism that is always and everywhere harmful to 'indigenous communities'. David Simon's chapter, 'Development revisited', treads a careful course through these debates. 'Development', as defined by Simon,[6] emerges in the new millennium as a broader, contested, and perhaps more humble term as a result, and this move away from teleology and narrow economism is a trend that we welcome. At the same time, we share with Simon his concern that post-development critiques need to be more realistic in their assessments of alternative(s) to development, and as a result must not neglect either the pressing material needs of poor people in the South or the potential for more enlightened forms of development to meet these.[7]

Within this book, we do not see our role as setting an editorial line on these debates, for individual authors' positions, and degree of engagement with them, differ greatly. Instead we have encouraged authors to reflect on these issues in the light of their own research topics. A critical engagement with development-as-practice forms the central core of many chapters of this book. Here the contributions by Vandana Desai, Sarah Jewitt and Yolanda Foster usefully show that the new 'sacred cows' of post-development writing (non-governmental organisations (NGOs), indigenous knowledge and local resistance) need careful empirical investigation. Elsewhere, Craig Jeffrey and Lakshman Yapa look at terms such as 'caste' and 'poverty' that have been central to the imaginings of South Asia by Western scholars, and examine both their discursive and material impact on South Asians. By addressing some of these theoretical concerns through concrete development issues, the relevance of, and limitations to, alternative and anti-developmental perspectives in a South Asian context can be seen more clearly.

If we are to take the theoretical changes of post-colonialism and post-modernism seriously, we need to cast a critical eye over the subject of our text, the region of 'South Asia' itself. Here, we have chosen to accept the definition of South Asia as a collection of nation-states recognised as 'South Asian' within the region itself: India, Pakistan, Bangladesh, Sri Lanka, Nepal, Bhutan and the Maldives (see Figures 1.1 and 1.2). Given the changing geographies of a globalising world, is the definition of South Asia used in this volume appropriate today? In a recent article, John Agnew reminds us that 'metageographies', or schemes of global division, should be open to critical review: 'Regional schemes are never simply intellectual. Neither are they simply political. They play with facts about the world at the same time they must reflect the biases, intellectual and political, of their originators' (Agnew 1999: 95). He argues that since 1989, the collapse of the Second World has thrown our existing regional schemes into disorder, and he questions whether the changing character of the world economy is changing the shape of regions, and suggesting new metageographies of ordering. Specifically, he asks whether 'inserting national units into compact regions' is valid today, given that 'an increasingly patchworked geography of economic development is less and less captured by such territorial containers as nation-states or world regions' (*ibid.*: 95).

These are important questions for this volume to address, notwithstanding the fact that the boundaries of South Asia given here are also those adopted by the members of the South Asian Association for Regional Co-operation (SAARC) in their own self-definition of the region. Our first response to Agnew would be to reject any claims of the 'naturalness' of South Asia, which is self-evidently to a significant degree a political construct. Despite the barrier of the Himalayas that divides 'the Indian subcontinent' from other parts of Asia, our definition is as much a badge of convenience as it is a geographical absolute. On the basis of common cultural and administrative histories, we could have equally easily included

Figure 1.1 South Asia in global perspective

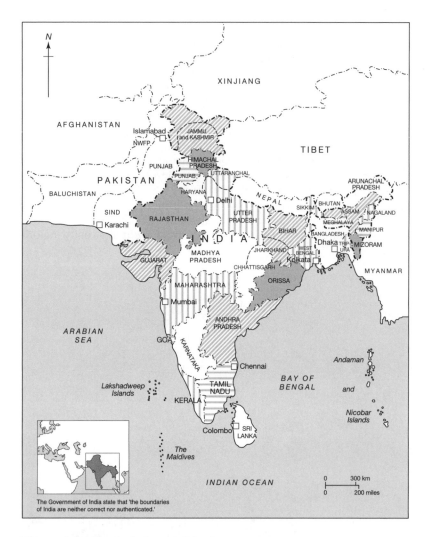

Figure 1.2   South Asia – political

neighbouring Afghanistan (or even Myanmar), or stretched the boundaries of the region to include diasporic South Asian communities from Nairobi to New York. Our definition is thus somewhat arbitrary, one of many possible ways of 'playing with facts about the world', but what does it tell us about 'the biases, intellectual and political, of [its] originators'?

We will freely admit to two 'biases' here, the first being the need to recognise the importance of history and culture in defining regions. 'South Asian civilisation' is neither a uniform whole, nor could it be made to fit unproblematically into the states mentioned, but there are sufficient elements of common history to make considering these places together a worthwhile exercise. Over five millennia, the region has seen a succession of overlapping kingdoms and empires, as well as the complex interplay of many different religions, of which four – Hinduism, Jainism, Buddhism and Sikhism – arose within South Asia, while Islam and Christianity also play vital roles in different parts of the region. The British Raj – which was itself an evolving patchwork of forms of rule, rather than a uniform whole – was an important, but by no means dominant, part of this history.[8] In this sense, an idea of 'South Asia' is as open-ended and internally divided as an idea of 'Europe'. Today, as for many centuries, various parts of South Asia are integrated into the world economy in different ways, but these differences are perhaps less important for the themes we are investigating in this collection than the region's shared histories and cultural traits.[9]

Our second 'bias' is that we consider that nation-states remain important territorial containers in contemporary South Asia, and hence we are willing to use their boundaries to define the region and key subregions. Despite the progress of neo-liberal agendas, national governments continue to play an important part in the region's economic development: indeed, as Stuart Corbridge and John Harriss's chapter in this volume illustrates, states are important actors in shaping the process of liberalisation itself. Our focus on nation-states does not imply that they or their boundaries are 'natural': nations are always imagined communities, to use Benedict Anderson's powerful descriptor (Anderson 1983), and South Asia provides numerous examples of where these imaginations are being reshaped. From Kashmir in the north to Jaffna in the south, the building and maintenance of state borders is a contested project throughout the region, as Emma Mawdsley's chapter clearly shows.

Beyond these definitional questions, John Agnew's questioning of regional schemes raises other challenges for an exercise of regional geography. Reflecting on 'metageographies', whether or not this is prompted by present preoccupations about globalisation, involves rethinking the use of spatial units throughout our research, and this has two important implications on our collection. The first is that giving a 'fair' representation of the region is an almost impossible task. As editors, we have attempted to ensure that all of the major countries of South Asia feature strongly within at least one chapter, and that all of the authors make reference to other areas and examples within South Asia wherever possible.[10] But, in many ways, the stories told in each chapter are unique: Arif Hasan's study of the informal housing market in Pakistan highlights particular actors and processes important within Karachi, and as such cannot represent similar issues in Colombo or Kolkata, although it can provoke interesting comparisons. The second is that all spatial units are social constructions, and this includes 'the local' as much as it does 'the region'. Within some versions of alternative development, 'the local' is accorded special status as the container of indigenous values and authentic cultures (Mohan and Stokke 2000), and we welcome the opportunity to challenge such 'wobbly romanticism' (Corbridge 1998) in this volume. In different ways, Craig Jeffrey and Yolanda Foster both highlight the particular configurations of power/knowledge (with their associated exclusions and even violence) that are needed to produce 'authentic' local cultures.

7

Themes for a reconstructed regional geography

Given these broader concerns, what are the issues that such a reconstructed regional geography should focus on? We have chosen three themes, *environmental challenges, the politics of development* and *social inequality and social change*, that allow us to address current debates within geography and development studies without ignoring the traditional concerns of either discipline. Perhaps more significantly, the themes we have selected also address questions that are of immediate practical importance to people living within the region – questions about ecological sustainability, the control of development projects and processes, and social transformation. Michael Edwards once argued that development studies can never be relevant in the abstract (Edwards 1989): we hope that the chapters of this collection 'ground' our themes within a South Asian context in a manner that is of interest to practitioners, and disciplinary

Figure 1.3 South Asia: location of places referred to within the text

and area specialists alike. (Figure 1.3 indicates the areas from which case study material has been taken).[11]

'Environmental challenges', the first section, recognises that 'human–environment' relationships have moved back to a central position in geographical research – in the 'South' as well as in the 'North' – over the last decade. In part this has come from the recognition of the degree to which human impact is now global in scale and pervasive in its regional and local effects. Economic development in South Asia is widely interpreted as impinging on the viability and sustainability of both rural and urban systems, and in this section the authors explore key rural and urban envir-onmental issues.

Although a reasonable claim could be made, contra Michael Lipton's urban bias theory, that the urban world of the South is underrepresented in South Asian development debates, the agricultural sector remains critically important to the livelihoods of the majority of South Asia's people. No sector of economic activity illustrates more clearly the critical debates which have erupted over the last three decades on development and anti-development, and in particular the wide range of gender- and environment-related issues associated with agricultural change. Sarah Jewitt's chapter addresses concerns that have been expressed across the global South about the displacement of 'traditional' farming methods by modern (Northern) agricultural practices and the implications of this for local livelihoods and environments. Her study of a predominantly rainfed rice-farming ecosystem in Jharkhand, an area of South Asia marginal to the 1960s 'Green Revolution', challenges the narrative of technology displacement, biodiversity loss and major environmental problems that has been described for areas such as the Punjab (Shiva 1991). As such, her chapter is an important reminder of the fact that technology does not automatically create environmental problems, and that proper attention needs to be given to its effects in context.

As Bradnock and Saunders' chapter shows, the study of environment, once focused at a distinctively regional and local scale, has been powerfully affected by the rise of global consciousness. They explore the implications of new global environmentalist agendas for understanding regional level environmental change in Bangladesh with particular reference to global warming, sea level rise, subsidence and development. They show how the global discourse of climate change fails to do justice to regional environmental processes, with potentially serious policy consequences. Throughout, they note that the challenges posed by economic development throw up a range of political responses, and that these in turn are important in shaping environmental agendas.

This theme is addressed by Chapman with reference to the crucial issue of water resource development. He shows how technology, science and social science play interlocking roles in shaping both perceptions of environmental opportunities and challenges for development. The scale of South Asia's need for water to meet basic food supply requirements leads not to a simplistic chain of cause and effect, but to a range of development needs and opportunities that open political as well as social and economic choices, sensitive to influences from the global to the most local of scales. As he shows, irrigation, which can be conceived in narrow engineering terms, has far wider ramifications when considered in a development context.

The following section, 'The politics of development', deals more explicitly with the processes of intentional development. Intentional development, whether state-led or under the agency of NGOs, is an important part of the lives of many South Asian citizens today. Issues that in the North would be classified under headings from macroeconomic policy to community self-help are still gathered together under the rubric of 'development' by many of those speaking for, and governing, South Asians. Here we take recent theoretical challenges seriously in that we see development as an inherently contested process, and one in which such labels matter. Just as the concept of 'environment' is increasingly contested, so 'development' is as much a political project as it is a process of economic change, and it is one that can pit states against their citizens at a variety of different levels. The individual authors have taken a wide range of planned interventions – which, for all their differences, fit within current definitions of 'development' concerns – and traced the political pressures and resistances that are brought to bear on states and their subjects in trying to realise these strategies.

Stuart Corbridge and John Harriss's chapter on economic reform in India addresses these concerns at a macro-scale, dealing with debates around liberalisation that clearly have relevance to a variety of developing countries inside and beyond South Asia. By taking three accounts of liberalisation – a 'free marketeer's' perspective, the politics of reform, and finally the political economy of reform – the chapter achieves a number of important tasks, of which two are particularly worthy of mention here. First, the chapter highlights the continued importance of paying attention to differences of class and economic interest. By questioning the 'inevitability' of reform and its socially regressive effects, it demonstrates that while socialist development is off the agenda (in India and elsewhere), Marxist-inspired accounts still raise important questions for the current development orthodoxy. Second, it draws attention to the importance of reform in redrawing the relationship between the developmental state and society more generally.

If class matters within Corbridge and Harriss's portrayal of the political economy of development, it is clear from Emma Mawdsley's chapter that the geography of the state is also a key factor. Given the vast differences of language, culture and religion within all of South Asia's larger states, nation-building is by no means an easy task. By teasing out the relationships between changing concepts of federalism, the internal borders of the Indian state, and the actions of political parties, Mawdsley shows how nation-building remains a contested task in India even over 50 years after Independence. Importantly, her account also questions the assumptions that decentralisation automatically leads to either improved participation in government, or to a weaker central state.

Elements of the themes raised in these two chapters are taken forward in the two that follow. The quickening changes in state–society relationships over the 1990s noted by Corbridge and Harriss are echoed in Vandana Desai's account of the changing role of NGOs in South Asia. One of the assumptions underlying the neo-liberal post-Cold War agenda was that governments were inherently incapable of delivering on key objectives of the economic development agenda. However, as a result of the subsequent reforms the state has become increasingly 'hollowed out', and it is NGOs that are often called upon (by the World Bank among others) to plug some of the gaps of social provision that result. As Desai notes, this often places NGOs in contradictory positions, transforming but ultimately embroiled in the difficulties of the state/society relationship they were supposed to transcend.

Finally, Yolanda Foster's chapter draws our attention to the spatiality of resistance to development projects, a theme of importance across South Asia, as shown by the Narmada Dam protests, among others. With the World Bank (and the Asian Development Bank) emphasising the import-ance of infrastructural development to growth within the region, economic reform may even lead to a proliferation of such large-scale projects. Foster's study of a phosphate mining project in Eppawela, Sri Lanka, shows here the political importance of such 'economic' development, not merely through provoking local reaction, but also through its contribution to the 'complex emergency' of the country's civil war.

The final section engages with one of the enduring themes of develop-ment geography, 'Social inequality and social change'. A concern with poverty and social change has been written into development theory and practice in many ways, perhaps most famously in President Truman's inaug-ural address. This speech, with its paternalistic overtones, has rapidly become an icon of the anti-developmentalist crusade against particular readings of developing countries and their people (see, *inter alia*, Rist 1997; Simon and Dodds 1998: 595–6). By focusing on social inequality in South Asia, we are not attempting to reproduce Truman's selective vision of developing societies by presenting South Asian citizens as helpless victims. However, inequality and growing social exclusion have such a dramatic influence on the length and quality of life of so many South Asians that to ignore these issues would be an unjustifiable omission within any con-cerned geography of the region. Again, our approach has been to consider questions of social inequality and social change in the light of recent theor-etical debates.

'Poverty' is, of course, a keyword in these debates. Since the end of the Second World War, various strategies of poverty alleviation have been deployed in South Asia as elsewhere in the developing world, and defin-itions and measures of poverty have themselves shifted from narrow eco-nomic concerns to encompass questions of empowerment and social exclusion. Lakshman Yapa looks at how these poverty alleviation strategies have played out in Sri Lanka, and argues that governmental poverty dis-course is itself disabling and disempowering for those outside the nation's elite. His critical reading of local Buddhist understandings of poverty shows both the possibilities and problems of stepping outside official development discourse, as economic reductionism becomes replaced with implicit ethnic and religious chauvinism.

Craig Jeffrey's chapter in some ways turns traditional developmental concerns with poverty upside down, by looking at social dominance and its reproduction within north Indian society. The chapter shows that caste differences are not disappearing in modern India – in contrast to the hopes and expectations expressed by the country's development theorists and political leaders at Independence. But nor is caste dominance an unchanging phenomenon: it is in part a product of changing patterns of material and cultural accumulation. Importantly, Jeffrey's chapter thus challenges both simplistic ideas of development as 'modernisation' and idealisations of the local implicit in some neo-populist development writing. His chapter carefully combines detailed empirical study with engagement in important theoretical debates: a theme that we hope runs through all three sections of this book.

By 2050 over half the total population of South Asia's 2 billion people will be living in urban environments. Issues of urban sustainability and the role of informal settlements in coping with urban environmental stresses can be expected to be of paramount significance throughout South Asia. Arif Hasan's chapter on experience of the informal housing sector in Pakistan illustrates the interaction of environmental issues with factors ranging from demographic change to intertribal relations. Urban issues are also addressed by Shahnaz Hussain in her work on gender and development in Bangladesh. Hussain's chapter draws our attention to the speed with which, as Bangladesh urbanises, traditional gender roles are undergoing radical change. Population growth, lack of rural opportunity, exposure to natural hazards and the newly perceived need to earn a cash income are transforming the role of women across Bangladeshi society. Despite Bangladesh's low position on the UNDP's Gender Development Index, including such variables as female literacy and share of total earned income, Shahnaz Hussain finds that women are playing the primary role in changing their own situation. Reporting on work among the female bustee populations of Dhaka, she shows how urban migration has rapidly been accompanied by transformed attitudes to work and home.

Using this book

Finally, we turn to the question of how we hope those interested in geography and development will use this book. It was never our intention to write a 'textbook' in the standard sense. As the rest of this introduction has hopefully made clear, we want to address contemporary theoretical concerns and their relevance to South Asia. We hope that the three themes we have selected allow a critical engagement with some important aspects of the contemporary development of the region. Each chapter has been specially commissioned for this volume, and throughout, the authors have provided a summary of current debates in their fields and indicated how their studies relate to others elsewhere within the region.

We recognise, however, that for many readers this may be the first collection of South Asia-specific material that they have encountered. As editors we have insisted throughout that the case studies presented by individual authors speak to wider debates, and also explain any locally specific and technical terms to the non-specialist. We hope the chapters are accessible as a result, and that the bibliographies for each chapter help to provide useful directions for further reading, both of related case study material and of theoretical debates. Those wanting to gain a more broad-based overview of South Asian geography and development should examine the list of introductory materials below, which points the reader towards a range of supporting textbooks and other information sources. The glossary of regional-specific terms at the end of the book should also be of use to those unfamiliar with the region.

Our intention has been to make the South Asia volume of the DARG series an accessible collection of cutting-edge research papers with a clear thematic focus, of interest to the student and specialist alike. We hope that it also demonstrates that writing a (reconstructed) regional geography is a vitally important intellectual project, especially in an era of 'globalisation'.

Notes

1. The end-of-millennium reports of the World Bank and the United Nations Development Programme (UNDP) provide two among many views of what globalisation might mean for world development (World Bank 1999; UNDP 1999). The World Bank's faith in the neo-liberal agenda contrasts with the more cautious approach of the UNDP, where the risks of globalisation are stressed and used to justify calls for new forms of global governance. For critical appraisals of the globalisation discourse, see among others Massey (1999) and Swyngedouw (2000). Allen (1995) and Crang (1999) provide useful entry points to some of these debates for students.
2. Longer-term income poverty trends suggest that while East Asia and North Africa have seen a fall in the numbers and percentage of the poor since the late 1980s, South Asia and sub-Saharan Africa have had far smaller decreases in their poverty rates combined with a growing absolute number of people in poverty.
3. Of course, these global consumption patterns have their historical antecedents, a point some simplistic accounts of a shrinking world tend to forget.
4. For students unfamiliar with these debates, useful points of entry include Sarah Radcliffe's and Stuart Corbridge's chapters in *Introducing Human Geographies* (Cloke *et al.* 1999), or Jonathan Crush's edited collection *Power of Development* (Crush 1995).
5. For particular South Asian examples, see Shrestha (1995) and Rangan (1996).
6. 'The process of enhancing individual and collective quality of life in a manner that satisfies basic needs (as a minimum), is environmentally, socially and economically sustainable, and is empowering in the sense that the people concerned have a substantial degree of control over the process through access to the means of accumulating social power' (Simon 1999: 21).
7. Stuart Corbridge's review essay (Corbridge 1998) provides a commentary on post-development debates that could be usefully read alongside Simon's account. The work he reviews leads Corbridge to suggest that some versions of post-development thinking are at least as ignorant of the real needs of people living in developing countries as the straw men they are looking to knock down. We would also note that economism and teleology are alive and well in neo-liberal readings of development, and it is vital that development academics continue to engage critically with such accounts.
8. Students wishing to gain an overview of the history of the region could consult the introductory texts by Farmer (1993) and Wolpert (2000).
9. Discussing South Asia as a 'civilisation' in this sense can only be justified if we stress histories and cultures in the plural. Attempts to repress this plurality are

common in contemporary South Asia, and can be seen in various contexts from the conflict over the filming of Deepak Metha's *Water* in Varanasi, to the state's repression of the Tamil language in Sri Lanka. The political consequences of such claims to singular cultural 'authenticity' are significant and disturbing, and lead to our unease with Rana P.B. Singh's somewhat uncritical equation of Hinduism with Indian culture in an earlier volume in this series (Singh 1999).

10. The terms on which we have tried to be 'representative' are themselves open to question: one could make equally logical claims to divide the coverage of chapters up proportionally to land area or population size – in which case Sri Lanka would be less 'essential' to the collection than many Indian states.

11. We could have chosen numerous other issues – from international security to demographic change – to similarly address theoretical and practical concerns. The omission of these issues is not intended to suggest their irrelevance: we merely wanted to address a restricted range of themes in depth.

References

Agnew, J. (1999) 'Regions on the mind does not equal regions of the mind', *Progress in Political Geography*, **23**(1), 91–6

Allen, J. (1995) 'Global worlds' in D. Massey and J. Allen (eds), *Geographical Worlds*. Oxford University Press, London.

Anderson, B. (1983) *Imagined Communities*. Verso, London.

Cline-Cole, R. (1999) 'Contextualising professional interaction in Anglo-(American) African(ist) Geographies', in Simon and Närman (eds), *op. cit.*

Cloke, P., Crang, P. and **Goodwin, M.** (1999) *Introducing Human Geographies*. Arnold, London.

Corbridge, S. (1998) '"Beneath the pavement, only soil": the poverty of post-development', *Journal of Development Studies*, **34**(6): 138–48.

Cowen, M.P. and **Shenton, R.W.** (1996) *Doctrines of Development*. Routledge, London.

Crang, P. (1999) 'Local–global', in Cloke *et al.*, *op. cit.*

Crush, J. (ed.) (1995) *Power of Development*. Routledge, London.

Edwards, M. (1989) 'The irrelevance of development studies', *Third World Quarterly*, **11**: 116–35.

Farmer, B.H. (1993) *An Introduction to South Asia*, 2nd edn. Routledge, London.

Guha, R. (1997) *A Subaltern Studies Reader, 1986–1995*. Oxford University Press, London.

Massey, D. (1999) 'Power-geometrics and the politics of space-time', Department of Geography, University of Heidelberg, Heidelberg.

Mohan, G. and **Stokke, K.** (2000) 'Participatory development and empowerment: the dangers of localism', *Third World Quarterly*, **21**(2): 247–68.

Rangan, H. (1996) 'From Chipko to Uttaranchal: development, environment, and social protest in the Garhwal Himalayas, India', in R. Peet and M. Watts (eds), *Liberation Ecologies: Environment, development and social movements*. Routledge, London.

Rist, G. (1997) *The History of Development: From western origins to global faith.* Zed Books, London.

Sachs, W. (ed.) (1992) *The Development Dictionary: A guide to knowledge as power.* Zed Books, London.

Sen, A. (1981) *Poverty and Famines.* Clarendon Press, Oxford.

Sen, A. (1989) 'Food and freedom', *World Development*, **17**: 769–81.

Sen, A. (1992) *Inequality Re-examined.* Clarendon Press, Oxford.

Shiva, V. (1988) *Staying Alive: Women, ecology and survival in India.* Kali for Women, New Delhi.

Shiva, V. (1991) *The Violence of the Green Revolution. Third World Agriculture, Ecology and Politics.* Zed Books, London.

Shrestha, N. (1995) 'Becoming a development category' in Crush (ed.), *op. cit.*

Simon, D. (1999) 'Development revisited: thinking about, practising and teaching development after the Cold War', in Simon and Närman (eds), *op. cit.*

Simon, D. and **Dodds, K.** (1998) 'Introduction: rethinking geographies of North–South development', *Third World Quarterly*, **19**(4): 595–606.

Simon, D. and **Närman, A.** (eds) (1999) *Development as Theory and Practice; Current perspectives on development and development co-operation.* Longman, Harlow.

Singh, R. (1999) 'Rethinking development in India: perspective, crisis and prospects', in Simon and Närman (eds), *op. cit.*

Swyngedouw, E. (2000) 'Authoritarian governance, power, and the politics of rescaling', *Environment and Planning D: Society and Space*, **18**(1), Feb.: 63–76.

UNDP (United Nations Development Programme) (1999) *Human Development Report 1999.* Oxford University Press, New York.

UNDP (United Nations Development Programme) (2000) *Human Development Report 2000.* Oxford University Press, Oxford.

Wolpert, S. (2000) *A New History of India,* 6th edn. Oxford University Press, London.

World Bank (1999) *World Development Report 1999/2000.* Oxford University Press, New York.

World Bank (2000) *Attacking Poverty: World Development Report 2000/2001.* Oxford University Press, New York.

Further reading

For those unfamiliar with the region, the following provide some general background information:

Bose, S. and Jalal, A. (1998) *Modern South Asia: History, Culture and Political Economy.* Routledge, London.

Bradnock, R.W. (2002) *The Political Geography of South Asia.* Christopher Hurst, London.

Chapman, G.P. (2000) *The Geopolitics of South Asia: from early empires to India, Pakistan and Bangladesh.* Ashgate, Aldershot.

Corbridge, S. and Harriss, J. (2000) *Reinventing India*. Polity, Cambridge.
Farmer, B.H. (1993) *An Introduction to South Asia,* 2nd edn. Routledge, London.
Shurmer-Smith, P. (2000) *India: Globalisation and change*. Arnold, London.
Stern, P. (1993) *Changing India*. Cambridge University Press, Cambridge.
Wolpert, S. (2000) *A New History of India,* 6th edn. Oxford University Press, London.

Web links

There are a range of good quality South Asian newspapers and journals online, and many networking sites. These include:

Dawn	www.dawn.com
Economic and Political Weekly	www.epw.org.in
Frontline	www.frontline.com
India Today	www.indiatoday.com
The Hindu	www.hinduonline.com/today/
The Statesman	www.thestatesman.org
The Telegraph	www.telegraphindia.com
Times of India	www.timesofindia.com Bangladesh Environment Network
The Economic Times	www.economictimes.com/today/ pagehome.htm

Institutional sites

Census of India	www.censusindia.net/results/index.html
Government of India	http://alfa.nic.in/
World Bank South Asia	wbln1018.worldbank.org/sar/sa.nsf

Environment sites

http://www.enviroweb.org/ben

Centre for Science and the Environment, India	www.cseindia.org
Bangladesh Environment Network	http://www.ben-center.org

Links pages

www.indiaconnect.com/
www.pakistanlink.com/

Part One Environmental challenges

Modern farming, socio-environmental disasters and the displacement of traditional agriculture?
A reassessment from Ranchi District, Jharkhand

Sarah Jewitt

Introduction

Since the late 1980s, 'grand theories' of development and externally imposed, 'top down' development programmes have attracted much criticism for their detrimental economic, social and environmental impacts and local insensitivity (Booth 1994; Corbridge 1990, 1995; Long and Long 1992; Schuurman 1993b). In particular, the transfer of capital-intensive technology from 'developed' to 'developing' countries and the implicit assumption that the West is the model for the rest of the world to follow has been questioned. A prime example is the Green Revolution agricultural technology which, despite raising agricultural productivity in many areas,[1] creating substantial extra employment and restraining food prices, has been criticised for its ecological impacts, potential for economic differentiation and promotion of mechanisation-induced landlessness (Johnston and Kilby 1975; Farmer 1983; Byres 1981, 1983; Shiva 1991a).

For a number of radical populist and (anti-)development theorists, these problems have been regarded as representative of more fundamental flaws within development itself (Esteva 1987; Banuri 1990; Marglin 1990a; Escobar 1992; Sachs 1992) and more recently Singh has addressed the question of Indian conceptualisations of development (Singh 1999).[2] Highlighting the damage caused to Third World eco- and social systems by transfers of Western technology and ideas, some anti-developmentist writers liken development to a Western disease that destroys its Third World hosts (Rahnema 1988; Shucking and Anderson 1991).[3] Of particular concern is the displacement of 'traditional' indigenous knowledge and culture which many view as the key to a more environmentally sensitive and locally appropriate development alternative. Consequently, they argue that the West should no longer be a role model for the rest of the world and propose a radical populist alternative for the Third World based on traditional technology, environmental knowledge and community organisation.

A problem that this type of 'supply side' populism (Richards 1990) shares with more traditional development discourses is its tendency to speak for others without consulting them. Yet the growing importance of development-oriented 'new social movements' underlines the fact that local people are often quite capable of articulating what they want (as well as what they do not want) from the development process (Vivian 1992; Escobar and Alvarez 1992; Schuurman 1993a; Kaufman 1997). Indeed, such 'demand-side' populism seems rather more promising as a means of achieving a more bespoke (and locally controlled) mixture of tradition and modernity as it comes from the grassroots.

With these different approaches to development in mind, this chapter will examine the interaction of traditional and 'modern' technology within the context of rice-based peasant farming systems in Jharkhand. Concern about the displacement of traditional farming methods by modern (Western) agricultural practices and the implications of this for local livelihoods and environments form the main research emphases. By focusing on one of the many predominantly rainfed areas 'left out' by the Green Revolution (Lipton with Longhurst 1989), the chapter seeks to present an alternative picture of Indian agricultural change to that associated with the more advanced irrigated areas.

Using census data, the information from Survey and Settlement Operations Reports 1902–10 and agricultural statistics dating from 1901, coupled with detailed ethnographic work on the same region published in 1915, the third part of the chapter charts the main agricultural changes that have occurred in Ranchi District over the last century. This is followed by more detailed empirical data on local agricultural practices from two predominantly tribal villages in Bero Block, Ranchi District, collected between 1993 and 1999. The chapter argues that although declining water table levels and pollution from agrochemicals and other sources will require careful monitoring (see Chapman's chapter in this book), agricultural intensification in this area has brought relatively few environmental problems to date. It also suggests that although several traditional technologies have been somewhat marginalised by modern agricultural practices, there is at present quite a healthy and harmonious mixture of both old and new techniques. Moreover, the widespread adoption of modern agricultural practices in one of the villages has been associated with substantially better food security, reduced levels of seasonal migration and greater village-wide co-operation. Before moving on to these empirical sections, however, the second part of the chapter will provide some background to the debates over modern versus traditional agricultural practices in the developing world.

'Modern' versus 'traditional' agriculture

For many development planners during the 1950s and 1960s, the rural sector was seen from rather an orientalist perspective (Said 1978). The view that Third World farmers were 'ignorant and needed to be taught how to farm' (Hecht 1987; 18–19) helped to justify the continued transfer of Western agricultural technology. In the mid-1960s, India developed a 'new agricultural strategy' (Farmer 1983) focused around the introduction of high-yielding varieties (HYVs) of wheat (with associated irrigation, chemical fertiliser and pesticide inputs) into the key agricultural districts of Punjab, Haryana and

Uttar Pradesh. Rice HYVs were also introduced but produced less marked yield increases and proved unsuitable for cultivation in over 75 per cent of India's rice-growing areas due to their inability to withstand flooding or drought. The adoption of HYV cereals was therefore much greater in areas with good irrigation (notably Punjab) than in predominantly rainfed or flood-prone States such as Madhya Pradesh, Maharashtra, Karnataka, Andhra Pradesh, Kerala, Orissa and West Bengal. Even in favourable areas, however, many farmers initially chose not to adopt the new technology. Much of the Marxist literature on the Green Revolution highlighted its lack of scale neutrality, inaccessibility to poor farmers and potential for agrarian class differentiation (Johnston and Kilby 1975; Byres 1981, 1983; Brass 1994), problems that were especially pronounced in capital-intensive wheat-based systems. Also problematic was the pest-susceptible nature of many early HYVs (Berry 1980; Lipton with Longhurst 1989).

Environmental problems and the Green Revolution

Although plant scientists were able to breed much greater pest and disease resistance into later modern varieties (MVs) and thereby cut the need for large applications of agro-chemicals (Greenland 1997), it nevertheless became apparent that intensive Green Revolution agriculture had significant environmental costs. Irrigation systems were often wasteful and poorly drained, causing waterlogging, salinisation and soil toxicity (Greenland 1997; Swaminathan 1996; WARDA 1997). Monocropped cereals such as rice and wheat often displaced less valuable crops and brought a reduction in varietal diversity as well as (in the case of double and triple cropping) causing soil exhaustion. Water resources, meanwhile, became polluted as the availability of heavily subsidised agrochemicals encouraged farmers to 'insurance spray' their crops against increased weed growth, disease and pest persistence. In addition to finding their way into water sources, these chemicals often killed natural predators and stimulated the development of resistance among surviving pests (International Rice Research Institute 1994; Swaminathan 1996, Greenland 1997; Pingali *et al.* 1997).

Even more worrying from the point of view of future food security has been the general decline in MV cereal output growth (Khush 1995; Greenland 1997; Pingali *et al.* 1997). Aggregate rice output growth has declined to 1.5 per cent per annum and there has been a steady decline in yields even at major research centres. The main cause of this is thought to be the intensive and continuous nature of wet rice cropping which reduces nitrogen availability by slowing organic matter mineralisation (Greenland 1997). Salinity build-up and increases in soil pests and micronutrient deficiencies are also important in some areas (Swaminathan 1996).

As a means of alleviating some of these problems, much recent emphasis has been placed on biotechnology and genetically modified (GM) plant research which, in theory at least, could improve yield sustainability and labour-intensive employment in less favourable rainfed areas not reached by the Green Revolution. The GM staples enriched with vitamin A, iron, zinc or iodine could also have important nutritional and health benefits for the poor and could be grown in currently unusable (often rainfed) environments if weed-inhibiting characteristics, salt tolerance and moisture stress resistance were introduced (Lipton 1999). Indeed, they have been promoted as the key to a 'new doubly green revolution' (Conway 1998) that could increase yields in an ecologically sustainable manner. In tandem with current declines in total fertility rates, they might also represent the means to eliminate world hunger if '"one more heave" from agricultural research…can raise yields over a wide area…providing income from labour-intensive farm work to just two more generations of the poor' (Lipton 1999: 6).

Given the anti-GM lobby's concern about health risks and existing evidence of environmental problems associated with the Green Revolution, however, biotechnology has attracted considerable scepticism. Moreover, GM research has attracted much bad press for its focus on the demands of multinational agribusiness and rich farmers rather than the needs of poor farmers and consumers. Particularly contentious has been the insistence by GM suppliers on germplasm patenting and intellectual property rights: rights that favour the final developers and ignore the farmers who have been selecting seed for generations and agricultural research institutions which have traditionally improved and distributed it free (Lipton 1999).

Interest in indigenous knowledge as the route to more appropriate development

From the early 1980s when the environmental impacts of Green Revolution agriculture started to be taken seriously, the tone of much development and agronomic literature shifted away from a desire to 'educate' supposedly 'tradition-bound' Third World farmers by exposing them to Western technology, towards a sense of disillusionment with the transfer of technology model. At around the same time, concern about previous failures to recognise the richness of many indigenous knowledge systems came to the fore (Brokensha *et al.* 1980; Chambers 1983; Richards 1985, 1986). In recent years, these issues have been taken on board by 'participatory' development approaches which emphasise empowering local people by working with them, learning from them and trying to be sensitive to their world views (Chambers 1992; Vivian 1992; Ágrawal 1995).[4] At the same time, traditional agro-ecological knowledge and environmental management practices have attracted attention for their potential to implement more appropriate and sustainable forms of rural development.[5]

Interest in indigenous knowledge before it is displaced

Another, not unrelated, reason for researching indigenous knowledge systems has been to ensure that information on traditional agro-ecological practices and crop varieties can be obtained before the 'dominating knowledge' of the West displaces them completely (Apffel Marglin and Marglin 1990). As Gupta (1992) points out, there is a tendency for biodiversity to be associated with the more disadvantaged areas of the world, and if this is lost through the impact of modernisation, the capacity of these areas to secure future autochthonous development could be seriously reduced. The psychological effects of modern technological developments can also be very damaging as indigenous people may feel that their own knowledge is worthless (Chapman 1983; IDS Workshop 1989a; Scoones and Thompson 1993, 1994; Agrawal 1995).[6] Even attempts to build upon and improve indigenous crops and agricultural practices run the risk that the end result will become 'delegitimised' in the eyes of its possessors, or worse, trivialised if removed from its 'cultural context and forced into the framework of western epistemology' (IDS Workshop 1989a: 38).

Similarly, Westernised or 'city based' food preferences can have an important impact on cropping patterns by encouraging demand for rice and wheat at the expense of more traditional (yet often nutritious and hardy) pulses that come to be perceived as 'poor people's crops' (Shiva 1988, 1991a,b). The rapid rate at which both new plant varieties become available and food preferences change has generated fears that valuable indigenous genetic material could be displaced permanently (IDS Workshop 1989b).

On the other hand, there is a danger that too much concern about such issues encourages populist romanticisation of traditional lifestyles which have little bearing on the aspirations of the people concerned. As Bebbington's research on indigenous groups in the central Andes of Ecuador shows, a 'return to the past' may be neither possible nor desirable. This is because land subdivision and the sale of cattle have caused a shift away from traditional food crop varieties grown with organic manure towards more market-oriented intensive farming and a dependence upon chemical fertilisers. Moreover, many local people have become very attracted to modernity and oppose the 'supply side' populist emphasis of local NGOs on the need to revive traditional agriculture as part of a 'cultural determination' strategy. Instead, they have demanded greater access to modern agricultural technology which will enable them to maintain family incomes through cash cropping, reduce their dependence upon migration for employment and thereby strengthen indigenous culture as a result of their ability to participate more in community life. They also associate this technology with a rise in sociocultural standing and the ability to 'claim the full rights of citizenship from which they have so long been excluded' (Bebbington 1990: 256).

In many ways, these responses reflect those of the research area where a desire for greater food security (and variety) in the face of population increase and land fragmentation has encouraged a shift towards more capital-intensive and market-oriented agriculture. Before examining the impacts of this on specific village environments, cultures and household survival strategies, however, it is necessary to provide some more general background on agriculture in Ranchi District (both traditional and modern) and the fieldwork undertaken.

Agricultural change in Ranchi District

Ranchi District is particularly appropriate as a case study for this chapter in part because it is typical of a rather 'backward' rainfed area, but also because its significant tribal (*adivasi*) population has a long history of detailed ethnographic investigation that gives insight into agricultural change over time (Roy 1912, 1915, 1928). An additional factor is that the dominant tribal group in the District, the Oraons, view themselves primarily as cultivators and have great pride in their agricultural prowess. In contrast to some of the more remote forest-based Jharkhandi tribes, Oraons have long preferred to clear forests for agriculture rather than hunt and gather in them and even claim to have introduced plough agriculture into Chota Nagpur. Also, unlike tribes such as the Mundas which own all village land as a corporate body, the Oraons maintained their forests and other public lands under community ownership but allowed individual family ownership of cultivated land: a situation that seems to have both allowed and favoured agricultural development (Roy 1915). In theory, therefore, one would expect to find the most dramatic agricultural changes, socio-environmental impacts (and displacement of traditional technology) in Oraon-dominated areas.

The research on which this chapter is based has taken place over a period of nine years. It draws heavily upon my doctoral research which involved a nine-month period of fieldwork during 1993 based in two *adivasi* (Oraon) dominated villages in Bero Block, Ranchi District. The village case studies draw upon in-depth agro-ecological information collected in 1993 and updated during numerous field visits since then. As the two study villages were characterised by rather different levels of agricultural development, they provided good case studies for an examination into the displacement (or otherwise) of traditional agriculture by modern farming methods.

Traditional agriculture in Ranchi District

When S.C. Roy was undertaking his ethnographic work on the *adivasis* of Chota Nagpur during the early part of the last century (Roy 1912, 1915, 1928), agriculture formed a major socio-religious focus for the Oraons as well as being their prime economic activity. Nevertheless, cultivated food crops were usually supplemented with game, fruit, flowers, leaves, leafy vegetables, mushrooms and tubers from nearby forests. The flower of the *mahua* tree (*Madhuca indica*), which can either be eaten or distilled to make country liquor, formed a particularly important relief mechanism in times of famine, sometimes forming the equivalent of two months' food supply (Hallett 1917).

The equipment used for cultivation was fairly simple. Cattle-drawn implements including wooden single furrow ploughs (with iron ploughshares), harrows and levellers were all made locally. Clods were broken with a wooden mallet and crops were harvested with a sickle and threshed by cattle walking over them. Winnowing was undertaken with a bamboo winnowing basket (*soop*) and husking was carried out with the use of a foot-operated wooden rice pounder known as a *dhenki*. The main form of transport used was a bullock cart made of wood and bamboo. Being expensive to build, wells were not particularly common and were used primarily for drinking and bathing water. The main systems used for drawing water were either a bucket on a rope or a *latha* which consisted of a long wooden lever pivoted between two posts and weighted at the end furthest from the well. A rope holding a metal pail was attached to the other end and could be drawn down to collect and lift water out of the well.

Table 2.1 Typical crops and productivity levels of different land types in Ranchi District

Land type	Yield (t/ha)*	Main crop
Don 1	1.9	*Aghani* paddy (harvested in December)
Don 2	1.9	*Aghani* paddy (harvested in December)
Don 3	1.5	*Khartika* paddy (harvested in November)
Don 4	0.9	*Bhadoi* paddy (harvested in October)
Tanr 1	0.8	*Gora* paddy, pulses, millets and oilseeds
Tanr 2	0.4	*Gora* paddy, pulses and millets
Tanr 3	0.2	*Gora* paddy and coarse millets

Source: Reid (1912).
* These figures have been converted from the original data, which were quoted in maunds. Although the value of the maund varies across India from between 25 and 90 lb (roughly 10–40 kg) in the Jharkhand region 1 maund is roughly equal to 40 kg.

As irrigation was limited to those Oraon households with wells on their uplands or fields close to streams and natural depressions, cultivation was restricted almost exclusively to the *kharif* (monsoon) season (June–November). Rice was the principal food crop and, depending on the rainfall, took between 84.4 and 61.2 per cent of Ranchi District's net cultivated area between 1901 and 1917 (Government of India 1911a; Reid 1912; Hallett 1917). The best paddy has traditionally been grown on the relatively wet and fertile terraced lowlands (*don*) which can be divided into four elevation-related categories with different harvesting times and productivity levels (see Table 2.1). At the time when S.C. Roy was writing, *don* lands were cultivated annually and manured every second or third year. In order to alleviate the monocropped nature of *don* paddy cultivation, farmers used to plant different varieties of paddy in their fields each year and alternate transplanting with broadcasting (Sinha 1976).

The drier and poorer quality uplands are known as *tanr* and have traditionally been planted with *gora* (upland) paddy in rotation with other crops (Reid 1912). Like *don* land, *tanr* has also been classified according to its elevation, quality and productivity. In more remote areas, *tanr* land was traditionally left fallow once every two or three years although Reid was of the opinion that as 'cultivation becomes more intense and the lands are more liberally manured and better cultivated, this class of *tanr* will, no doubt, greatly improve' (Reid 1912: 4). In Bero Block, however, cultivation was fairly intensive even in 1912 and the 'average uplands' were 'cultivated approximately every year' (Reid 1912: 62). Most *tanr* in this area has traditionally been cultivated in a four-year rotation consisting of *marua* millet, *gora* paddy, *urad* pulse and a coarse millet called *gondli* which yields even on the poorest land (see Table 2.2). By the end of the rotation, the soil was usually friable and ready for ploughing and manuring for the next crop cycle (Reid 1912; Hallett 1917; Sinha 1976).

Those Oraons who could afford the time and labour to dig wells on their uplands often grew irrigated *rabi* vegetables or made nurseries for paddy

Table 2.2 The traditional four-year rotation

	Main kharif *crop*	*Intercrop*	*Late-*kharif *pulses (if residual moisture available)*
Year 1	*Marua* millet (*Eleusine corocana*)	*Bodi* pulse (*Vigna catiang*)	
Year 2	*Gora* paddy	*Arhar* millet (*Cajanus indicus*)	
Year 3	*Urad* pulse (*Phaseolus roxburghii*)		Pulses including lentils, gram and *kurthi* (*Dolichos biflorus*); oilseeds including mustard, linseed and *surguja* (*Guizota abyssinica*)
Year 4	*Gondli* millet (*Panicum milaire*)		*kurthi* or *surguja*

Sources: Reid (1912), Hallett (1917) and Sinha (1976).

seedlings which would later be transplanted on to *don* land. Many of the better-off Oraons also owned a plot of *bari* or homestead land which they could supervise closely and cultivate intensively. Such land was commonly enclosed and planted with kitchen vegetables such as gourds, okra, beans, maize, potatoes and aubergines (Roy 1915).

Apart from the cultivation of hardy crops such as *bodi, arhar*, lentil, *kurthi* and oilseeds on the more moisture-retaining *tanr* or garden vegetables on intensively managed *bari* land, *rabi* crops were not important until relatively recently.[7] The main limitations were the waterlogging of the lowlands, the dryness of the uplands and the difficulty of digging wells in an area with hard base rock and a low water table (Sinha 1976). As a result, only 0.2 per cent of Ranchi District's gross cultivated area was irrigated in 1911 (Government of India 1911a). Significantly, however, the District Gazetteers report that famines were strongly linked to areas with a high proportion of *tanr* lands relative to *don* as 'water drains away from the uplands (*tanr*) and even the higher low lands (*don*) with great rapidity; and unless the crops growing on these lands receive a plentiful and continuous supply of rainfall, they wither and die very quickly' (Reid 1912: 5).[8] Indeed, according to Hallett (1917), 'Famines due to deficiency of rainfall may be said to be the only natural calamity from which the inhabitants of Ranchi can suffer.... When famine or scarcity has occurred, it has been due almost invariably to the failure of the rains in the latter part of August and in September and October' (Hallett 1917: 134).

In areas where tanks had been constructed for irrigation, however, Hallett found the impact of drought to be much less severe. As a result, he bemoaned the fact that well irrigation was 'an expensive labour which the aboriginal cultivator will not make' (Hallett 1917: 116) and emphasised that given the prevalence of rivers and streams in the district, much 'could be done in this respect by the judicious expenditure of capital' (Reid 1912: 5). The District's Survey and Settlement Officer, Reid, agreed with this analysis, arguing that without irrigation, '*rabi* crops are of little importance and can never be relied on as affording even a partial substitute for the main crop of the year' (Reid

1912: 5). Foreseeing the land fragmentation that rapid population increase would bring, Hallett predicted that farmers would be 'compelled to learn what are to them new methods of cultivation, to improve their uplands by the use of manure, and to make good crops a certainty in their lowlands by means of irrigation' (Hallett 1917: 114).

Agricultural intensification and modern farming practices

Hallett was, of course, right about land subdivision and the role of population increase in forcing agricultural change. Over time, landholdings became more fragmented and household survival strategies became more tenuous. The average farm size of 3.5 ha recorded in the 1927–35 Revisional Survey and Settlement Report fell to the current average of around 2 ha per household, though the data are too sparse to tell whether this change was gradual or uneven. In addition, widespread forest decline in Jharkhand, as elsewhere, has reduced the availability of many forest-based subsistence items including wood, fuel, fruits, nuts, leaves, herbs, mushrooms, wild vegetables, tubers, liquor-making ingredients (notably *mahua* flowers) and game. Many households were therefore forced to diversify their economic activities, with poorer farmers seeking work as agricultural or casual labourers while the better-educated tried to move into off-farm employment such as business or government service. Seasonal migration to brickfields in Uttar Pradesh and West Bengal also became quite widespread after the late 1970s when bad harvests coupled with widespread forest loss put existing household survival strategies under severe strain.

Most families that retained farming as their primary economic activity were forced to consider how to increase the productivity of their fragmented holdings. Farmers with land close to degraded forests sometimes came to an 'arrangement' with the forest guard to extend their area of cultivation. For other owner-cultivators, intensification was the obvious option. Between 1930 and 1960, the net sown area increased by around 7 per cent whereas the number of working animals increased by about 22 per cent and ploughs by 40 per cent (Bose and Ghosh 1976). Nowadays, therefore, very little agricultural land remains uncultivated during the *kharif* season and the broadcasting of *don* paddy has largely given way to transplanting.

More changes have occurred on the uplands, many of which are now far more intensively farmed than before. The upland fallows common in remote areas a century ago have all but disappeared as villagers use annual manure (and increasingly chemical fertiliser) applications to enable these lands to be cultivated at least once in a year. In areas like Bero Block where annual upland cultivation has a long history, a shift to yearly manuring has caused a decline of the traditional four-year rotation. Reflecting wider Indian food preferences that favour rice and wheat and regard coarse millets as inferior 'poor people's crops' (Shiva 1988, 1991a,b), *gora* paddy has increased at the expense of *gondli*, although the latter is still frequently grown on poor land that other crops will not tolerate.

Marua, which is used to make both *roti* and beer, is grown more widely than before, possibly reflecting the reduced availability of *mahua* flower liquor as a result of forest decline. To maximise *marua* yields, some farmers make seedling beds in June and transplant the young plants on to *tanr* land in July. Most, however, still broadcast their *marua* and *gora* paddy in June, often inter-cropping it with small amounts of *arhar*, *bodi* and *urad* as they believe this to be less 'exhausting' for the soil. On fields with good residual moisture, these crops are often still followed by late-*kharif* pulses and oilseeds.

By far the most important source of intensification, however, has been the spread of irrigation which has made *kharif* crops less vulnerable in dry years, increased the importance of *rabi* season *tanr* cultivation and opened up opportunities for using MVs along with chemical fertilisers and pesticides. While the environmental implications of this may become more pronounced in future, they have to date been limited and drastically outweighed by the greater freedom from famine and food security that irrigated cropping offers in the event of a failure in the *kharif* crop.

As villagers made efforts to dam streams, create irrigation tanks and dig wells on their uplands, the first winter crops that they grew consisted mainly of *rabi* wheat, pulses and oilseeds. By 1966, vegetables were grown on around 13,000 ha in Ranchi District and in 1969–70, 2.68 per cent of the gross cultivated area was irrigated (Kumar 1970; Tripathy 1989). Wheat cultivation increased significantly when MV seeds became widely available. As awareness about cash crop markets increased, villagers also experimented with commercially valuable vegetables such as potatoes and peas alongside kitchen vegetables for their own use. In time, most of the good quality irrigated *tanr* and *bari* land came to be planted with wheat, potatoes, sweet potatoes, tomatoes, peas, beans, okra, maize, aubergines, gourds, pumpkins, radishes, onions, chilli and garlic. The best fields are now usually planted with improved varieties of wheat and vegetables which are well supervised and heavily fertilised with farmyard manure, compost and, increasingly, chemical fertiliser.

Benefiting from Bero Block's higher than average irrigated acreage (12.6 per cent of the 1981 net cropped area compared to 4.3 per cent for Ranchi District as a whole),[9] local farmers responded to the demand by both local and distant urban centres (including Kolkata) for winter vegetables. Initially, the dependence of these crops on timely, regular watering and (to a lesser extent) chemical fertiliser and pesticide meant that villagers with favourable land and capital assets were often in a better position to cultivate them than others: a situation that has been associated with the spread of modern agricultural technology throughout the developing world (Byres 1981, 1983; Shiva 1988, 1991a; Conway and Barbier 1990). Nevertheless, rural Jharkhand's relatively low levels of initial economic inequality, restrictions on the sale of land to non-tribals and limited scope for large-scale irrigation have largely enabled it to avoid the types of agrarian class differentiation experienced in more advanced agricultural regions (Johnston and Kilby 1975; Byres 1981, 1983; Brass 1994). In addition, opportunities for resource-poor villagers to take up irrigated *rabi* cropping have been increased by the availability of Block-level loans for pumpsets (costing Rs18,500) and irrigation wells. To maximise the use of their new irrigation facilities some farmers have also started to grow summer vegetables as well as new paddy varieties on their most productive *bari* land (Sinha 1976).

Table 2.3 Changes in the proportion of irrigated land in Ranchi District (%)

Year	Ranchi District
1900–11	0.2
1916–17	0.19
1969–70	2.68
1980–81	4.53

Sources: Government of India (1911b), Hallett (1917), Kumar (1970) and Government of India (1981, 1991).

The recent increase in dry season vegetable growing owes much to the fact that it can be both profitable and high yielding. It can also be carried out by resource-poor farmers as it requires little land and is more labour- than capital-intensive. A study of Bero Block by the Xavier Institute of Social Services (XISS) between 1979 and 1984 (XISS 1984) found that potatoes had the highest average productivity (3.64 t/ha) of all crops. The category 'other vegetables' also had a high average yield (1.74 t/ha) and profitability compared to more traditional *rabi* crops such as wheat (14.4 t/ha). It seems no wonder, therefore, that irrigated *rabi* cultivation focusing on high-yielding and commercially valuable vegetables like potatoes has been such a widespread intensification strategy in this area. By reducing pressure to reclaim declining areas of waste and forest land for cultivation, the increase in double cropping has also had significant environmental benefits.

From the point of view of rural livelihoods, meanwhile, intensification has been significant in helping to increase agricultural productivity and food security in the face of uncertain monsoon rainfall, rapidly increasing population, fragmented landholdings and declining opportunities for forest-based subsistence. The option of supplementary irrigation for the staple *kharif* paddy crop has been a major factor in reducing the risk of crop failure in dry years. It also benefits both owner-cultivators and agricultural labourers by providing the option of growing labour-intensive *rabi* crops to compensate for paddy shortfalls. As a result, it has helped resource-poor farmers to meet at least some of their subsistence needs while landless villagers have benefited from more year-round agricultural employment. In Bero Block, for example, the proportion of villagers (as a percentage of the total working population) classified as 'cultivators' changed little between 1981 and 1991 (from 82.9 to 82.4 per cent), indicating that intensification has not been accompanied by increased landlessness during this period (see Table 2.4). The percentage of agricultural labourers declined very slightly between 1981 and 1991 (from 10.9 to 9.9 per cent of the total working population), but the numbers of people involved increased significantly (of cultivators by 29.4 per cent and of agricultural labourers by 18.2 per cent), approximately in line with the percentage increase in population.

Table 2.4 Changes in the number and proportion of cultivators and agricultural workers in Bero Block between 1981 and 1991

	All workers	Number of cultivators	Proportion of cultivators as a percentage of all workers	Number of agricultural labourers	Proportion of agricultural labourers as a percentage of all workers	Proportion of cultivators and agricultural labourers as a percentage of all workers
Bero 1981	22,955	19,037	82.93	2,500	10.89	93.82
Bero 1991	29,912	24,638	82.37	2,956	9.88	92.25

Sources: Government of India (1981, 1991).

Tradition within modernity

In spite of these changes, agriculture in Chota Nagpur still bears many similarities to how it was 100 years ago with rice as the major crop, dominating almost all of the *don* and much of the *tanr* land during the monsoon season. Indeed, *kharif* season cultivation in areas like Bero Block that have no recent tradition of *tanr* fallows remains remarkably similar to the descriptions given by Reid (1912), Hallett (1917) and Roy (1915). Even average rice yields have changed little. The XISS study carried out between 1979 and 1984, for example, reported average *don* paddy yields of 1.1 t/ha and *gora* paddy yields of 0.7 t/ha (XISS 1984). These correspond closely and not particularly favourably with the working average of '1.25 t/ha of *dhan don*, and 0.58 t/ha *tanr* (net cropped)' (Reid 1912: 115) noted in the Settlement and Survey report of 1902–10: a situation that highlights the importance of irrigated *rabi* cropping to the District's food security.

Also, and in significant contrast to the main wheat-based Green Revolution areas like Punjab, there has been little change in the agricultural equipment used in the region. Far from making a shift to labour-displacing machinery, most farmers still use cattle-driven single furrow wooden ploughs and harrows, arguing, as they did in colonial times (Reid 1912) that their cattle are not strong enough to manage 'improved' implements (Kumar 1970). Tractors are rarely feasible due to the small size of the plots and the limited capital of most farmers, so bullock carts are still the primary means of transporting bulky goods. Sickles are still used for harvesting, cattle for threshing, *soops* for winnowing and *dhenkis* for husking. Although kerosene-driven pumpsets are now commonly used for irrigating *rabi* crops, many wells still have the traditional *latha* system in place.

Overall, therefore, statistical and secondary source data on Ranchi District and Bero Block displays relatively little evidence either that traditional agro-ecological knowledge is being displaced by modern agricultural methods or that local labourers are losing work as a result of mechanisation. Nor is there any significant evidence to date that agricultural intensification in the area has been associated with either increased landlessness or biodiversity loss. Indeed, all of the crops grown a century ago are still grown today and many of the intercropping systems remain the same. To examine these changes more closely, along with their ecological impacts, however, it is necessary to turn to the case study villages.

The fieldwork villages

In this section, a more detailed investigation will be made into the environmental problems associated with modern farming methods and the extent to which they are displacing or delegitimising traditional farming techniques and cropping patterns. An attempt will be made to investigate whether local people see modern agriculture as the key to greater livelihood security and cultural stability, or whether they see traditional knowledge and local agricultural expertise as central to a more populist form of development (Shiva 1988; Apffel Marglin and Marglin 1990) or as an alternative to development (Esteva 1987; Escobar 1992; Sachs 1992). An additional area for investigation is the extent to which Jharkhandi agricultural institutions have tried to challenge the 'one-way transfer of technology from the industrial world to the Third World' (Altieri *et al.* 1987: 196) by attempting to unite local and 'scientific' agricultural knowledge.

Ambatoli

Ambatoli is the larger of the two villages, with seven hamlets and a total population of 1,751 people in 1991. The majority (76.5 per cent) of villagers are Scheduled Tribes (mostly Oraons with some Mahlis, Mundas and Lohras) although there is a significant population of Scheduled Caste (17.5 per cent) and Backward Caste (6 per cent) households. Levels of prosperity in the village are quite mixed, but even the 'wealthiest' villagers live in predominantly mud-built houses and have no electricity or piped water. Average landholdings are small (1.9 ha) but irrigation has increased significantly from 4.1 ha in 1981 (Government of India 1981) to around 18 ha in 1993. Nevertheless, only 41 per cent of Ambatoli's sample households regularly grow enough food to meet their annual subsistence requirements and most have to find additional sources of income to make ends meet. Of the village sample population, 61 per cent undertake some form of casual labour and 10.8 per cent undertake regular seasonal migration. According to the 1991 census, 706 members of the village's working population were classified as 'cultivators' and 116 were listed as 'agricultural labourers': taken together, 93.73 per cent of the village's working population who depended on agriculture as their primary economic activity (see Table 2.7).[10]

In Ambatoli, agricultural practices should, in theory, be more modern because it was adopted as a 'project village' under the Hindustan Fertiliser Corporation's (HFC's) 'Rainfed Farming Project' in the late 1980s and has had a full-time 'village motivator' living there for much of the time since then.[11] The aim of the Rainfed Farming Project is to undertake a 'sustainable cropping system approach' to 'develop risk management strategies for resource poor farmers...based on their total participation' (Hindustan

Fertiliser Corporation 1990). At the cutting edge of the project are village motivators based in 'project' villages who identify the main agricultural constraints faced by local people and introduce new crop varieties, intercropping techniques and alternative income-generating strategies to be tried out in the village.

Significantly, Ambatoli's patterns of experimentation with, and successful adoption of, HFC-distributed MV paddy correspond well with the classic model of Green Revolution technology diffusion: the wealthier farmers adopting first as they can afford the risks and inputs associated with the new technology, and the poorer farmer adopting later when the advantages of the new seeds have been demonstrated (Chambers *et al.* 1989). In 1993, MV paddy varieties were planted by only 12 of the 100 households that I sampled and took up 4.51 per cent of the net cropped area. All of these households were relatively well-off with favourable (often irrigable) landholdings. The rest of the net cropped area was planted with traditional varieties (TVs) of paddy, pulses, millets and oilseeds. As a result of farmers' preference for maintaining TVs and growing small quantities of MVs alongside them, the number of crops and crop varieties grown today is significantly greater than 100 years ago (see Table 2.5): a situation that contrasts with fears about biodiversity loss in more 'developed' regions (Shiva 1988, 1991a,b; Gupta 1992).

The 12 families who grew MV paddy plus a handful of the wealthier *adivasis* were also the only sample villagers who applied chemical fertilisers (in addition to farmyard manure) to their *don* paddy and chemical pesticides to their *don* and *gora* paddy. Nevertheless, the average MV paddy yields were not significantly greater than the average TV paddy yields (see Table 2.6). The variation in yields reflects both rainfall and the quality of the land, with MVs tending to be planted only on the better quality land. Intercropping seemed to make little difference to *gora* paddy yields although farmers felt that it improved soil quality and allowed a degree of risk minimisation.

Rabi season vegetable cultivation was practised by 34 per cent of the village sample population in 1993 and, because of its dependence on irrigation,[12] was dominated by villagers with access to wells and pumpsets or with landholdings close to the village's two ponds. Interestingly, the increased availability of pumpsets in the village (around a third of the sample households owned one in 1993) has not brought about a decline in the use of *lathas*. Indeed, most villagers use their pumpsets only during the dry season, whereas *lathas* are used to extract bathing and drinking water throughout the year. They also act as substitutes for pumpsets when mechanical problems arise or kerosene shortages occur.

The amount of land planted with *rabi* season crops is usually determined by the success of the previous paddy harvest. In poor years, villagers grow more vegetables which are used both for home consumption and for cash sales. In 1993, 111 ha of irrigated *rabi* crops were planted by the sample villagers, most of whom aimed to grow at least a few kilograms of vegetables as cash crops. Most villagers planted potatoes, peas and cauliflowers in September or October, to enable them to be sold as early vegetables for the Kolkata market. Potatoes were particularly popular as they yielded well (up to 8 t/ha on good irrigable land) and often fetched up to Rs12,000 per ha.

Table 2.5 Crops commonly grown in Ranchi District in 1915 compared to 1993

Year	Cereals	Millets	Pulses	Oilseeds	Vegetables	Spices and condiments
1915	*Don* paddy *Gora* paddy Summer paddy Wheat Barley	*Marua* *Gondli* Bajra Jowar	*Arhar* *Bodi* Horsegram *Kurthi* *Urad* Gram	Linseed *Surguja* Mustard	Aubergine Bean Potato Onion Pumpkin Bottle gourd Okra Maize Radish Jhingi (*Luffa acutangula*)	Chilli Turmeric Garlic
1993	*Don* paddy *Gora* paddy Summer paddy Wheat BarleyHYV *don* paddy HYV *gora* paddy HYV wheat	*Marua* *Gondli* Bajra Jowar	*Arhar* *Bodi* Horsegram *Kurthi* *Urad* Gram	Linseed *Surguja* Mustard	Aubergine Bean Potato Onion Pumpkin Bottle gourd Okra Maize Radish Jhingi Bitter gourd Cauliflower Tomato Pea Spinach Bitter gourd Sweet potato	Chilli Turmeric Garlic Ginger

Sources: Roy (1915) and Jewitt (1996).

Table 2.6 Yields for traditional and modern paddy varieties (t/ha)

Crop	Average yield	Yield range
MV *don* paddy	2.1	0.8–4.0
TV *don* paddy	2.0	0.5–4.0
MV *gora* paddy	1.4	1.2–1.5
TV *gora* paddy	0.8	0.5–1.5
Gora paddy/*marua/bodi/arhar* intercrop	0.8	0.5–1.5

Source: Jewitt (1996).

Rabi wheat was cultivated by only 14 per cent of Ambatoli's sample population in 1993. It is usually sown a little later, often on good quality *tanr*, to reduce the amount of fertiliser required. A few households also grew oilseeds (linseed, *surguja* and mustard) as *rabi* season cash crops to supplement household budgets during the lean (monsoon) season. Others grow summer vegetables (harvested in July) and reduce the amount of *kharif* season *gora* paddy that they planted in favour of late-planted *urad* with *bodi* and/or *arhar* combinations.

As irrigated *rabi* and summer crops do not benefit from the nutrients carried in the monsoon rainfall and runoff, it is common practice to apply chemical fertilisers. In 1993, 30 of Ambatoli's 34 *rabi* cultivators used them, but often applied them in less than optimal doses (heavily supplemented with manure and compost) to minimise capital expenditure. The same villagers who used chemical fertilisers also applied small quantities of chemical pesticides (but not herbicides) to their *rabi* crops.[13]

Primarily as a result of the small amounts of agro-chemicals applied plus the use of organic alternatives, water pollution from pesticides and chemical fertilisers is not (yet, at least) perceived to be a problem in the village. A number of villagers with wells on elevated land, however, were becoming concerned about falling water table levels as a result of increased levels of irrigation. Although this is not a problem for villagers with better quality, less hilly *tanr* land, it may, given the environmental problems associated with intensive irrigated cultivation elsewhere (Greenland 1997), be something that requires close monitoring in future. From a socio-economic perspective, meanwhile, the influx of more modern farming practices seems to have been fairly benign to date. Intensification, especially through irrigated *rabi* cropping, has enabled villagers to cope with population increase and land fragmentation without a significant increase in landlessness or unemployment. Although the percentage of cultivators (as a proportion of the total working population) declined between 1981 and 1991, the proportion of agricultural labourers increased (see Table 2.7).[14] Most local people see

Table 2.7 Changes in the number and proportion of cultivators and agricultural workers in Ambatoli between 1981 and 1991

	All workers	Number of cultivators	Proportion of cultivators as a percentage of all workers	Number of agricultural labourers	Proportion of agricultural labourers as a percentage of all workers	Proportion of cultivators and agricultural labourers as a percentage of all workers
Ambatoli 1981	525	465	88.00	18	3.40	92.00
Ambatoli 1991	877	706	80.95	116	13.23	93.73

Sources: Government of India (1981, 1991).

land fragmentation and not land sales as the main cause of this shift; a situation which suggests that cultivators losing out from this have gained from the employment that intensification has offered. Indeed, the percentage of villagers involved in agriculture (cultivators plus agricultural labourers) increased marginally over this period (Table 2.7).

Jamtoli

Jamtoli is a smaller, hillier village with only three hamlets and a total population of around 1,330 people. It is more ethnically and economically homogeneous than Ambatoli with an *adivasi* population of 82 per cent (of which around 73 per cent are Oraons) and a Backward Caste population of 18 per cent. The average landholding is 1.66 ha (of which 0.55 ha are irrigated), and between 1981 and 1993 the village's irrigable land increased from 23.25 ha (Government of India 1981) to around 500 ha. According to the 1991 census, 478 members (68.29 per cent) of the village's working population were listed as 'cultivators' and 205 (29.29 per cent) were classified as 'agricultural labourers' with the result that agriculture was the main economic activity for 97.57 per cent of the village's working population. Income differentials (as well as poverty levels) have been substantially reduced by a very successful irrigation scheme initiated by an important village elder, Simon Oraon.

The adoption of modern farming practices in Jamtoli

In contrast to Ambatoli, Jamtoli is not a Rainfed Farming Project village yet it is characterised by a significantly higher degree of modern farming coupled with active and ongoing experimentation with new seeds and farming techniques. Although its topography is slightly more favourable than Ambatoli's, the difference in cultivation reflects, in large part, the agricultural expertise, ability to innovate and, for want of a better expression, 'charismatic leadership' of Simon Oraon.

Although Simon had to withdraw from primary school after a year to help on the family farm, he was nevertheless able to conceptualise, plan and organise the construction of five check dams which irrigate a substantial portion of the village's *don* and *tanr* land. When he became *Parha Raja* (leader of an association of villages) almost 40 years ago, he took advantage of a stream running through the village from a nearby hill to build a shallow (2.5 m deep) dam. His success with *rabi* vegetables and summer paddy encouraged him to expand Jamtoli's irrigable potential, so he gathered together over 100 local people to work as daily labourers on the construction of a bigger dam with 3 km of irrigation channels. Benefiting from the irrigation provided by this dam, Jamtoli villagers were able to survive the very poor monsoons in 1977–8. This encouraged them to start building an even bigger dam in 1978 which enabled over 60 ha of land to be irrigated. Two further ponds have been built since 1990 to irrigate nearby land as well as providing bathing facilities for both people and cattle.

As a result, food security has increased dramatically as villagers can provide supplementary irrigation for their paddy when rainfall is poor and maximise cropping intensities by growing *kharif*, *rabi* and even summer crops. Indeed, higher paddy yields plus the profits made from (and employment generated by) winter vegetables has almost halted the need for seasonal migration.[15] Nowadays, over 75 per cent of Jamtoli's households grow enough food to last them throughout the year: a factor that has helped to significantly increase the village's sense of unity as well as a culture of co-operation for further development work and agricultural experimentation.

Interestingly, most Jamtoli villagers cultivate local rather than high-yielding varieties of *don* and *gora* paddy, although around 60 per cent regularly use chemical fertiliser on their *don* paddy (to maximise the yield of their staple crop) and 28 per cent use chemical fertiliser on their *gora* paddy.[16] On fields close to check dams, canals and wells, average paddy yields are higher than elsewhere, with irrigable *don* regularly yielding 5 t/ha and irrigable *tanr* yielding over 1.5 t/ha. Yields on non-irrigable *don* usually range from 1.5 to 3.0 t/ha, while those for non-irrigable *tanr* range from 0.5 to 1.2 t/ha.

Another advantage of irrigation in Jamtoli is that it has made the cultivation of late-*kharif tanr* crops significantly more widespread than in Ambatoli, with around 35 per cent of the village population cultivating vegetables and 24 per cent cultivating oilseeds. Some farmers also grow late-*kharif* pumpkins on their *don* land but the fields must be elevated enough not to be waterlogged but within reach of irrigation if the need arises.

In the *rabi* season, the major factor influencing villagers' adoption of wheat and vegetable cultivation has been not wealth, as in Ambatoli, but the proximity of their land to a source of irrigation. By 1993, Simon's assistance in obtaining Block-funded well and pumpset loans for villagers with no irrigable land had made *rabi* season cultivation possible for over 75 per cent of Jamtoli's households.[17] On well-drained *don* land, wheat and early vegetables are planted after the paddy is cut and harvested. Wheat, onions and garlic are usually planted in November and harvested in February or March. Some farmers then plant a second crop of vegetables or even summer paddy on their lowest *don* land which is harvested in late May or June. Such triple cropping is not as cash intensive or environmentally demanding as it sounds, however, as *don* land (being more fertile) requires less fertiliser than *tanr* and rarely needs irrigating until January or February.

Rabi season *tanr* cultivation is carried out by approximately 75 per cent of Jamtoli's households but is somewhat more intensive than in Ambatoli with almost all villagers using chemical fertilisers to supplement their applications of manure and compost. The other main inputs are also applied rather more widely in Jamtoli with around half of the households using pesticide and a fifth using herbicide.

The reasons behind Jamtoli's adoption of modern agriculture

Although Jamtoli is not a Rainfed Farming Project village, agriculture there is more intensive and less differentiated by wealth than is the case in Ambatoli. The primary reasons for this are Simon Oraon's emphasis on village development and the success of his assistance to the most 'resource poor' farmers. The proximity of Bero market also helps villagers to keep abreast of any changes to the prices and demand structures of both crops and agricultural inputs.

Table 2.8 Changes in the number and proportion of cultivators and agricultural workers in Jamtoli between 1981 and 1991

	All workers	Number of cultivators	Proportion of cultivators as a percentage of all workers	Number of agricultural labourers	Proportion of agricultural labourers as a percentage of all workers	Proportion of cultivators and agricultural labourers as a percentage of all workers
Jamtoli 1981	346	291	84.10	42	12.14	96.24
Jamtoli 1991	700	478	68.29	205	29.29	97.57

Sources: Government of India (1981, 1991).

Another important factor is the manner in which Jamtoli's success has fed upon itself. A prime example is Simon Oraon's flow irrigation system which gave the village significantly greater food security as well as generating substantial local employment both during its construction and afterwards, from more intensive agriculture. Indeed the increase in the percentage of agricultural labourers (as a proportion of the total working population) between 1981 and 1991 indicates the importance of intensification in softening the impact of population increase, land fragmentation and the loss of agricultural land to irrigation tanks (see Table 2.8). Although between 1981 and 1991 there was a decline in the percentage of cultivators in Jamtoli, the overall proportion of villagers employed primarily in agriculture (cultivators plus agricultural labourers) increased slightly (Table 2.8).

A major sociocultural benefit of Jamtoli's modern farming practices that is greatly envied elsewhere in the area is its contribution to reducing the need for seasonal migration. In addition to strengthening village 'unity', it has provided a strong foundation for co-operative development work including a very successful forest protection system (Jewitt 1995a). Simon has also used his influence with Block Development Officers to obtain government-owned common land for cultivation by the poorest villagers in Jamtoli: a situation that has significantly improved levels of support for his work.

Are 'modern' farming practices displacing 'traditional' agro-ecological knowledge?

Although it cannot be denied that modern agriculture, especially irrigated cultivation, has been significant in terms of increasing food security and profits, it is nevertheless important to assess these benefits in the light of potential environmental problems and the risk of them displacing traditional technology. As was noted above, the last 100 years have brought little real change in the farm equipment used in the region except for the introduction of kerosene pumpsets. Far from displacing labour or traditional technology, however, these pumpsets have created employment by allowing winter cropping and are largely used in conjunction with traditional *lathas*. Unlike the large agricultural machinery found in areas like Punjab, the relative cheapness (coupled with the availability of Block loans) and land-augmenting nature of pumpsets makes them accessible even to resource-poor farmers or farmer co-operatives.

It is important to acknowledge, however, that the introduction of MVs coupled with more intensive farming and a widespread adoption of irrigated *rabi* cropping has brought quite dramatic change, the implications of which may not be fully apparent for many years. Chemical fertiliser use, for example, has increased significantly over the last 20 years and has brought a decline in the quantity and range of organic nutrients used by farmers. Formerly, cattle manure and household compost were used in conjunction with other nutrients including ash, *parsa* (*Butea monosperma*) flowers and leaves and *karanj* (*Pongamia glabra*) flowers, leaves and seed cake. Now it is often supplemented only with household compost and chemical fertiliser. Likewise, pesticide use has increased in conjunction with *rabi* vegetable cultivation and the traditional practice of using ash on vegetable foliage to repel or kill insects has become less widespread. Herbicide use is as yet relatively uncommon, although the labour-saving implications of its spread could be serious.

Similarly, the environmental problems associated with intensive irrigated cultivation coupled with large applications of agrochemicals are well known (Greenland 1997; Pingali *et al.* 1997) and could provide grounds for future concern. To date, however, the research area seems to be coping quite well environmentally. Being a predominantly rainfed area, irrigated well- and tank-based *rabi* cropping is too small scale to cause the types of waterlogging, salinisation and soil toxicity that have been associated with major irrigation schemes (Shiva 1988, 1991a). Nor are yield declines resulting from a degradation of the paddy ecosystem likely to be a problem as the dry winter season prevents continuous wetland cropping (Greenland 1997). Groundwater pollution, meanwhile, has been limited by the relatively small amounts of agrochemicals currently used, although this and falling water table levels may both become more widespread in future.

The MV seeds are another aspect of modern farming that have attracted much criticism in the populist and anti-developmentist literature for their environmental damage and limited overall benefits (Shiva 1988, 1991a,b; Apffel Marglin and Marglin 1990). In the study villages, however, most farmers continue to use traditional varieties of paddy alongside MVs: a situation that reflects the greater success that plant breeders have had with irrigated compared to rainfed MVs (Greenland 1997). Moreover, like many other farmers in the developing world (Shiva 1988, 1991a; Lipton with Longhurst 1989) villagers in Ambatoli and Jamtoli favour TV paddy because they taste better and often yield as well as MVs on poor land with fewer inputs. They are also liked for their ability to produce good quality, long straw that can be used as cattle fodder, thatch, bedding and for making items such as grain storage baskets. In contrast to fears about the biodiversity loss resulting from MV cropping (IDS Workshop 1989b; Conway and Barbier 1990; Shiva 1991a,b), therefore, evidence from the study area suggests that MVs can help to increase overall varietal diversity in rainfed environments where risk minimisation is essential for food security.

The fact that winter crops must be irrigated, however, reduces the risk of losses caused by uncertain rainfall and makes it more economic for villagers to invest in MV wheat and vegetables accompanied by modern agrochemicals. A profit-oriented emphasis on irrigated *rabi* vegetable cultivation has also played a major role in enabling resource-poor farmers to buy MV seeds and associated inputs for intensive cultivation. Indeed, the lucrative nature of early vegetable cultivation encourages even effectively landless farmers to grow a few winter vegetables as cash crops on their homestead land. And as the Jamtoli case study has shown, the financial returns to irrigated *rabi* cultivation have enabled many villagers to meet their food requirements without having to migrate. They have also allowed local people to eat a more varied and balanced diet which, although perhaps signalling a change in traditional food preferences (Shiva 1988, 1991a), may simultaneously represent a shift to better nutrition. Overall, therefore, there appears to be relatively little evidence *to date* that the introduction of modern farming practices into the study villages has created significant landlessness, unemployment or a loss of traditional farming practices. On the environmental side, groundwater pollution may become more widespread in future if the use of agrochemicals continues to increase and water table levels need to be carefully monitored.

Populist concerns that the displacement of traditional technology may reduce options to secure future autochthonous development (Appadurai 1990), meanwhile, seem to have little relevance to the study area. Although some traditional practices such as the use of ash as a pesticide and a wide variety of composting materials as fertiliser have been displaced somewhat by the increased use of agrochemicals, they have certainly not disappeared. In contrast to Gupta's (1992) concern that it is usually the poorest farmers who suffer most from a loss of traditional technology or increased prices of modern alternatives, evidence from Ambatoli and Jamtoli suggest that the smallest landholders are the most likely to use a judicious mixture of agrochemicals with organic manures and pesticides. Almost everybody that I spoke to in both Jamtoli and Ambatoli could name at least four or five people who still use these technologies and most people said that they would readopt them if agrochemicals increased greatly in price. Indeed, during a series of in depth interviews in both Jamtoli and Ambatoli, nobody that I spoke to could think of any agro-ecological techniques used by their parents that had been lost completely during their lifetimes.

While this is not particularly surprising given the continuity of agriculture in the research area over the last 100 years, it is nevertheless necessary to step back from conventional orientalist assumptions about indigenous knowledge systems being static because they contain so-called traditional knowledge. Farmers in Ambatoli and Jamtoli have always conducted experiments to test the suitability of new seeds and agricultural techniques in a manner that is not dissimilar to (although less refined than) the work of agricultural research stations. It is essential for farmers to constantly adjust their agricultural practices because this is the normal way in which 'indigenous' agro-ecological knowledge has always developed and adapted to change. The main difference in recent years is that local people's subsistence strategies have been squeezed by population increase and land fragmentation so they have seized opportunities to integrate more modern practices into their agro-ecological knowledge bases. Rather than focusing on the displacement (or otherwise) of traditional technology, then, it is perhaps more pertinent to ask whether current rates of change in farming practices are so unlike anything that has occurred in the past that a future coexistence of traditional and modern farming techniques and knowledge will be impossible.

Certainly in the short term, this seems unlikely because the current mixture of agro-ecological knowledge is a very positive and practical one for an area where agriculture itself is far from predictable. Contrary to Nandy and Visvanathan's claim that the adherents of Western *episteme* are totally unable 'to regard competing systems with anything but contempt' (Marglin 1990a: 25), it is also a mixture that receives a substantial amount of support from KRIBHCO's Rainfed Farming Project as well as the main agricultural university in Chota Nagpur, the Birsa Agricultural University (BAU) in Ranchi.

Indeed, researchers from the BAU seem to have taken heed of 'bottom up' emphases on learning from (not talking at) local farmers (Brokensha *et al.* 1980; Chambers *et al.* 1989) as they work closely with local people on their own farms and aim to improve 'farmers' technologies rather than to import completely new technologies' (Birsa Agricultural University 1992: 145). They have also taken issue with the idea that local and scientific knowledge are incompatible (Banuri and Apffel Marglin 1993a; Scoones and Thompson 1994; Agrawal 1995) and have been very successful in their efforts to build on farmers' experiments and to combine local and modern farming techniques in a cost-effective and ecologically sustainable manner. Moreover, they have tried to show that this can be done in ways that, far from disturbing or delegitimising local knowledge, integrate well with wider development aspirations: not least the need to adjust to changing economic circumstances and to meet household subsistence requirements.

One example of this is the BAU's efforts to build upon and expand the ways in which local farmers use a combination of organic nutrients and chemical fertilisers to cut down cultivation costs (Professor Mohsin, personal communication, March 1993). Another is the trialling undertaken to test the effectiveness of locally available *neem* (*Azadirachta indica*) oil as a foliar insect repellent spray, *neem* leaves to reduce moth damage during potato storage and *kusum* (*Schleichera oleosa*) cake as a nutrient-rich cattle feed. The university also conducts frequent experiments on pest-repellent and nitrogen-fixing intercrops, zero tillage, optimum crop sowing times and standard varietal trials in different farming environments (Birsa Agricultural University 1992, 1995).

So in addition to taking steps to minimise the environmental impacts of intensive agriculture, the BAU also seeks to encourage the use of traditional technologies that have environmental or financial benefits for local farmers. An important example of this are the efforts of BAU researchers to raise farmers' awareness about the problems of monocropping and the advantages of maintaining varietal diversity. Indeed, a significant underlying principle that the BAU shares with many local people is that if water pollution becomes a problem or if the agricultural input price rises envisaged by writers such as Appadurai (1990) do occur, farmers would be able to readopt a predominance of labour-intensive irrigation methods coupled with organic fertilisers and pest-repellent techniques.

Conclusion

In this particular region it seems that much has been learned from the more advanced agricultural areas about the potential of modern agriculture as well as its possible socio-economic and environmental impacts. Populist critiques of the transfer of technology model and development assistance have been tremendously valuable in publicising such problems and in helping to stimulate a shift towards more participatory approaches based on working with farmers and building upon indigenous agro-ecological knowledge. They have also raised awareness of the psychological impacts that modern technology can have on indigenous social and belief systems as well as highlighting the dangers of biodiversity loss in a world that is changing rapidly both physically and culturally.

As the BAU's work indicates, the idea of participatory development rooted in indigenous knowledge and practices has been taken quite seriously in Jharkhand, but most people are keen not to throw out the baby with the bath water. Unlike many Green Revolution critics who downplay the fact that there has never before been an 'increase in food production that was remotely comparable in scale, speed and spread' (Lipton with Longhurst 1989: 1), most Jharkhandis are well aware of the benefits of modern agriculture. They also recognise that without the intensification that has taken place in the region, many people alive now would be dead. Faced with rapid population increase and fragmented landholdings, a shift to double cropping, particularly irrigated *rabi* cultivation, literally represented a life-saving option for some and a substantial increase in annual food security (not to mention better nutrition) for many more.

What the future holds for agriculture in this region, or for Asia as a whole, is by no means clear, especially given current declining trends in staples yield growth. In order to feed the 4,000 million people who will depend on rice as their major staple by 2015, paddy production will have to increase by 50 per cent to maintain current nutritional standards or by 70 per cent to achieve satisfactory nutritional levels (Greenland 1997). Given populist fears about and criticism of biotechnology and GM foods (not to mention patenting issues – Shiva 1991b, 1993), public sector research in this direction has been severely hampered. Private sector research, meanwhile, caters to wealthy farmers' demands for traits such as prolonged shelf life rather than poor people's requirements for plants that can raise yields, resist moisture stress, fix nitrogen and improve the health of millions (Lipton 1999). Indeed, Conway (1998) argues that GM crops have potential for increasing agricultural productivity in just those unfavourable, rainfed environments that benefited little from the Green Revolution. According to Lipton, however, the tendency of the anti-GM lobby to exaggerate the risk from GM crops 'belittles the far greater risk of worsening nutrition in their absence' (Lipton 1999: 2). Moreover, 'if the duck we are shooting at is the third great breakthrough of the century in tropical staples – after maize hybrids and wheat and rice semidwarfs – then GM plants quack plausibly, and are the only duck on the block' (Lipton 1999: 10).

Given a harsh choice between the ability to feed themselves and the risk of future environmental problems (or even increased agrarian class differentiation – Byres 1981, 1983; Brass 1994), there is little doubt that most Third World farmers would choose the former. But as professional risk minimisers, they would probably find ways, in conjunction with local knowledge and indigenous practices, of avoiding or dealing with environmental problems when they arose. In the past, and in spite of their practical experience (and Marxist, populist and anti-developmentist critiques) of the Green Revolution's side effects, farmers in India's advanced agricultural areas have overwhelmingly voted with their pockets for modern farming methods. In relatively backward, rainfed, rice-based ecosystems with less scope for economic polarisation of the type experienced in Punjab's wheat-based systems, meanwhile, most farmers are keen to continue growing MVs and using agrochemicals because they have found them to be effective for meeting household subsistence needs.

In other words, Third World farmers are not throwing away perfectly good indigenous technology because they are told to adopt modern farming methods by state authorities that cannot resist the influence of multinational agrochemical, seed and biotechnology corporations.[18] Rather, they adopt new techniques because they believe that it is worthwhile for them to do so (Nanda 1991). Similarly, if MVs really are so 'resource wasteful', vulnerable to pests and diseases and 'compared to the cropping systems they displace...are not "high yielding" or "improved" at all' (Shiva 1988: 122), surely local people would revert to growing TVs. As the Ambatoli and Jamtoli case studies show, many farmers are very skilled in maintaining a judicious mix of modern and traditional seeds and technologies to minimise risk, environmental damage and unnecessary capital expenditure.

Given local farmers' clearly articulated desire for increased food security (and profits if possible), it seems prudent to be somewhat wary of the radical populist and anti-developmentist assumption that local people should and could 'go back' to simpler existences based on traditional agricultural practices (Esteva 1987; Shiva 1988, 1991a,b; Apffel Marglin and Marglin 1990; Sachs 1992). Although traditional societies may appear 'affluent' (Sahlins 1972) in terms of their 'satisfaction of basic and vital needs' (Shiva 1988: 12), other studies suggest that they often encounter difficulties in meeting their requirements for carbohydrates and water (Kelkar and Nathan 1991), not to mention basic health care and education.

Many supply side populist views also ignore the fact that local people are not silent and helpless 'others', but are quite capable of choosing to what extent they want to access the things that development has to offer. As Bebbington's (1990, 1992) work in Ecuador illustrates, local people are often keen to modernise their indigenous farming techniques when this offers them greater food security, cultural cohesiveness and a way out of undesirable household survival strategies like migration.

In many ways, Jamtoli's approach has been similar as intensive irrigated farming has enabled villagers to give up seasonal migration and has helped to increase village unity to a much greater degree than in Ambatoli where food security is poor and seasonal migration common. As a means of achieving local self-determination, Simon Oraon's strategy has been very effective as it combines indigenous and modern agro-ecological knowledge in a way that helps to strengthen local culture. Yet it is a strategy that has been based on working *with* the state rather than resisting state-imposed development. Indeed, it has been the evident success of Simon's astute combination of indigenous and modern farming methods that has enabled him to obtain state funding for a form of locally controlled development that has improved the lives of many people in both material and sociocultural terms. Perhaps it is a strategy that could work effectively in other marginal rainfed or otherwise risk-prone environments that have largely been left out of India's Green Revolution.

Notes

1. Among other things, the Green Revolution raised Asian per capita food production by 27 per cent and enabled India to become self-sufficient in food (Lipton with Longhurst 1989).
2. Escobar (1992) argues that development has, like 'the orientalist discourses examined by Said...functioned as a mechanism for the production and management of the Third World in the post-war period' (p. 413).
3. According to Shucking and Anderson (1991), development assistance 'can be likened to the AIDS virus; a pathogen that destroys the ability of the host country to resist the invasion of a foreign socio-economic system' (p. 21).
4. In an attempt to make such approaches more 'actor oriented' and politically astute, 'beyond farmer first' writers have emphasised that local knowledge occurs within 'socially and politically constituted networks of different actors, organisations and institutions' (Long and Villareal 1992 quoted in Scoones and Thompson 1993: 16). Any attempt to encourage a meaningful dialogue between local and 'scientific' knowledge must therefore take these socio-political differentiations into account. Similarly, participatory approaches to development are likely to achieve little if they attempt to build upon local knowledge or resource management strategies without considering the uneven distribution of knowledge and inequalities in landownership and control over resources.
5. In many ways, the dichotomy of traditional versus modern agriculture is misleading as indigenous agriculture is constantly adapting to change. Moreover, farmers are constantly experimenting with both new and old varieties (and combinations of both) that can be cultivated in different places at different times with various inputs.
6. Chapman (1983) notes how indigenous farmers are often very embarrassed about their farming practices and strive hard to reduce 'dissonance' in conversations with extension workers or agricultural scientists. Examples of how they do this include hiding traditional crop varieties from visiting extension workers, giving stock answers that they do not understand but which they have learned will make extension workers go away, and rejecting indigenous farming practices altogether (pretending that they never practised them) in favour of 'all modern' methods (Chapman 1983; IDS Workshop 1989b).
7. In the *Ranchi District Gazetteer* of 1917, however, it is noted that market gardening was practised near to Ranchi and Lohardaga towns (Hallett 1917).

8. Ranchi District suffered severe droughts associated with food scarcity in 1820, 1823, 1827, 1837, 1866, 1873, 1920–1, 1927–8, 1945–6, 1948–9, 1952–3 and 1957–9 and declared famines in 1896–7, 1899–1900 and 1907–8 (Reid 1912; Hallett 1917; Kumar 1970).

9. It has not been possible to update this information from the 1991 census as the Ranchi District level handbook is not yet available and landuse data are not contained on the 1991 census disks.

10. Most of the remaining 6.27 per cent were artisans (bamboo weavers, blacksmiths, etc.). The rest were either businessfolk or had waged jobs outside the village.

11. The HFC's Rainfed Farming Project has since been taken over by Krishak Bharati Co-operative Ltd (KRIBHCO) as part of the British Department for International Development (DfID) sponsored Eastern India Rainfed Farming Project.

12. I came across only four families who owned irrigated land and did not grow vegetables at all.

13. Of Ambatoli's sample households, 42 per cent used pesticide on either their *kharif, rabi* or summer crops.

14. Interestingly, the proportion of female workers, cultivators and agricultural labourers all increased significantly in both Ambatoli and Jamtoli between 1981 and 1991. Although some of this change may reflect an increase in female-headed households (resulting from male seasonal migration), in-depth interviews on agricultural change suggest that it has more to do with how women's work was classified in the two censuses.

15. Only 2.25 per cent of the sample population undertook seasonal migration in 1993 and most of these villagers migrated out of curiosity (as a means of earning money to buy consumer goods or visit major cities like Kolkata) rather than necessity.

16. The only person experimenting with IR36 on his *don* land was Simon Oraon.

17. At this time, Jamtoli's sample villagers had an average of 0.38 pumpsets, 1.43 wells and 0.55 ha of irrigated land per household compared to 0.34 pumpsets, 0.58 wells and 0.2 ha of irrigated land for Ambatoli's sample households.

18. See, for example, Shiva's (1989, 1991a) accounts of the detrimental impacts of the Green Revolution.

References

Agrawal, A. (1995) 'Dismantling the divide between indigenous and scientific knowledge', *Development and Change*, **26**: 413–39.

Altieri, M.A. with contributions by Norgaard, R.B., Hecht, S.B., Farell, J.G. and Liebman, M. (1987) *Agroecology. The Scientific Basis of Alternative Agriculture*. Westview Press, Boulder, CO, and Intermediate Technology Publications, London.

Apffel Marglin, F. and Marglin, S.A. (eds) (1990) *Dominating Knowledge. Development, Culture and Resistance*. Clarendon Press, Oxford.

Appadurai, A. (1990) 'Technology and the reproduction of values in rural western India', in Apffel Marglin and Marglin (eds), *op. cit.*: 185–216.

Banuri, T. (1990) 'Development and the politics of knowledge: a critical interpretation of the social role of modernization', in Apffel Marglin and Marglin (eds), *op. cit.*: 29–72.

Banuri, T. and Apffel Marglin, F. (eds) (1993a) *Who Will Save the Forests? Knowledge, Power and Environmental Destruction*. Zed Books, London.

Banuri, T. and **Apffel Marglin, F.** (1993b) 'The environmental crisis and the space for alternatives: India, Finland and Maine', in Banuri and Apffel Marglin, (eds), *op. cit.*: 24–52.

Bebbington, A.J. (1990) 'Indigenous agriculture in the Central Ecuadorian Andes. The culture, ecology and institutional conditions of its construction and its change', PhD Dissertation, Graduate School of Geography, Clark University, Worcester.

Bebbington, A.J. (1992) 'Searching for an, "Indigenous" agricultural development: Indian organisations and NGOs in the central Andes of Ecuador', Centre of Latin American Studies, University of Cambridge, Cambridge, Working Paper No. 45.

Berkes, F. (ed.) (1989) *Common Property Resources. Ecology and Community-Based Sustainable Development*. Belhaven Press, London.

Berkes, F. and **Taghi Farvar, M.** (1989) 'Introduction and Overview', in Berkes (ed.), *op. cit.*: 1–17.

Berry, S.A. (1980) 'Decision making and policymaking in rural development', in Barlett, P.F. (ed.), *Agricultural Decision Making. Anthropological Contributions to Rural Development*. Academic Press, New York: 321–36.

Birsa Agricultural University (1992) *Annual Report 1991–1992*. Birsa Agricultural University, Ranchi.

Birsa Agricultural University (1995) *Annual Report 1994–1995*. Birsa Agricultural University, Ranchi.

Birsa Agricultural University (1996) *Indian Council of Agricultural Research. National Agricultural Research Project for Zone IV, V and VI in Bihar. 'Impact'*. Birsa Agricultural University, Ranchi.

Blaikie, P.M. (1985) *The Political Economy of Soil Erosion in Developing Countries*. Longman Scientific and Technical, New York.

Blaikie, P.M. and **Brookfield, H.** (1987) *Land Degradation and Society*. Methuen, London.

Booth, D. (ed.) (1994) *Rethinking Social Development: Theory, Research and Practice*. Longman Scientific and Technical, Harlow.

Bose, S.R. and **Ghosh, P.P.** (1976) *Agro Economic Survey of Bihar. A Pilot Study*. B.K. Enterprise, Patna.

Boserup, E. (1970) *Woman's Role in Economic Development*. St Martin's Press, New York.

Brass, T. (1994) 'Some observations on unfree labor, capitalist restructuring, and de-proletarianisation', *International Review of Social History*, **39**(2).

Brokensha, D.W., Warren, D.M. and **Werner, O.** (eds) (1980) *Indigenous Knowledge Systems and Development*. University Press of America, Lanham, MD.

Byres, T.J. (1981) 'Agrarian structure, the new technology and class action in India', *Journal of Peasant Studies*, **(8)**: 322–42.

Byres, T.J. (1983) *Green Revolution in India*. Open University Press, Milton Keynes.

Chambers, R. (1979) 'Rural development: whose knowledge counts? *IDS Bulletin*, Special Issue, **10**(2), Institute of Development Studies, University of Sussex.

Chambers, R. (1983) *Rural Development: Putting the last first*. Longman, New York.

Chambers, R. (1992) 'Rural appraisal: rapid, relaxed and participatory', IDS Discussion Paper No. 311. Institute of Development Studies/International Institute of Environment and Development, London.

Chambers, R. (1994) 'Paradigm shifts and the practice of participatory development', *IDS Working paper* No. 2. Institute of Development Studies/International Institute of Envionment and Development, London.

Chambers, R., Pacey, A. and **Thrupp, L.A.** (eds) (1989) *Farmer First. Farmer Innovation and Agricultural Research*. Intermediate Technology Publications, London.

Chapman, G.P. (1983) 'The folklore of the perceived environment in Bihar', *Environment and Planning A*, **15**(7): 945–68.

Conway, G.R. (1998) *The Doubly Green Revolution: Food for all in the 21st century*. Cornell University Press, Ithaca, New York.

Conway, G.R. and **Barbier, E.B.** (1990) *After the Green Revolution. Sustainable Agriculture for Development*. Earthscan, London.

Corbridge, S.E. (1986) 'State tribe and religion: Policy and politics in India's Jharkhand'. 1900–1980, Ph.D. Dissertation, University of Cambridge, Cambridge.

Corbridge, S.E. (1990) 'Post-Marxism and development studies. Beyond the impasse', *World Development*, **18**(5): 623–39.

Corbridge, S.E. (1995) 'Thinking about development. Editor's introduction', in Corbridge, S.E. (ed.), *Development Studies: A reader*. Edward Arnold, London: 1–16.

Escobar, A. (1992) 'Reflections on "Development"', *Futures*, **24**: 411–36.

Escobar, A. (1995) *Encountering Development. The Making and Unmaking of the Third World*. Princeton University Press, Princeton, NJ.

Escobar, A. and **Alvarez, S.** (eds) (1992) *The Making of Social Movements in Latin America*. See especially the Introduction by Escobar and Alverez and the chapters by Garcia and Schneider. Oxford University Press, Delhi.

Esteva, G. (1987) 'Regenerating people's space', *Alternatives*, **XII**: 125–52.

Farmer, B.H. (1977) *Green Revolution*. Westview, Boulder, CO.

Farmer, B.H. (1983) *An Introduction to South Asia*. Methuen, London.

Government of India (1907) *Judicial and Administrative Statitstics of British India for 1905–1906 and Preceding Years*. Superintendent, Government Printing, Calcutta.

Government of India (1911a) *Census of India, 1911*, Vol. 1. Superintendent, Government Printing, Calcutta.

Government of India (1911b) *Census of India, 1911*. Vol V. *Bengal, Bihar and Orissa and Sikkim. Part 1. Report*. Superintendent, Government Printing, Calcutta.

Government of India (1971) *Census of India, 1971*. Government of India, New Delhi.

Government of India (1981) *Census of India, 1981*. Government of India, New Delhi.

Government of India (1991) *Census of India, 1991*. Government of India, New Delhi.

Greenland, D.J. (1997) *The Sustainability of Rice Farming*. CAB International, IRRI, Wallingford.

Gupta, A.K. (1992) 'Sustainability through biodiversity: designing a crucible of culture, creativity and conscience', Indian Institute of Management, Ahmedabad, Working Paper No. 1005.

Hallett, M.G. (1917) *Bihar and Orissa Gazetteers. Ranchi.* Superintendent, Government Printing, Bihar and Orissa, Patna.

Hecht, S.B. (1987) 'The evolution of agroecological thought', in **Altieri, M.A.** with contributions by **Norgaard, R.B., Farell, J.G.** and **Liebman, M.** (1987) *Agroecology. The Scientific Basis of Alternative Agriculture.* Westview Press, Boulder, CO, and Intermediate Technology Publications, London: 1–20.

Hindustan Fertiliser Corporation (1990) 'Towards sustainable agriculture', *Prayas Rainfed Farming Project News Letter*, **1**(3), December.

IDS Workshop (1989a) 'Farmers' knowledge, innovations and relation to science', in Chambers, Pacey and Thrupp (eds), *op. cit.*: 31–8.

IDS Workshop (1989b) 'Interactions for local innovation', in Chambers, Pacey and Thrupp, (eds), *op. cit.*: 43–50.

International Rice Research Institute (1983) *Women in Rice Farming.* Proceedings of a Conference on Women in Rice Farming Systems. Gower, IRRI, Aldershot.

International Rice Research Institute (1994) *Integrated Pest Management: The IRRI perspective.* IRRI Information Series No. 3. IRRI, Los Banos.

Janzen, D.H. (1973) 'Tropical agroecosystems', *Science*, **182**: 1212–19.

Jewitt, S. (1995a) 'Voluntary and "Official" forest protection committees in Bihar: solutions to India's deforestation?', *Journal of Biogeography*, **22**: 1003–21.

Jewitt, S. (1995b) 'Europe's "Others"? Forestry policy and practices in colonial and post-colonial India', *Environment and Planning D: Society and Space*, **13**: 67–90.

Jewitt, S. (1996) 'Agro-ecological knowledges and forest management in the Jharkhand, India: tribal development or populist impasse?', PhD dissertation, University of Cambridge, Cambridge.

Johnston, B. and **Kilby, P.** (1975) *Agriculture and Structural Transformation.* Hutchinson, London.

Kalland, A. and **Persoon, G.** (1997) *Environmental Movements in Asia.* Curzon, London.

Kaufman, M. (1997) *Community Power and Grassroots Democracy. The Transformation of Social Life.* Zed Books, London.

Kelkar, G. and **Nathan, D.** (1991) *Gender and Tribe. Women, Land and Forests in Jharkhand.* New Kali for Women, Delhi.

Khush, G.S. (1995) 'Breaking the yield frontier of rice', *Geojournal*, **35**: 329–32.

Kumar, N. (1970) *Bihar District Gazetteers. Ranchi.* Government of Bihar Gazetteers Branch, Patna.

Lightfoot, C., De Guia, J.R. and **Ocado, F.** (1988) 'A participatory method for systems-problem research: rehabilitating marginal uplands in the Philippines', *Experimental Agriculture*, **24**: 301–9.

Lipton, M. (1999) 'Reviving the stalled momentum of global poverty reduction: what role for genetically modified plants?' Crawford Memorial Lecture, Washington, DC, October.

Lipton, M. with **Longhurst, R.** (1989) *New Seeds and Poor People.* Unwin and Hyman, London.

Long, N. and **Long, A.** (eds) (1992) *Battlefields of Knowledge. The Interlocking of Theory and Practice in Social Research and Development.* Routledge, London.

Marglin, S.A. (1990a) 'Towards the decolonization of the mind', in Apffel Marglin and Marglin (eds), *op. cit.*: 1–28.

Marglin, S.A. (1990b) 'Losing touch: the cultural conditions of worker accommodation and resistance', in Apffel Marglin and Marglin (eds), *op. cit.*: 217–282.

Mishra, S. (1988) *Irrigation Development and Economic Growth (A Case Study of Bihar Economy).* Capital Publishing House, Delhi.

Nanda, M. (1991) 'Is modern science a western patriarchal myth? A critique of the populist orthodoxy', *South Asia Bulletin,* **XI**: 36–61.

Okali, C., Sumberg, J. and **Farrington, J.** (1994) *Farmer Participatory Research: Rhetoric and Reality.* Intermediate Technology Publications, London.

Pingali, P.L., Hossain, M. and **Gerpacio, R.V.** (1997) *Asian Rice Bowls. The Returning Crisis?* International Rice Research Institute, Manila.

Pretty, J. and **Chambers, R.** (1993) 'Towards a learning paradigm: new professionalism and institutions for agriculture', Institute of Development Studies Discussion Paper 334. Institute of Development Studies/International Institute of Environment and Development, London.

Pretty, J. and **Chambers, R.** (1994) 'Towards a new learning paradigm: new professionalism and institutions for agriculture', in Scoones, I. and Thompson, J. (eds), *Beyond Farmer First. Rural People's Knowledge, Agricultural Research and Extension Practice.* Intermediate Technology Publications, London: 182–203.

Rahnema, M. (1988) 'On a new variety of AIDS and its pathogens: homo economicus, development and aid', *Alternatives,* **13**(1): 117–36.

Ranchi District Gazetteer (1905) *Statistics 1901–02.* The Bengal Secretariat Book Depot, Calcutta.

Ranchi District Gazetteer (1915) *Statistics 1900–1901 to 1910–1911.* Government Press, Calcutta.

Reid, J. (1912) *Final Report on the Survey and Settlement Operations in the District of Ranchi, 1902–1910.* Bengal Secretariat Book Depot, Calcutta.

Rhoades, R.E. and **Bebbington, A.J.** (1988) 'Farmers who experiment: an untapped resource for agricultural research and development'. Paper presented at the International Congress on Plant Physiology, New Delhi, 15–20 February.

Rhoades, R.E. and **Bebbington, A.J.** (1995) 'Farmers who experiment: an untapped resource for agricultural research and development', in Warren, D.M., Slikkerveer, L.J. and Brokensha, D. (eds), *The Cultural Dimension of Development. Indigenous Knowledge Systems.* Intermediate Technology Publications, London: 296–307.

Richards, P. (1985) *Indigenous Agricultural Revolution: Ecology and rice production in West Africa.* Hutchinson, London.

Richards, P. (1986) *Coping with Hunger: Hazard and experiment in an African rice-farming system.* Allen and Unwin, London.

Richards, P. (1990) 'Local strategies for coping with hunger: northern Nigeria and central Sierra Leone compared', *African Affairs*, **89**: 265–75.

Roy, S.C. (1912) *The Mundas and Their Country.* Reprinted in 1970. Asia Publishing House, London.

Roy, S.C. (1915) *The Oraons of Chota Nagpur: Their history, economic life and social organisation.* Reprinted in 1984. Ranchi.

Roy, S.C. (1928) *Oraons Religion and Customs.* Reprinted in 1985. Gian, New Delhi.

Roy, S.C. (1985) *Oraons Religion and Customs.* Gian, New Delhi.

Sachs, W. (ed.) (1992) *The Development Dictionary.* Zed Books, London.

Sahlins, M. (1972) *Stone Age Economics.* Chicago University Press, Chicago.

Said, E.W. (1978) *Orientalism.* Penguin, Harmondsworth.

Schuurman, F.J. (1993a) 'Modernity, post-modernity and the new social movements', in Schuurman, F. (ed.), *Beyond the Impasse. New Directions in Development Theory.* Zed Books, London.

Schuurman, F.J. (ed.) (1993b) *Beyond the Impasse. New Directions in Development Theory.* Zed Books, London.

Scoones, I. and **Thompson, J.** (1993) 'Challenging the populist perspective: rural people's knowledge, agricultural research and extension practice', Institute of Development Studies Discussion Paper 332. Institute of Development Studies/International Institute of Environment and Development, London.

Scoones, I. and **Thompson, J.** (eds) (1994) *Beyond Farmer First. Rural People's Knowledge, Agricultural Research and Extension Practice.* Intermediate Technology Publications, London.

Scott, A. (1990) *Ideology and the New Social Movements.* Unwin Hyman, London.

Scott, J.C. (1985) *Weapons of the Weak. Everyday Forms of Peasant Resistance.* Yale University Press, New Haven, Connecticut.

Shiva, V. (1988) *Staying Alive: Women, ecology and survival in India.* Kali for Women, New Delhi.

Shiva, V. (1991a) *The Violence of the Green Revolution. Third World Agriculture, Ecology and Politics.* Zed Books, London.

Shiva, V. (ed.) (1991b) *Biodiversity: Social and ecological perspectives.* Zed Books, London.

Shiva, V. (1993) *Monocultures of the Mind. Perspectives on Biodiversity and Biotechnology,* Zed Books, London.

Shucking, H. and **Anderson, P.** (1991) 'Voices unheeded and unheard', in Shiva, V. (ed.), *Biodiversity: Social and ecological perspectives.* Zed Books, London: 13–41.

Singh, R.P.B. (1999) 'Rethinking development in India: perspective, crisis and prospects', in Simon, D. and Närman, A. (eds), *Development as Theory and Practice.* DARG Longman, London: Ch. 3.

Sinha, B.B. (1979) *Socio-Economic Life in Chotanagpur.* B. R. Publishing, Delhi.

Sinha, V.N.P. (1976) *Chota Nagpur Plateau. A Study in Settlement Geography.* K. B. Publications, New Delhi.

Stolzenbach, A. (1994) 'Learning by improvisation: farmer experimentation in Mali', in Scoones, I. and Thompson, J. (eds), *Beyond Farmer First. Rural People's Knowledge, Agricultural Research and Extension Practice.* Intermediate Technology Publications, London: 155–9.

Swaminathan, M.S. (1996) *Sustainable Agriculture. Towards Food Security.* Konark, New Delhi.

Tripathy, R.N. (1989) *Technology, Farm Output and Employment in a Tribal Region.* Mittal Publications, New Delhi.

Vivian, J.M. (1992) 'Foundations for sustainable development: participation, empowerment and local resource management', in Ghai, D. and Vivian, J.M. (eds), *Grassroots Environmental Action. People's Participation in Sustainable Development.* Routledge, London: 50–77.

WARDA (1997) *Salt of the Earth.* WARDA Annual Report, 1997. WARDA, Bouake.

Warren, D.M. and **McKiernan, G.** (1995) 'CIKARD: a global approach to documenting indigenous knowledge for development', in Warren, D.M., Slikkerveer, L.J. and Brokensha. D. (eds), *The Cultural Dimension of Development. Indigenous Knowledge Systems.* Intermediate Technology Publications, London: 426–34.

Wignaraja, P. (1993) *New Social Movements in the South.* Zed Books, London.

XISS (1984) *Bero, A Longitudinal Report from 1979–1984.* Xavier Institute of Social Services, Ranchi.

CHAPTER 3

Rising waters, sinking land?
Environmental change and development in Bangladesh

Robert W. Bradnock and Patricia Saunders

Introduction

As we argued in the introduction to this book, 'academic and geopolitical shifts have destroyed the certainties that underpinned the geographical imaginations of development in the mid-twentieth century'. Just as our understandings of development have been challenged, so geographical conceptualisations of the environment have also been transformed in the last three decades. In part this re-evaluation has been stimulated by the emergence of 'the environment' as a subject of political concern. Geographers have long been concerned with human–environment relationships. Lowenthal's 1965 edition of Marsh's *Man and Nature* highlighted over a century of scholarly interest, and in the latter part of the twentieth century environmental change remained a core area of geographical research (Lowenthal 1965; Goudie 1979, 1981). Early recognition of a wholly new conceptualisation of the environment as a field of political action was given shape by O'Riordan (1976), and in academic geography there is now a strong focus on exploring conceptualisations of the environment from both cultural and physical perspectives (Redford *et al.* 2001).

South Asia too has long been a field of intensive environmental research, from the eighteenth- and nineteenth-century foundations of scientific study of the Bengal delta of Rennell (1781, 1792) and Fergusson (1863) to the wide-ranging twentieth-century studies of Geddes (1925, 1960, 1982), Bagchi (1944), Spate and Learmonth (1968) and Johnson (1982). However, beyond the academic discipline of geography the 'environment' has only recently come to be seen as a formative 'context' of human activity and as a central factor in development. In the last 30 years, human-induced environmental change has increasingly been presented as one of the major obstacles to 'sustainable development', with a strong predisposition to oppose human intervention in 'the environment'. Recent reconceptualisation has been stimulated by concern with the environment as a global phenomenon and subject to global processes. The sense that global scale processes control regional and local human responses has a vital effect on the role of the regional and the local as domains of research and practice.

Perhaps nowhere is currently more widely cited as an example of the assumed catastrophic effects of global environmental change than

Bangladesh. The cyclone and associated storm damage of November–December 1970, with an estimated loss of life of over 250,000 people (Houghton 1997: 113), brought what was then East Pakistan to world attention, and the new country of Bangladesh's apparently crushing poverty lent credence to Henry Kissinger's often quoted description of Bangladesh as a 'basket case'. In that context, the vicious circle of global warming, sea level rise and coastal inundation has such common-sense logic and apparently severe human consequences that the name of Bangladesh has become synonymous with flooding, mass death and migration on a globally threatening scale. In the process Bangladesh has become one of the most potent symbols in the arguments surrounding climate change and global warming. This chapter examines this construction and explores the extent to which global constructions of environmental change with respect to sea level help to interpret and predict regional scale change, and questions their role as a basis for development planning.

The environment and global climate change

The language of the environment has been taken up at national and international levels through the UN Conventions on Climate Change and Biodiversity at Rio and Kyoto in 1992 and 1997 respectively, though the language is often polemical rather than scientific. Running through the debate is the message that without drastic action to control human interference with the natural world both the environment and human development itself will be severely compromised. In the last decade concern has focused primarily on climate change. From the 1990 report of the Intergovernmental Panel on Climate Change (IPCC), world policy-making bodies have accepted the arguments of those who have proclaimed the disbenefits of global climate change. That view has stimulated an extraordinary range of international research and policy initiatives. Yet, in view of the acceptance which political fora have now placed on the phenomenon of global warming and its assumed associated destructive impacts, it is remarkable how little success there has yet been in understanding exactly how any putative global level climate change may impact on climate itself, still less on wider characteristics of environmental change and development at a regional level.

Models of climate change designed at the global scale have no direct or necessary relationship with specific changes in climate at a regional or local level. Warrick *et al.* (1994: 16) have argued that even where predictions of climate change at a global scale are secure – which currently they are not – there are three obstacles to predicting climate change for a region such as Bangladesh:

first, the regional details of the Global Climate Change Models are suspect; second, most GCM results are for equilibrium, not time dependent changes; and third, the simpler models (which provide time dependent changes at the global scale) do not provide geographic detail.

Under the terms of the Framework Convention on Climate Change (1992), to which Bangladesh is a signatory, Bangladesh is obliged to formulate

programmes containing measures to mitigate climate change and to facilitate adaptation (Warrick *et al.* 1994: 26). Considerable research effort is
now going into evaluating the possible regional scale implications of global
climate change. Yet the uncertainties characteristic of the global projections
themselves are magnified many fold when applied to regional patterns of
wider environmental change and its implications for development.

Such a regional level approach is part of a global concern expressed by
the IPCC and the United Nations Environment Programme (UNEP) for
analysing the vulnerability of coastal zones around the world, which are
held to be particularly susceptible to climate change and sea level rise. As
a result both the IPCC and UNEP have sponsored research into vulnerability, which has led to the publication of technical guidelines for the assessment of vulnerability and an elaboration of the guidelines for nine
physiographic systems and coastal zones (Klein and Nicholls 1999: 182).

However desirable predictions in these two areas may appear, the ability to model links between environmental change and development with
any degree of reliability has yet to be demonstrated. In a dynamic where
technology, political and societal values, and human adaptability all shape
the relationships between physical environment and development, global
climate change is just one, often very minor, factor influencing wider
patterns of development. Many of these non-climate factors are subject to
long-term but inherently unpredictable changes, as well as to major short-
term shocks – in both the physical and human environments – which are
even less susceptible to accurate modelling than climate change. In arguing
for the implementation of a series of 'fast track' regional climate impact
studies based on uniform assumptions, Parry *et al.* recognised that action to
control the emission of greenhouse gases was hampered by uncertainty
about the effects of climate change. They accepted that in part this was due
to 'the sheer complexity of the issue and the state of science in impact
assessment that is less well developed than in climate change modelling'
(Parry *et al.* 1999: S1). This chapter explores the extent to which either the
conceptual or the evidential basis exists for regional scale modelling of
climate impacts on development.

Understanding global and regional environments

The contemporary environmental debate offers plenty of scope for confu-
sion. The term 'environment' itself is so broad that it can legitimately
include such diverse phenomena as plate tectonics, the management of
small-scale 'family' farms, climate and hydrology or the geopolitics of
interstate relations. As geographers have long recognised, much of the
earth's environment has been radically affected by human activity. In South
Asia, and in Bangladesh in particular, tens of millions of people depend on
the success of such fundamental modifications to the physical environment
as irrigation, forest clearance for agriculture, and the use of inanimate
sources of energy.

From its late nineteenth-century flirtation with environmental determin-
ism, twentieth-century academic geography developed more subtle and
nuanced views of the complexity of the relationships between human society

and the environment (Goudie 1979; Messerli *et al.* 2000; Haggett 2001). Goudie (1979, 1981) pointed to the importance of innate variability in the natural environment and the problems of distinguishing human induced from natural causes of variability. The neo-determinism of today (though rarely recognised as such), sees human activity as critically dependent on the stability of a wide range of attributes of that environment, while portraying human activity itself as the major destabilising force in a world otherwise held in equilibrium.

From the global to the regional: narratives of environmental degradation

Climate change is just one facet of a wider debate about environmental change. In the process of conceptualising environmental degradation, 'science' has sometimes become simply a tool for 'the assertion of a number of propositional truths' rather than 'analysis and potential falsification through the praxis of science' (Sullivan 2000: 16). The emphasis on the degradation of the 'environment' has often been accompanied by a reduction of the inherent complexity of variables which together comprise the interlocking physical and human environments to highly simplified models of cause and effect.

The 1990s emphasis on global warming had its immediate antecedents in 1960s and 1970s debates on global environmental degradation. Concepts of 'desertification' and 'deforestation' became central to public debate on environmental change, change itself being increasingly interpreted as the one-directional and degraded result of inappropriate human activity. The environmental degradation discourse became a dominant feature of public policy debates, particularly with respect to the developing world in the 1980s, and has remained prominent today. One effect of such a shift has been to value the global far more highly than the regional as an interpretative scale of reference.

There has been an increasingly sharp academic critique of this process, although to date the critique has largely been ignored and overridden in the global policy-making debate. Sullivan (2000: 15) argues that many such contemporary environmental narratives of 'environmental degradation':

> have become accepted as 'fact' in the absence of what most natural scientists would today acknowledge as the praxis of science, that is, the standardised and 'transparent' collection of data to explore propositional or 'testable' statements, and the interpretation of such data within a defendable, albeit changing and contentious, theoretical framework.

Sullivan points to the widening body of critical literature which has not only shed doubt on the scientific validity of the concepts of desertification (Thomas 1993, 1994) and deforestation (Grainger 1993) but demonstrated how the promotion of the desertification discourse has served the interests of severely repressive policies (Homewood and Rogers 1987; Fairhead and Leach 1996; Homewood and Brockington 1999). Sullivan has argued that one factor contributing to regional and national level support for this agenda is the effect 'on national knowledge production of the transmission of a reactive, "northern" environmentalist worldview, informed by concerns of pending ecological collapse as formulated by "deep ecology", "ecophilosophy" and in some calls for "political ecology"' (Sullivan 2000: 26).

The prevalence of this global discourse is evident across South Asia. Quoting the nineteenth-century British agronomist Alfred Howard, Vandana Shiva (1988), one of South Asia's leading environmentalist critics of 'economism', argued *inter alia* that traditional Indian agriculture had achieved 'a perfect balance between the manurial requirements of crops harvested and natural processes which recuperate fertility' (Shiva 1988: 106). She went on to claim that Green Revolution agriculture, an 'exploitative masculine technology', 'was destroying the very capacity of ecosystems and people to produce food' (1988: 124). Far from achieving improved agricultural productivity, Shiva argued that the Green Revolution was responsible for ills ranging from desertification and soil degradation (p. 140), rural impoverishment, 'breeding out the feminine principle' and increased female infanticide (p. 119) and a dramatic reduction in the nutritional value of food (p. 131).

In a powerful and detailed rebuttal of Shiva's thesis Jewitt and Kumar (2000: 92–5; see also Chapter 2 of this volume) have shown that the evidential basis of many such claims is weak. None the less, they continue to be widely repeated, justifying Jewitt's conclusion that:

> what is urgently needed is a shift away from an undifferentiated and largely instrumental 'add women and stir' WED [Women Environment and Development] approach (Braidotti *et al.* 1994), towards a deeper, more place- and people-sensitive form of gender analysis that can take on board the highly political yet ever changing socially constructed nature of people–environment interactions (Jewitt and Kumar 2000: 109).

Global warming

The primary position in the currently perceived chain of human cause and environmental effect is widely accorded to global warming which has become the point of reference for an extraordinarily wide range of features loosely described as 'environmental'. Its ascendancy is evident in policy fora and in popular conceptions of environmental change, and in the scale of research funding devoted to climate change and related research, embracing geographical scales from the global to the local and temporal scales from the daily to that of millennia. In its 2001 report the IPCC recognised that while there were still many difficulties with interpreting global models, it was still impossible to trust any predictions at the regional scale. Yet many have sensed that climate change necessarily spells disaster. George Monbiot, a leading British environmental journalist, has expressed the sense of desperation of those campaigning about climate change. Writing in *The Guardian* he claimed:

> Climate change is perhaps the gravest calamity our species has ever encountered. Its impact dwarfs that of any war, any plague, any famine we have confronted so far. It makes genocide and ethnic cleansing look like sideshows at the circus of human suffering. A car is now more dangerous

than a gun; flying across the Atlantic is as unacceptable, in terms of its impact on human well-being, as child abuse. The rich are at play in the world's killing fields (Monbiot 1999).

In the politics (though not necessarily in the science and still less in the social science) of the environment, climate change is thus increasingly being taken as a primary cause or driver of both the environment and of development.

Climate change and development in Bangladesh

Bangladesh is often quoted as a key example of the several destructive effects of climate change. Warming in itself has been argued to lead directly to reduced agricultural productivity, although the picture is far from straightforward. Warrick *et al*. (1994: 9) suggest that

> laboratory experiments show that increased carbon dioxide concentrations in the atmosphere increase photosynthesis rates by which plants absorb carbon. They also decrease transpiration, resulting in greater efficiency of water use by plants. Thus, if other factors are not limiting, expected increases in carbon dioxide levels could significantly increase plant growth rates and yields. Crops such as rice, wheat, jute and pulses would benefit more than maize and sugar cane.

More recently Parry *et al*. have argued that by 2080, assuming no climate change, world cereal production will be 4,012 million tonnes (Mt) compared with 1,800 Mt in 1990. Climate change will be responsible for a modest increase in mid- and high-latitude cereal production and a decrease in low-latitude production. Under the Hadley Climate Research Centre's HadCM3 model there will be a reduction of about 4 per cent in total cereal production, though the developing world will experience the largest negative changes – as much as 16.5 per cent. Such a conclusion may be deemed by some to be heroic in view of the stated assumptions, which include 'no major changes in the political or economic context of world food trade', and population growth as projected by the World Bank (1994). Parry *et al*. also state that 'for the most part the results of this study assume optimum nutrient levels' (Parry *et al*. 1999: 554). There are further, unstated, assumptions, notably that trends will increase in a linear fashion, there will be no major shocks nor any major breakthroughs.

Sea level rise

Sea level rise in Bangladesh has come to play an important part in the global discourse of global warming, and the Bengal delta is widely seen as a region where the human costs of sea level rise are among the most severe in the world. Its role as an exemplar raises many of the fundamental issues underlying the transfer of global concepts to the regional scale. In part these difficulties are embedded in the problem of scale itself. Ives and Messerli (1988) and Hofer (1998) have demonstrated how processes of environmental change dominant at the individual catchment level in the Himalayas may be insignificant at the scale of the mountain range and the downstream catchment as a whole. As an example they showed how processes influential in stimulating soil erosion in one catchment area had no discernible effect at a larger regional scale.

The wider environmental changes predicted as accompanying global warming have led to forecasts of a catalogue of environmental disasters, all ultimately attributable to increased CO_2 levels in the atmosphere. McKie and Morris painted a typical portrait of the scale of the environmental damage to which rising CO_2 levels would lead: in 2050, they claimed, Alaska would be 'turned into a quagmire as permafrost and glaciers disintegrate...the remnants of Florida are declared America's second island state...Africa's coastline from Cairo to Lagos is completely flooded and many of the major cities have been abandoned...and Bangladesh is almost totally inundated. Tens of millions stand on the brink of death' (McKie and Morris 2001).

Bangladesh and the environmental disaster scenarios

A view of the imminence of environmental disaster in South Asia pre-dates the current fixation on global warming. Saunders (2000: 224–5) has shown that in the 1960s there was a strong view that India and East Pakistan's failure to produce enough food to cope with their growing overpopulation, meant, in the words of the Paddock brothers in *Famine 1975!*, that they 'could not be saved'. The policy implications were that the West should cut off all development assistance to such countries and allow them to go to the wall.

Today, however, when India has moved into large-scale food surplus and Bangladesh is self-sufficient, concern focuses less on food shortages than on the 'inevitability' of mass migrations of 'environmental refugees', perceived as a major threat to global security (Saunders 2000: 227–31). Milliman *et al.* (1989) 'predicted that millions would be displaced by a combination of subsidence and sea level rise in Egypt and Bangladesh' (Saunders 2000: 237). The theme was picked up by Jacobsen, writing for the World Watch Institute, who wrote of Bangladesh that 'eventually, the combination of rising seas, harsher storms, and degradation of the Bengal delta may wreak so much havoc that Bangladesh as we know it may virtually cease to exist' (Jacobsen 1988: 35, quoted in Saunders 2000: 237).

Thus today, it is claimed, increased CO_2 levels in the atmosphere are leading to unparalleled disasters. The Bengal delta offers a particularly significant region in which ideas of environmental change and development can be examined. In its relatively small area – at 125,000 km^2 approximately the size of England and Wales – and over the seemingly almost uniform plain of the Ganges–Brahmaputra–Meghna (GBM) delta of which it is largely comprised, Bangladesh is the most densely populated and one of the poorest countries in the world, with a world ranking in terms of GNP per capita of 167 (World Bank 2001: 274). The recipient of over $1.5bn of foreign aid in 1998–9, it is a country in which links between current economic activity and the physical environment are still highly visible (Government of Bangladesh Bureau of Statistics 2000: 304).

Bangladesh's environment is now frequently constructed as an environment in a state of what might be termed 'hyper-risk', in an equilibrium poised on the edge of catastrophic disaster. In 2001 the Bangladesh Environment Network carried the report of an international conference on Bangladesh's environment which listed 21 areas of severe environmental degradation, all related causally to human-induced environmental change. These ranged from the reported drying up of Bangladesh's rivers to the severe problem of arsenic contamination of groundwater in Bengal. Sea level rise, an increase in storm frequency, flood and cyclonic damage and desertification are most commonly referred to as the direct result of global warming, and the remainder of this chapter explores the role of global warming and sea level rise in the debate over environmental change in Bangladesh.

The physical environment of the Bengal delta

Recent research on the tectonic origins of the Bengal delta have provided new insights into the geomorphological processes operating today. They demonstrate one dimension of the delta's environmental uniqueness. At the same time, Brammer's work has demonstrated the interlocking complexity of the various components of Bangladesh's physical and human geography which have played and continue to play a major role in the nature and scope of development opportunities taken up by Bangladesh's people. In exploring the significance of global processes of environmental change for Bangladesh, this chapter analyses key features of its geological origins, the geomorphological characteristics of the Bengal delta, and the evidence of sea level rise, before moving on to consider the implications of projected changes in the environment for development.

The tectonic origins of the Bengal delta

The key geological characteristics of the contemporary Bengal delta can be traced to its origins in the collision of the Indian plate with the Asian plate approximately 45 Myr ago (Acharyya 1998). Uddin and Lundberg (1998b) argue that 'the Bengal foredeep has been formed between the stable but northwards moving peninsular shelf and the building ranges of the Indo-Burman ranges to the east and the Himalaya to the north and northwest'. They suggest that sedimentary and metasedimentary rocks beneath the shelf, which owe their origin to the creation and erosion of the Himalayas, are as much as 22 km thick.

Throughout the period of Himalayan uplift, subsidence characterised most of the substrata of the Bengal delta. Studies confirm the astonishing sedimentation rates through much of this period, as high as 1.2 mm/yr in the Surma basin – 'one of the highest sedimentation rates in Earth history, that was sustained for millions of years' (Worm *et al.* 1998: 487). The detail, however, was and remains complex, and recent trends have witnessed uplift of the Madhopur tract in central Bangladesh and the Barind tract in the north-west, while the Sylhet trough has subsided at more than average rates. See Figure 3.1.

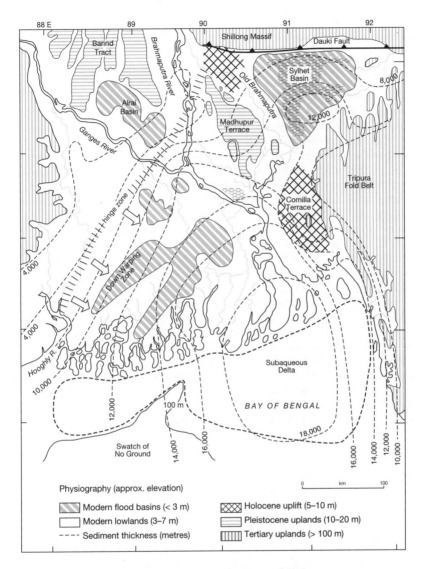

Figure 3.1 Uplift and subsidence in the Bengal delta
Source: after Goodbred and Kuehl (1999).

Late Pleistocene and Holocene sea level change

On to the long-term and continuing trend of tectonically induced subsidence have been superimposed the geologically very recent patterns of sea level rise during the last 20,000 years. Around 18,000 years ago global sea level was approximately 140 m lower than at present, and 120 m of the rise occurred between 18,000 BP and 6000 BP, with a rise of only 20 m in the last 6,000 years. However, it is clear that the long-term trends were marked by dramatic variations, such as the rise of 16 m within 300 years between 14.6 and 14.3 thousand years BP (Hanebith *et al.* 2000: 1033). This represents an annual rate of over 53 mm – more than 1 m every 20 years.

The key factor in the development of the delta at this period lay not in sea level rise or in tectonic subsidence, both of which were strong (though, it must be stressed, not uniform across Bangladesh) but in the establishment of the modern monsoon system between 15,000 and 10,000 BP. By 11,500 BP the humid monsoon was stronger than now (Gasse 1991; Goodbred and Kuehl 2000), 'forcing tremendous sediment discharge of approx 2.5×10^9 tyr^{-1}' (Goodbred and Kuehl 2000). It is this extraordinarily high sediment discharge of the precipitation-fed rivers, pouring down from the Himalayan ranges continuously forced up by tectonic movements that was responsible for the creation of the delta. Milliman and Syvitski have estimated that at the mouth of the GBM the rivers have an estimated seaward-directed suspended sediment load of 1.06×10^9 t/yr (Hoshina 1998; Milliman and Syvitski 1992).

The unique geological and sedimentary character of the Bengal delta is demonstrated in part by the speed of its adjustment to the most rapid period of global sea level rise in recent world geomorphological history. In Bangladesh subsidence caused greater relative sea level rise throughout this period than did eustatic sea level rise. This uniqueness is further demonstrated by the tectonic and sedimentary characteristics outlined above, but it is noteworthy that parts of the Bengal delta continued both to prograde and to infill during a period in which eustatic sea level rose at a rate two and a half times the rate predicted in the worst case global warming scenarios, and parts of Bangladesh subsided at double that rate. Thus had sedimentation not occurred parts of the surface of the Bengal delta would already have been over 300 m under the sea 10,000 years ago, instead of extending southwards into the Bay of Bengal.

Translating eustatic sea level rise predictions to the regional scale

Mörner (1976: 125), a critic of the concept of eustatic sea level rise as a uniform global change in relative sea level, proposed redefining the concept as 'ocean level changes'. Estimates of contemporary eustatic sea level change as a result of global warming continue to vary widely in both scale and speed. Successive IPCC reports have lowered estimates of the likely sea level rise directly attributed to global warming and to emphasise the wide margin of error in the forecasts. In their 1990 report the IPCC estimated a sea level rise of 20 cm by 2030 and 1 m by 2140. In their Second Assessment Report estimates of global mean sea level rise by the year 2100 were put at between 20 and 86 cm, with a best estimate of 46 cm (Klein and Nicholls 1999). Some geomorphologists and marine scientists have considered that such a rise in sea level still cannot be demonstrated to be taking place (Bird 1993: 3), although some climate scientists assert that an annual sea level rise of 1–2.5 mm is already under way (Klein and Nicholls 1999).

Although there is abundant evidence from which post-Pleistocene rises in sea level can be inferred, global change has often correlated imperfectly with regional and local changes in sea level. Furthermore, changes through time are slow in human generational terms, though they may be very rapid in terms of geological timescales. Evidence of sea level change in the short term is riddled with uncertainty.

Even defining sea level is problematic. Jelgersma and Tooley have highlighted the divergence of evidence for global sea level change through the pronounced oscillations of the Quaternary period, the last 2.47 Myr (Jelgersma and Tooley 2000). Bird points out that 'statements about present sea level usually refer to mean sea level, which can be estimated as the long-term average of high and low tide levels, i.e. mean tide level. Strictly, mean sea level is defined as the arithmetic mean of the height of calm sea surface measured at hourly intervals over a period of at least 19 years' (Bird 1993: 9, Carter 1988).

The relationship between local sea level and eustatic sea level is complex (Carter 1988), as local and temporal variables may have far greater effects on sea level than are discernible from long-term means and global trends. In addition to variations in tidal ranges and tidal cycles corresponding to the lunar cycle, there is a long-term 18.6 year nodal tide, based on the precession of the lunar orbit. Atmospheric pressure may account for variations in sea level of several metres, a rise in sea level of approximately 1 cm being the result of every millibar fall in pressure, and seasonal temperature, pressure and wind regimes all play a part alongside the specific characteristics of the coastline. Furthermore, global sea level is also known to be uneven. The geoid – the surface of the undisturbed sea – has been known for 40 years to be 'anything but level' (Carter 1988: 245). Mörner has pointed to a difference of 240 m between the highest and lowest points of the ocean surface (Bird 1993: 13), with a gradient of approximately 15 mm/km between the central Indian Ocean and the eastern Pacific (Carter 1988: 245). Geoidal variations may migrate, and could be responsible for local scale variations in sea level far outweighing eustatic sea level changes on specific coastlines, but these movements are as yet little understood (Tooley 1993: 88).

In addition the problems of measurement of mean sea level change are severe. Tide gauges are often located on port structures which may be subject to local effects, including local subsidence of their foundations, which will be recorded as a rise in sea level. These problems of evaluation are acute in Bangladesh. They make precise evaluation of data from the Bengal coast impossible. The difficulties are starkly illustrated in the paper of an influential SAARC study which claims to have distinguished a rise in sea level in coastal Bangladesh directly attributable to global warming. The conclusions, however, are marred by a failure to explain how any of the above problems have been disposed of (Khan *et al.* 2000). It remains impossible to say anything definitive about the contribution of eustatic sea level change to contemporary relative sea level in Bangladesh. However, from the geological record it is becoming increasingly possible to show how the Bengal delta has reacted to past changes in relative sea level, including the eustatic sea level change since approximately 18,000 BP.

Delta formation in the Bengal regional environment

The recent geological history of the Bengal delta has important implications for understanding the possible consequences of eustatic sea level rise. According to Khan *et al.* (2000: 1), during the last 100 years global sea level has risen by 10–20 cm, a conclusion which, as has been shown above, is still highly contested. If for the sake of illustration Haq's (1994) estimates of subsidence are accepted, as much as one-third of Bangladesh has subsided during the same period by 2.5 m, dwarfing the significance of global sea level rise. Under the IPCC's 'other things being equal scenario', under which there is a global rise in sea level but no local subsidence or sedimentation, most of the land shown within the area less than 3 m above sea level on the often reproduced maps of the areas of Bangladesh at risk would by now already have been inundated as a result of the combination of eustatic sea level rise and subsidence.

Not only has this submersion failed to materialise, recent remote sensing imagery confirms the continuing extension of Bangladesh's surface area into the Bay of Bengal. The main factor responsible for the failure of the models to account for the actual pattern of geomorphological development of the Bengal delta lies in the specific regional pattern of subsidence and uplift, and in Bangladesh's unique sedimentation history. Weber *et al.* argue that detailed analysis of data from the Bengal fan – the world's largest – shows active growth during the most recent sea level rise and highstand (extended period of high sea level). They argue that this is a unique phenomenon that contradicts current models of delta formation (Weber *et al.* 1997: 318).

The regional history of the Bengal delta demonstrates the crucial significance of its unique features which continue to distinguish it from other great world deltas such as the Mississippi and Nile, with which comparisons are frequently sought. As a result, despite the continuing rapidity of subsidence in parts of the delta, the changes in relative sea level are quite different from those which are predicted under the global warming scenarios commonly assumed as the basis for impact assessment. The western section of the delta, close to the tectonic hinge line which runs broadly from south-west to north-east from Kolkata to Sylhet, is subsiding slowly or not at all, yet is the only part of the delta to have witnessed coastal recession, albeit on a marginal scale. The central and eastern part of the delta, notably the current mouths of the GBM rivers themselves, which, along with the tectonic trough of the Sylhet basin, is experiencing the most rapid subsidence in Bangladesh, is also experiencing the significant expansion of newly formed land as islands push out into the Bay of Bengal.

In a recent study of *char* lands, constantly forming islands of silt formed in the post-monsoon receding waters of the GBM system, de Wilde (2000) states that there is a net accretion of *char* land of approximately 1,600 ha per annum along the delta. The Dutch and Bangladesh governments inaugurated a joint development project on *char* land in 1978. 'Being convinced that ultimately social and economic benefits can be accrued from this accretion of land, the Governments of Bangladesh and the Netherlands initiated the Land Reclamation Project (LRP).' Subsequent publications from the Meghna Estuary Study project and Char Development and Settlement Project (CDSP, started in 1994) demonstrate at the local scale the agroeconomic significance of developing this new resource, created out of the annual supply of silt.

Analysis of the geomorphological history of Bangladesh demonstrates that eustatic sea level is unlikely to play more than a peripheral role in the foreseeable future of Bangladesh's coastal regions. Much of the coastline has remained stable or prograded with considerably higher rates of relative sea level change than those currently predicted under the global warming scenarios. The regional complexity of human adaptations to the delta environment complicates still further attempts to establish regional climate impacts from global climate change.

From environment to development

Brammer (1993, 1997) has shown how detailed variations in microrelief, soils and flooding in Bangladesh all have a bearing on patterns of land use in a highly dynamic relationship between human use and the environment. The significance of the dynamic nature of this relationship cannot be overemphasised, just as the search for single cause and effect relationships cannot do justice to the processes of change. Conway has argued that 'successful agricultural and rural development require the genuine integration of a wide range of skills and knowledge, ranging from anthropology to entomology. Bringing such varied disciplines together in an efficient and productive way to produce a common agreement on worthwhile action is an enormous challenge' (Conway 1990: 12).

It is here that the greatest problem lies in making global climate modelling relevant to understanding regional agricultural change. The last decade has seen enormous continuing interest in agricultural development in Bangladesh, but although the environment has come on to the agenda of the World Bank, the Asian Development Bank and other international agencies, the links between the global scale agenda and regional environmental issues have not yet been successfully established. Economists, anthropologists, political scientists and geographers have all brought different perspectives to bear on what is widely interpreted as 'the development challenge'. In the global development context Simon has recently pointed to the surprising lack of coverage on environment–development issues in texts devoted to development in the South, especially given the importance of the environment as one of the leading development topics of the 1980s and 1990s (Simon 1999: 41). Yet while the term 'environment' has played an increasingly prominent part in some of South Asian literature, there is no agreed basis on which the environment or environmental change can be modelled into alternative conceptualisations of agricultural development. Indeed, for much of the literature, 'the environment' is simply not an issue that enters the development agenda at all.

Fundamental to this problem is the recognition that environments are socio-cultural–economic constructs as much as they are objectively verifiable assemblages of scientifically observable phenomena. The problems of modelling are not simply those of complexity, data inadequacy and data inaccuracy, but of predicting the social, political and economic constructions and values which will be placed on 'the environment' through time.

Controls on agricultural productivity: regional diversity in Bangladesh

In view of the complexity of the environmental variables alone, discussed in detail by Brammer (2000), it is hardly surprising that much of the development literature pays scant regard to those features of global environmental change which are the centre of climate change scenario building. One example has to suffice to make the point that the regional and scale diversity of Bangladesh's agricultural system defies broad generalisation derived from global models.

In the mid-1990s Bangladeshi agriculture suffered an apparent stagnation in agricultural production, after a boom through the 1980s. However, as Brammer shows, broad trends have to be analysed at varying scales to be properly understood. Perhaps the most remarkable single feature of Bangladesh's agricultural system over the years since 1971 is the growth of *boro* season cropping. From under 2 per cent of the surface area cultivated in the dry winter season – the season of *boro* rice – the *boro* rice cropped area had risen to 2.8 million ha and winter wheat to nearly 810,000 ha, nearly 15 per cent of the total cultivated area. The *boro* is largely dependent on groundwater irrigation, and according to the Bangladesh Bureau of Statistics (2001) the yields of *boro* rice and wheat (2.82 and 2.25 t/ha in 1997–8 respectively) outstripped those of the *aman* rice (1.48 t/ha) and *aus* rice (1.19 t/ha). Something of this complexity is caught in Figure 3.3 which shows the relationship between the height of land below or above normal flood levels, the seasonality and the type of crop sown (Ahmad 2000: 8).

As Brammer (1997) has shown, the environmental constraints on each of the crops vary fundamentally, affecting every aspect of cultivation from crop types to use of high-yielding variety seeds, fertiliser and pesticide. Each of these questions involves farmer choices, mediated through the complexities of the social system, landownership and tenancy relations under pressures from continuing population growth, landlessness and fragmentation. Some of the complexity is captured in Brammer's maps of land suitable for different categories of crops, which take into account such variables as normal depth of flooding, soil type and degree of salinity (Figure 3.2).

Figure 3.3 shows that there is no straightforward relationship between the depth of flooding and productivity. Farmers in Bangladesh have adapted to regional and seasonal contrasts in flooding by a wide variety of different measures, including the recent rapid adoption of new technology such as high-yielding varieties of seed and both shallow and deep wells for irrigation. The factors which condition groundwater use range from their economic costs and benefits, through patterns of landholding and ownership to access to credit.

One of the greatest changes experienced in rural Bangladesh has been the opening up of new lines of development micro-credit pioneered by the Grameen Bank, which has sought thereby to stimulate a pattern of development though which the poorest section of society can be empowered. But credit itself is only one of many variables which have had a bearing on access to and development of key resources, including such macro-level

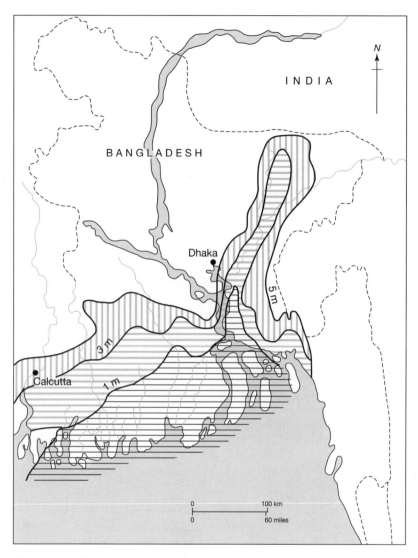

Figure 3.2 Predicted inundation of Bangladesh under varying sea
level rise scenarios
Source: after Houghton, J. (1997).

strategic issues as the geopolitics of resource sharing and development
between India, Bangladesh and Nepal (Bradnock and Saunders 2000).
Each of these variables may be expected to have a far more significant
impact on agricultural productivity, cropping patterns, sustainable development and the reduction of poverty than putative long-term changes in
eustatic sea level.

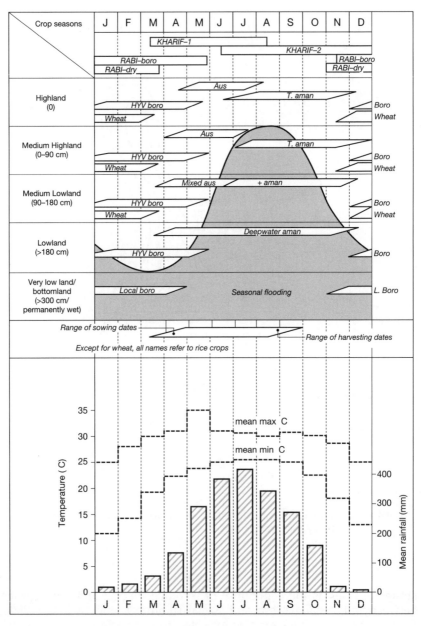

Figure 3.3 Land flooding, seasonality and cereal cropping patterns
Source: Brammen (2000: 24).

Demographic change: growth and urbanisation

Into the decision-making framework through which agricultural change takes place, comes the fundamental importance of demographic change. Population continues to grow. Despite a fall in the crude birth rate (CBR) from 36.9 to 21.4 per thousand between 1963 and 1998 the fall in the crude death rate (CDR) from 18.5 to 4.8 per thousand over the same period ensured that the overall population growth remained at 1.8 per cent per annum in 2001.

Meanwhile the distribution of the population is undergoing fundamental change. By 1999 urbanisation for Bangladesh as a whole had reached 25 per cent of the total population, having been under 8 per cent in 1974, while Dhaka reached over 9 million by 1998, and was expected to be as much as 16 million by 2015 (Islam 1999). The degree to which such demographic processes expose the population to changed climatic risk depends crucially on factors other than climate change per se, including long-term changes in technology, living patterns, housing and economic activity.

In this great complex of variables it is impossible to discern any measurable or predictable impact of global environmental change on the potential for further agricultural development. One illustration has come from the joint Dutch and Bangladesh governments' project on *char* land settlers, which has highlighted the importance of the qualities of the *char* settlers themselves as a key variable in the development prospects of the newly formed islands. They have found that the settlers are not homogeneous, that they have a low level of education and strong patron/client relationships. They are caught in an 'interlocking pattern of powerlessness, vulnerability, isolation, physical weakness and poverty' all characteristic of 'the poor' but in greater measure among *char* settlers. Unsurprisingly, these social and economic characteristics are considered to play a vital role in the prospects of *char* land development. It may seem surprising that susceptibility to storm surges, cyclones and sea level rise play so little part in their analysis of vulnerability, yet in the short term at least the experience of the last 20 years suggests that infrastructural measures including raising protected refuges and early warning can dramatically reduce the loss of life which has been the traditional hallmark of such events.

The political economy of agricultural change

It is striking that to date despite continuing extensive research into the agro-economics and political economy of development in Bangladesh no serious attempts have been made to build in the impact of sea level rise in Bangladesh to these analyses. Rogaly, Harriss-White and Bose's analysis of the political economy of agricultural growth in Bangladesh, for example, focuses entirely on the role of political economy – landownership, agrarian structure, local political institutions, migration, markets – as a key factor in shaping the pattern and nature of agro-economic development (Rogaly *et al.* 1999). Through an exploration of the continuing role of land, class, institutional politics and agrarian structures, they demonstrate the wide range of development paths which flow from alternative policy decisions and make only one fleeting reference to environmental 'threats', an indication that in the view of contemporary political science the environment is low down in the order of variables influencing development.

The World Bank has explored some of the economic dimensions of modelling environmental change into policy planning measures. In approaching the question 'What constitutes sustainable agricultural development?' Hazell and Lutz (1998: 9) have argued that 'continued agricultural growth is a necessity, not an option, for most developing countries...', but they then propose economic indicators that make no reference to changes in the physical environment itself, except in so far as this affects the value of the capital stock. They argue that 'Empirical valuation work in developing countries at the conjunction of agriculture and the environment is still quite limited' (p. 11). In the absence of both the concepts and the tools for such environmental analysis, they highlight issues of macroeconomic policy, and the design of incentive schemes to induce farmers and communities to factor important externality costs or benefits into their own resource management decisions, concluding that 'as a general finding, undistorted macroeconomic and sectoral policies tend to provide better incentives from an environmental perspective than highly distorted policies. Subsidies for agricultural inputs tend to be particularly costly in terms of economic efficiency, government budgets and the environment' (World Bank 1997: Chapter 4: 11).

Agricultural productivity changes in Bangladesh

While the proclamations of desertification and soil degradation as a direct result of the adoption of Green Revolution technologies, the impoverishment of soil fertility, the drying up of rivers and the incidence of devastating floods and storms are a recurrent theme of environmental commentators, the statistics of agricultural production in Bangladesh suggest that, in the face of all expectations over the last 30 years, agricultural productivity has more than kept pace with growing demand for food and agricultural products.

In its comprehensive review of agricultural development in Bangladesh in the twenty-first century, the World Bank picks on four key areas in which reform is needed for the momentum of agricultural development to be maintained:

(i) strengthening private sector trade by permanently abolishing relevant restrictions; (ii) making the price stabilization mechanism more efficient by relying on international trade (both public and private) in place of public storage; (iii) targeting the food insecurity and removing untargeted subsidies that are now captured, at least partly, by the affluent; and (iv) making public food operations more efficient by withdrawing the government from commercial activities (Faroque 1998).

While this prescription makes scant reference to environmental constraints to agricultural growth, elsewhere the report focuses extensively on what it terms the 'costs of resource degradation'. Deforestation, water depletion and degradation, salinisation, and fish stock depletion are among the chief variables highlighted. Yet all of these ills can be explained far more readily in terms of current trends in resource use within the region than in terms of global climate or sea level change outside it.

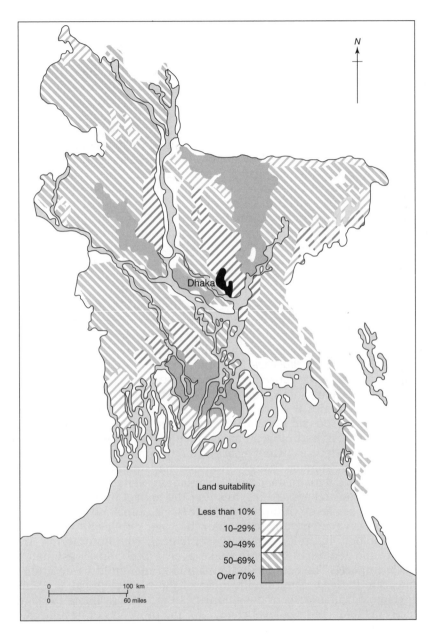

Figure 3.4 Bangladesh – land suitability for IRRI (HYV) *boro* and
transplanted *aus* rice

The difficulty of establishing clear relationships between environmental pressure and productivity is recognised by the report which highlights the regional contrasts in agricultural performance experienced in Bangladesh. 'The growth rate of rice production at the national level conceals strikingly large variations across regions. More importantly, the growth points have shifted from one period to another. While all the regions have had at least medium growth in one period or another, the early-starters have generally lagged behind other regions in the later periods. The exhaustion of easy sources of irrigation is the likely reason why production growth at the regional level has not been sustained over prolonged periods. But the explanation of this may also lie partly in the hypotheses mentioned above regarding the agronomic constraints to intensified rice cultivation.

Agricultural productivity has shown remarkable increases since independence in 1971, far higher than was believed achievable at the time. It is striking that in his evaluation of environmental stresses on agriculture in Bangladesh, Brandon identifies a series of environmental issues which are impacting adversely on both actual and prospective agricultural productivity. Yet every one of the environmental constraints identified is operating at the regional or local scale, and can be best addressed through policy decisions at those levels (Brandon 2001).

Conclusions

The emphasis on global scale processes has transformed much regional and national scale study from a hypothesis-testing scale in its own right to a scale from which 'such as' examples can be drawn to illustrate the damage assumed to be caused by changes at the global level. The effects of global level change in climate may be expected to vary very significantly at the regional level. It is clear that underlying the apparent simplicity of the Bengal delta's surface lies an enormous complexity of geomorphological, hydrological and geological variables, each of which has a role in the delta's current agricultural dynamics, yet many of which are unrelated to climate change. Brammer has demonstrated how 'the complexity and dynamism of the Bangladesh delta present immense problems to researchers seeking to study the impact in Bangladesh of a slowly rising sea level' (Brammer 1993: 261). He points out the extremely high cost of even attempting to monitor the range of relevant variables of geomorphology, hydrology, salinity, settlement, communications, land use and land tenure in order to make even a preliminary estimate of physical impacts. The complexity is magnified manyfold by an attempt to integrate these dynamics with those of the socio-political, economic and technological aspects of development in order to evaluate the significance of climate and sea level change for the prospects of development in Bangladesh.

Changes in the objective characteristics of regional climates are still inherently wholly unpredictable and unpredicted. But the value put on those characteristics is subject to a still wider range of influences that are themselves not susceptible to modelling. It is therefore impossible to determine what the costs or benefits of those changes at a regional scale will be. The Bangladeshi experience demonstrates that even apparently 'simple' environments are immensely complex. They show a constant pattern of adjustment to a highly dynamic context. The human use of that environment is also subject to its own complex dynamics. The definition of the 'relevant environment' is in part at least a political project, and conceptualisations of environmental degradation at the global level have to be interpreted as a part of that global political project rather than as a reflection of specified and understood relationships between scientifically observed links between global scale environmental phenomena and regional or national scale environments or the human activity so interdependent on those environments.

In the field of climate change alone, these three factors have contributed to creating a cascade of uncertainty in extending global climate change models both from the global to the regional, a cascade which becomes a deafening roar of 'noise' when attempts are made to extend the model beyond climate change to other environmental variables. There is simply no evidence that global scale climatic change or related environmental changes have any predictable implications for Bangladesh's environment and for the challenges posed by the enormous complexity of those environments for development. Klein and Nicholls (1999: 185) have argued that 'The future is, by definition, uncertain, so plausible scenarios must be constructed to assess these potential impacts. It is important to consider uncertainty by constructing more than one scenario for each relevant parameter, thus allowing for sensitivity analysis.' *A fortiori*, attempts to draw global development policy implications from highly simplified assumptions of the presumed regional effects of climate change in Bangladesh have to be rejected.

References

Acharyya, S.K. (1998) 'Break-up of the Greater Indo-Australian continent and accretion of blocks framing South and East Asia', *Journal of Geodynamics*, **26**(1): 149–70.

Ahmad, Q.K. (ed) *et al.* (2000) *Bangladesh Water Vision 2025: Towards a Sustainable Water World*. Bangladesh Water Partnership, Dhaka.

Alam, Mustafa and **Koudstaal, R.** (eds) (2000) *Riverine Chars in Bangladesh: Environmental dynamics and management issues*. University Press, Dhaka.

Alam, M. and **Samad, A.** (1996) 'Subsidence of the Ganges–Brahmaputra delta and impact of possible sea level rise on the coastal area of Bangladesh', *Dhaka University Journal of Science*, **44**(2): 179–91.

Allison, M.A. (1998) 'Historical changes in the Ganges–Brahmaputra delta', *Journal of Coastal Research*, **14**(4): 1269–75.

Bagchi, K. (1944) *The Ganges Delta*. Calcutta.

Bangladesh Bureau of Statistics (2001).

Bird, E.C.F. (1993) *Submerging Coasts: The effects of rising sea level on coastal environments*. John Wiley, Chichester.

Bradnock, R.W. (1989) *Agricultural Change in South Asia*. John Murray, London.

Bradnock, R.W. and **Saunders, P.** (2000) 'Sea-level rise, subsidence and submergence: the political ecology of environmental change in the Bengal delta', in P.A. Stott, and S. Sullivan, (eds), *Political Ecology: Science, myth and power*. Edward Arnold, London, Chapter 3.

Brammer, H. (1990a) 'Floods in Bangladesh 1: Geographical background to the 1987 and 1988 floods', *The Geographical Journal*, **156**(1), March: 12–22.

Brammer, H. (1990b) 'Floods in Bangladesh 2: Flood mitigation and environmental aspects', *The Geographical Journal*, **156**(2), July: 158–65.

Brammer, H. (1993) 'Geographical complexities of detailed impact assessments for the Ganges–Brahmaputra–Meghna delta of Bangladesh', in R.A. Warrick, E.M. Barrow and T.M.L. Wigley (eds) *Climate and Sea Level Change*. Cambridge University Press, Cambridge: 246–62.

Brammer, H. (1997) *Agricultural Development in Bangladesh*. Dhaka, University Press.

Brammer, H. (2000) *Agroecological Aspects of Agricultural Research in Bangladesh*. Cambridge University Press, Cambridge.

Brammer, H., Asaduzzaman, M. and **Sultana, P.** (1996) 'Effects of climate and sea-level changes on the natural resources of Bangladesh', in R.A. Warrick and Q.K. Ahmad (eds), *The Implications of Climate and Sea-level Change for Bangladesh*. Kluwer Academic, Dordrecht, Chapter 4.

Brandon, C. (2001) *Bangladesh: climate change and sustainable development*. World Bank, Washington, D.C.

Broadus, J.M. (1993) 'Possible impacts of, and adjustments to, sea level rise: the cases of Bangladesh and Egypt', in R.A. Warrick, E.M. Barrow and T.M.L. Wigley (eds) *Climate and Sea Level Change*. Cambridge University Press, Cambridge.

Carter, R.W.G. (1988) *Coastal Environments*. Academic Press, London.

Carter, R.W.G. and **Woodroffe, C.D.** (eds) (1994) *Coastal Evolution: Late Quaternary shoreline morphodynamics*. Cambridge University Press, Cambridge.

Carter, T.R., Parry, M.C., Nishioka, S. and **Harasawa, H.** (eds) (1994) *Technical Guidelines for Assessing Climate Change Impacts and Adaptations*. Report of Working Group II of the Intergovernmental Panel on Climate Change. University College London and Centre for Global Environmental Research, London and Tsukuba.

Conway, G.R. (1990) 'Agriculture and the environment: concepts and issues', in S. Huq., A. Rahman and G.R. Conway (eds) *Environmental aspects of agricultural development in Bangladesh*. University Press, Dhaka.

Eckholm, E.P. (1976) *Losing Ground: Environmental stress and world food prospects*. Norton for World Watch Institute, New York.

Fairhead, J. and **Leach, M.** (1996) *Misreading the African Landscape. Society and ecology in a forest–savanna mosaic*. Cambridge University Press, Cambridge.

Faroque, R. (ed) (1998) *Bangladesh agriculture in the 21st Century*. University Press, Dhaka.

Fergusson, J. (1863) 'Recent changes in the delta of the Ganges', *Proceedings of the Geological Society*, London: 322–54.

Gasse, F. *et al.* (1991) 'A 13,000 year climate record from Western Tibet', *Nature*, **353**: 742–5.

Geddes, A. (1925) *Au Pays de Tagore: La civilisation rurale de Bengale occidentale et ses facteurs géographiques.* Colin, Paris.

Geddes, A. (1960) 'The alluvial morphology of the Indo-Gangetic Plain: its mapping and geographical significance', *Transactions and Papers of Institute of British Geographers*, **28**: 253–76.

Geddes, A. (1982) *Man and Land in South Asia.* Concept, New Delhi.

Goodbred, S.L. Jr and **Kuehl, S.A.** (1998) 'Floodplain accretion and sedimentary processes in the Bengal Basin: new application of 137Cs and 210Pb geochronologies in heterogeneous environments', Poster summary from 1988 Fall meeting of the American Geophysical Union published as a supplement to *Eos, Transactions AGU*, **79**(45), 10 November.

Goodbred, S.L. Jr and **Kuehl, S.A.** (1999) 'Holocene and modern sediment budgets for the Ganges–Brahmaputra river system: evidence for highstand dispersal to flood-plain, shelf and deep-sea depocenters', *Geology*, **27**: 559–62.

Goodbred, S.L. Jr and **Kuehl, S.A.** (2000) 'The significance of large sediment supply, active tectonism, and eustasy on margin sequence development: Late Quaternary stratigraphy and evolution of the Ganges–Brahmaputra delta', *Sedimentary Geology* **133**: 227–48.

Goudie, A.S. (1979) *Environmental Change.* Oxford University Press, London.

Goudie, A.S. (1981) *The Human Impact.* Basil Blackwell, Oxford.

Government of Bangladesh (1999) *1998 Statistical Yearbook of Bangladesh.* Bangladesh Bureau of Statistics, Statistics Division, Ministry of Planning, Dhaka.

Government of Bangladesh (2000) *Statistical Pocketbook of Bangladesh.* Bangladesh Bureau of Statistics, Dhaka.

Government of Bangladesh Bureau of Statistics (2000) *1998 Statistical Yearbook of Agricultural Statistics of Bangladesh.* Statistics Division, Ministry of Planning, Dhaka.

Grainger, A. (1993) 'Rates of deforestation in the humid tropics: estimates and measurements', *Geographical Journal*, **159**(1): 33–44.

Haggett, P. (2001) *Geography: A global synthesis.* Pearson, London.

Hanebith, T., Stattegger, K. and **Grootes, P.M.** (2000) 'Rapid flooding of the Sunda shelf: a late glacial sea-level record', *Science*, **288**: 1033–5.

Haq, B.U. (1994) *Sea Level Rise and Coastal Subsidence: Rates and threats.* The World Bank, Washington.

Hazell, P. and **Lutz, E.** (1998) 'Integrating environmental and sustainability concerns into rural development policies', in E. Lutz, (ed.) *Agriculture and the Environment.* The World Bank, Washington, D.C.: 9–21.

Hofer, T. (1998) *Floods in Bangladesh: A highland–lowland interaction?* Geographia Bernensia for the University of Berne, Berne, G 48.

Hofer, T. and **Messerli, B.** (eds) (1998) *Floods in Bangladesh, History Processes and Impacts.* Geographia Bernensia for the University of Berne, Berne.

Homewood, K.M. and **Brockington, D.** (1999) 'Biodiversity, conservation and development in Mkomazi, Tanzania', *Global Ecology and Biogeography Letters.*

Homewood, K.M. and **Rogers, W.A.** (1987) 'Pastoralism, conservation and the overgrazing controversy', in D. Anderson and R. Grove (eds), *Conservation in Africa: People, policies and practice*. Cambridge University Press, Cambridge.

Hoshina, M. (1998) *The Expanding Earth: Evidence, causes and effects*. Tokyo University Press, Tokyo.

Houghton, J.T. (1997) *Global warming: The complete briefing*, 2nd edn. Cambridge University Press, Cambridge.

Huq, S., Rahman, A.A. and **Conway, G.R.** (eds) (1990) *Environmental Aspects of Agricultural Development in Bangladesh*. Dhaka University Press, Dhaka.

Intergovernmental Panel on Climate Change (1990) *Global Climate Change – the IPCC scientific assessment*. Cambridge University Press, Cambridge.

Intergovernmental Panel on Climate Change (1992) *Global Climate Change and the Rising Challenge of the Sea*. Coastal Zone Management Subgroup of the IPCC Response Strategies Working Group, The Hague.

IPCC CZMS (1992) 'A common methodology for assessing vulnerability to sea-level rise', 2nd revision in *Global Climate Change and the Rising Challenge of the Sea*. IPCC CZMS, Ministry of Transport, Public Works and Water Management, The Hague, The Netherlands, Appendix C.

Islam, N. (1999) 'Dhaka City: some general concerns', in N. Hamid and S. Huq (eds), *Asian Cities in the 21st Century: Contemporary approaches to municipal management*, vol. 3, Asian Development Bank, Manila: 71–82.

Ives, J.D. and **Messerli, B.** (1989) *The Himalayan Dilemma: Reconciling development and conservation*. Routledge, London and New York.

Jacobsen, J.L. (1988) *Environmental Refugees: A yardstick of habitability*. Worldwatch Paper 86, World Watch Institute, Washington, DC.

Jelgersma, S. and **Tooley, M.J.** (2000) 'Sea level changes during the recent past', Chapter 19 in *Holocene Cycles: Climate, sea levels, and sedimentation*, **17**: 123–39.

Jewitt, S. and **Kumar, S.** (2000) 'A political ecology of forest management: gender and silviculture knowledge in the Jharkhand, India', in P.A. Stott and S. Sullivan (eds), *Political Ecology: Science, myth and power*. Edward Arnold, London, Chapter 4.

Johnson, B.L.C. (1982) *Bangladesh*, 2nd edn. Heinemann, London.

Kausher, A., Kay, R.C., Asaduzzaman, M. and **Paul, S.** (1996) 'Climate change and sea-level rise: the case of the coast', in R.A. Warrick and Q.K. Ahmad *The Implications of Climate and Sea-level Change for Bangladesh*. Kluwer Academic Publishers, Dordrecht, Chapter 7.

Khan, T.M.A., Singh, O.P. and **Rahman, Md. S.** (2000) 'Recent sea level and sea surface temperature trends along the Bangladesh coast in relation to the frequency of intense cyclones', *Marine Geodesy*, **32**: 103–16.

Klein, R.J.T. and **Nicholls, R.J.** (1999) 'Assessment of coastal vulnerability to climate change', *Ambio*, **28**(2), March: 182–7.

Kudrass, H.R., Michels, K.H., Wiedicke, M. and **Suckow, A.** (1998) 'Cyclones and tides as feeders of a submarine canyon off Bangladesh', *Geology*, **26**(8): 715–18.

Kuehl, S.A. and **Goodbred, S.L. Jr** (1998) 'Sequence development along the Bengal margin during rapid sea-level rise: evidence for extreme high sediment discharge from the Ganges–Brahmaputra', Poster summary from 1988 Fall meeting of the American Geophysical Union published as a supplement to *Eos, Transactions AGU*, **79**(45), 10 November.

Kuehl, S. A., Levy, B.M., Moore, W.S. and **Allison, M.A.** (1997) 'Subaqueous delta of the Ganges–Brahmaputra river system', *Marine Geology*, **144**: 81–96.

Lowenthal, D. (ed) (1965) *Man and Nature*. Bell Knap Press, Cambridge, MA.

Lutz, E. (ed.) (1998) *Agriculture and the Environment*. World Bank, Washington, D.C.

McKie, R. and **Morris, P.** (2001) 'The world in 2050', *The Observer*, 1 April: 25.

Messerli, B., Grosjean, M., Hofer, T., Nunez, L. and **Pfister, C.** (2000) 'From nature dominated to human dominated environmental changes', *Quarternary Science Reviews*, **19**: 459–79.

Milliman, J.D., Broadus, J.M. and **Gable, F.** (1989) 'Environmental and economic implications of rising sea level and subsiding deltas: the Nile and Bengal examples', *Ambio*, **18**(6): 340–5.

Milliman, J.D. and **Syvitski, P.M.** (1992) 'Geomorphic/tectonic control of sediment discharge to the ocean: the importance of small mountainous rivers', *Journal of Geology*, **100**: 525–44.

Monbiot, G. (1999) 'Apocalypse now', *The Guardian*, 29 July.

Morgan, J.P. and **McIntyre, W.G.** (1959) 'Quaternary geology of the Bengal Basin, East Pakistan and India', *Bulletin of the Geological Society of America*, **70**: 319–42.

Mörner, N.-A. (1976) 'Eustasy and sea-level changes', *Journal of Geology*, **84**: 123–51.

O'Riordan, T. (1976) *Environmentalism*. Pion, London.

Paddock, W. and **Paddock, P.** (1967) *Famine 1975! America's decision: Who will survive?* Little, Brown and Co, Boston and Toronto.

Parry, M., Rosenzweig, C. Iglesias, A. Fischer, G. and **Livermore, M.** (1999) 'Climate change and world food security: a new assessment', *Global Environmental Change*, **9**: S51–67.

Rennell, J. (1781) 'An account of the Ganges and Burrampooter rivers', *Philosophical Transactions*, **71**, 87–114 (Subsequently added as an appendix to successive editions of the Memoir).

Rennell, J. (1792) *Memoir of a Map of Hindoostan, or the Mogul Empire*, 2nd edn. Nichol and Richardson, London.

Rogaly, B., Harriss-White, B. and **Bose, S.** (eds) (1999) *Sonar Bangla? Agricultural growth and agrarian change in West Bengal and Bangladesh*. Sage, New Delhi.

Saunders, P.L. (2000) 'Environmental refugees: the origins of a construct', in P.A. Stott and S. Sullivan, (eds), *Political Ecology: Science, myth and power*. Edward Arnold, London, Chapter 10.

Shiva, V. (1988) *Staying Alive: Women, ecology and survival in India*. Kali for Women, New Delhi.

Simon, D. (1999) 'Development revisited: Thinking about, practising and teaching development after the Cold War', in Simon and Närman (eds) *Development as Theory and Practice*. Longman/DARG, London.

Spate O.H.K. and **Learmonth A.T.A.** (1968) *India and Pakistan*, 2nd edn. Methuen, London.

Stanley, D.J. and Warne, A.G. (1994) 'Worldwide initiation of Holocene marine deltas by deceleration of sea-level rise', *Science*, **265**, 8 July.

Stott, P.A. and Sullivan, S. (eds) (2000) *Political Ecology: Science, myth and power.* Edward Arnold, London.

Sullivan, S. (2000) 'Getting the science right, or introducing science in the first place? Local "facts", global discourse – "desertification" in north-west Namibia', in P.A. Stott and S. Sullivan (eds), *Political Ecology: Science, myth and power.* Edward Arnold, London, Chapter 1.

Suter, J.A. (1994) 'Deltaic coasts', in R.W.G. Carter and C.D. Woodroffe (eds), *Coastal Evolution.* Cambridge University Press, Cambridge.

Thomas, D.S.G. (1993) 'Storm in a tea cup? Understanding desertification', *Geographical Journal*, **159**(3): 318–31.

Thomas, D.S.G. (1994) *Desertification: Exploding the myth.* Wiley, London.

Tooley, M.J. (1993) 'Long term changes in eustatic sea level', in R.A. Warrick, E.M. Barrow and T.M.L. Wigley (eds), *Climate and Sea Level Change: Observations, projections and implications.* Cambridge Univeristy Press, Cambridge, Chapter 6.

Uddin, A. and Lundberg, N. (1998a) 'Cenozoic history of the Himalayan–Bengal system: sand composition in the Bengal Basin, Bangladesh', *GSA Bulletin*, **110**(4): 497–511.

Uddin, A. and Lundberg, N. (1998b) 'Unroofing history of the eastern Himalaya and the Indo-Burman ranges: heavy mineral study of Cenozoic sediments from the Bengal Basin, Bangladesh', *Journal of Sedimentary Research*, **68**(3): 465–72.

Uddin, A. and Lundberg, N. (1999) 'A Paleo-Brahmaputra? Subsurface lithofacies analysis of Miocene deltaic sediments in the Himalayan–Bengal system, Bangladesh', *Sedimentary Geology*, **123**: 239–54.

Umitsu, M. (1987) 'Late Quaternary sedimentary environment and landform evolution in the Bengal lowland', *Geographical Review of Japan*, Series **B**(2): 164–78.

Umitsu, M. (1993) 'Late Quaternary sedimentary environment and landforms in the Ganges Delta', *Sedimentary Geology*, **83**: 177–86.

Warrick, R.A. (1993) 'Sea level rise', *Earthquest*, Spring, Science Capsule, Office of Interdisciplinary Earth Studies. Boulder, CO.

Warrick, R.A. and Ahmad, Q.K. (1996) *The Implications of Climate and Sea-level Change for Bangladesh.* Kluwer Academic Publishers, Dordrecht.

Warrick, R.A., Barrow, E.M. and Wigley, T.M.L. (eds) (1993) *Climate and Sea Level Change.* Cambridge University Press, Cambridge: 263–75.

Warrick, R.A., Bhuiya, A.H. and Mirza, M.Q. (1994) *The Greenhouse Effect and Climate Change.* Briefing document No. 1, Bangladesh Unnayan Parishad, Dhaka.

Warrick, R.A. and Rahman, A. (1992) 'Future sea level rise: environmental and socio-political considerations', in I.M. Mintzer (ed.), *Confronting Climate Change: Risks, implications and responses.* Cambridge University Press, Cambridge: 97–112.

Weber, M.E., Wiedicke, M.H., Kudrass, H.R. Hübscher, C. and Erlenkeuser, H. (1997) 'Active growth of the Bengal Fan during sea-level rise and highstand', *Geology*, **25**(4), April: 315–18.

Wigley, T.M.L. and **Raper, S.C.B.** (1992) 'Implications for climate and sea-level of revised IPCC emission scenarios', *Nature*, **357**: 293–300.

Wilde, K. de (ed.) (2000) *Out of the Periphery: Development of coastal chars in southeastern Bangladesh*. University Press, Dhaka.

Williams, G. (1999) 'Panchayati Raj and the changing micro-politics of West Bengal', in B. Rogaly, B. Harriss-White and S. Bose (eds), *Sonar Bangla? Agricultural growth and agrarian change in West Bengal and Bangladesh*. Sage, New Delhi, Chapter 8.

World Bank (1994) *World Development Report*. World Bank, Washington, D.C.

World Bank (1997) *World Development Report*. World Bank, Washington, D.C.

Worm, H.-U., Ahmed, A.M.M., Ahmed, N.U., Islam, H.O.O., Haq, M.M., Hamback, H.U. and **Lietz, J.** (1998) 'Large sedimentation rate in the Bengal Delta: magnetostratigraphic dating of Cenozoic sediments from northeastern Bangladesh', *Geology*, **26**(6), June: 487–90.

CHAPTER 4

Changing places
The roles of science and social science in the development of large-scale irrigation in South Asia

Graham P. Chapman

Introduction

Society and water are inextricably intertwined. It is even possible to conceive of society in terms of how it modifies and adapts the hydrological cycle. There is no substitute for water, and without it human beings die in days, and, after varying periods of deprivation, so do their crops and animals. In the industrialising state water provides energy for mills and turbines, the means of sanitation for the inhabitants, and the principal medium of countless industrial applications. As the level of development of nations has increased, so has their demand on water, and so has the scale of the harnessing of local and regional water resources.

This topic is so huge that whole institutions, vast quantities of legislation, innumerable books and journals are dedicated to it. Exactly the same can be said of the topic of this chapter, which is just is one aspect of water management – namely large-scale irrigation in South Asia. This chapter is bound, therefore, to be sketchy and incomplete. However, what I hope to do is to show how the development and management of water resources on a large scale are bound up not just with the political but also the intellectual climate of each era.

Pre- and post-Darwinian world views

Colonel Probey Cautley began the construction of the Ganges Canal (see below) some 160 years ago, 19 years before the publication of Darwin's *Origin of Species*. In Cautley's day the dominant world view was one of design. God was the greatest designer, who had designed the species of plants and animals. But He had also endowed the men of advanced societies with the capacity themselves to design – new machines and new weapons being among the products of such civilisations.

The idea of evolutionary change was not part of this world view. The explanation of how these things worked was mechanistic, and based on deterministic formulae. In the management of water and irrigation, accuracy was taken to its extreme:

> Other branches of engineering admit of some latitude, and even of errors, without disaster. Thus a mistake of 10 feet in 100 miles of railway levels was recently detected in Bengal, though no injury had resulted from the blunder. With water the mistake would at least have been registered with the utmost nicety to the fraction of an inch the first day the canal was opened (Deakin 1893: 228).

We accept now that mechanistic explanation is not the only means of analysis and association, because of the work of Darwin's relative, Francis Galton, and his pupil, Karl Pearson. One of the results of their work on the statistical analysis of hereditary characteristics was the formulation of the product moment correlation coefficient. From then on things could be shown to be at least related in some degree, though without saying anything about the arrow of causation. Pearson's claim can be seen in these days of bovine spongiform encephalopathy (BSE) and Creutzfeldt–Jakob disease (CJD) crises to have been quite prophetic:

> Formerly the quantitative scientist could only think in terms of causation, now he can think also in terms of correlation. This has not only enormously widened the field in which quantitative and therefore mathematical methods can be applied, but it has at the same time modified our philosophy of science and even of life itself (Forrest 1974: 199 citing Pearson 1930: III, 24).

The Gaussian – or 'normal' bell-shaped – distribution had been known for centuries – but its purpose was to aid in a proper estimation of a 'true' parameter. Deviation was error. For many engineers deviation is still error. But for many social scientists and evolutionary theoreticians, deviation is not error but needed and valued variation.

There are many more ways with many more elaborations that can be used to show the difference between thinking 'then' and 'now' – but these two strands are good enough for the moment, and underlie many modern disputes.

On the one hand there is the engineers' world of design, in which evolution plays no part, and in which mechanistic and reductionist principles apply. Historically, in colonial India this view fitted hand in glove with the background of the military engineers who started the first big schemes. Their *modus operandi* was command and control.

On the other hand there is the world of evolutionary change and adaptation, and a world in which the evidence of probabilities is admissible. It was a world which increasingly began to concern the colonial district officers, who tried to raise land revenue for the canal schemes, and it is the world of the modern anti-dam protesters, who show that over many schemes the correlation between the improvements predicted and those achieved are poor.

We have to consider how the relationship between these two approaches is worked out. The engineers' mechanistic world tends to work through correspondence – the theories and formulae are tested against empirical circumstance in replicable experiments. By contrast, the historical worlds of correlation tend to work more with coherence – checking belief among a community of believers to see how reasonable or not the views may be.[1] The level of coherence possible is to some extent dependent on the capacity of communications to distribute ideas to groups of people. In the nineteenth century communications capacities were small, and the information transmitted instantly was usually of commercial or military value. Scientific information transmitted intergenerationally by education was slanted more to the inculcation of general principles than the exceptions and failures in their application. But now the capacity of members of society to communicate with each other through vernacular press, through radio, through telephones, by post, and through the Internet, is such that there are communities of critics who can point to the many cases in which schemes do not do what they are supposed to do – sufficient that the variation cannot simply be dismissed as a few mistakes or errors.

The significance of the mass media in the age of the risk society has been taken up by Allan *et al.* (2000) among others, citing heavily the work of Beck (1995), who writes a foreword to their book. The thesis of the risk society is often illustrated with reference to the risks modern society manufactures, which have no known ultimate impact, and perhaps no ultimately knowable impact. BSE and genetically modified crops are cited as examples. But the scale of large irrigation projects in South Asia has also grown to the point where they too seem to have neither ultimate known nor ultimately knowable impacts. The institutions which take risky decisions bear the hallmarks of Beck's analysis. Risks are uncovered not within the institutions, but outside them, typically by the media as the eye of the public. Even when exposed, the institutions engage in the politics of denial, or of small incremental acknowledgements only. Human error is usually the only explanation – leaving open the possibility that things could have and should have gone (and will next time go) right, and no one is made liable. The degree of exposure by the media is, of course, variable across cultures and types of states. The Soviet system is often cited as having lost credibility over issues such as Chernobyl, precisely because the media were muzzled, thereby permitting excessive risk taking, with suicidal implications for the political system.

A fable of design and enactment

In theory – but in theory only – it is possible to conceive of a large irrigation system with the following component parts. This description is deliberately top-down, starting with the water source.

The system is designed and built according to scientific and engineering principles, which are ahistorical and of universal applicability, and using quantitative field data. River discharges are measured. The discharge may be stored behind large dams, and the capacity of the reservoir can be calculated in advance. Both in such schemes and in cases where there is no dam, a barrage (weir) of calculated height on the river may divert calculated quantities of water through the headwork regulators into the main canal. The canal is designed with a slope (fall) resulting in a velocity of flow such that there is neither siltation nor erosion (this is known as a mean stable regime). Fixed cross-structures can control the fall where necessary. The water is let through regulators into branch canals; from branch canals into distributary canals, and from those into minors. From these it passes through further regulators (in some cases fixed, in some variable gates) to the *chak*, or field command. Here a group of farmers use field channels to water their fields in rotation. Large systems may have main canals hundreds of miles long, and command areas covering hundreds of thousands of farmers, and millions of people. But in principle the total size of command area is known in advance, and the impact of water on agricultural productivity within it is also predictable. The beneficial results are calculated in terms of gains in crop productivity, of change in crop type from lower to higher value, of increasing cropping intensity (growing in more seasons). The building of such large schemes really took off under British administration in the nineteenth century particularly in Punjab in north-west India (now an area divided between India and Pakistan), and has continued with massive government support since Independence in India, Pakistan and Sri Lanka, but to a lesser extent in Bangladesh and Nepal. Rarely, if ever, do these schemes fulfil their promises. In particular, in scheme after scheme, the distribution system is never quite finished as intended, and the water supply at the tail end is much less and more irregular than intended. In consequence, the farmers in the tail reaches invest less in field levelling, etc. The zonation of crops is overruled – and water-intensive crops ('wet-irrigated', like rice) may be grown near the head when the plans called for 'dry-irrigated' crops like wheat.

How these divergences come about can be illustrated by means of a story (see Figure 4.1) about another problem of rural development. In Villagestan, the government was aware that the farmers were not able to maximise their wealth and their contribution to the GDP because rural infrastructure – particularly transport – was inadequate. They summoned a transport expert. He realised that the farmers would benefit greatly from a vehicle he called Co-op Lorry, that the farmers would be able to use in rotation to take their goods to market. He showed a drawing (picture 1) of his Co-op Lorry to the government, with his cost estimates. The government was impressed and sanctioned the project. The farmers in the countryside heard rumours about some improvements to transport – but they had heard rumours before and were not greatly expectant. The expert set up a project shed, and his team set to work on producing the complex and sophisticated core – the engine. Having finished it (picture 2), they put it outside while the shed was cleared to make the chassis and body. A nearby farmer (picture 3) saw the engine, and thinking that the government was for once really trying to help, he put it on his bicycle, turning it into a motorbike, which gave him speedy access to distant markets (picture 4). He became richer, and bought a trailer, with which he could pull even more goods to market. Meanwhile the expert finished the chassis and pushed it out of the shed, ready to install the engine: but that had been misappropriated (picture 5). So the offending farmer was called back (picture 6). It was explained to him that if he gave the engine back, he could share the Co-opLorry with all the other farmers. But the farmer did not like the idea of losing his privileged access, so, by now being a rich man, he suggested to the expert that he would buy the spare wheel

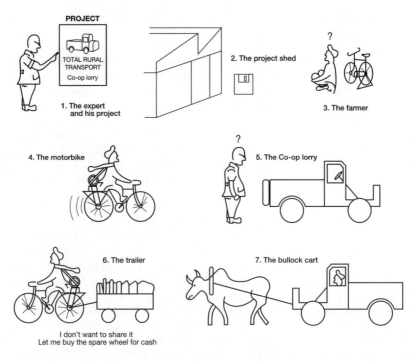

Figure 4.1 Development in Villagestan – the Project

from the chassis for a very good price, since his own bicycle wheel was breaking under the strain. The expert reported the 'theft' of the spare wheel and the engine, and then secretly built himself a fine retirement home in his native village. The other farmers heard rumours that something odd was happening, and when they came and saw the chassis, they hitched a bullock to it, and were quite pleased. Here was a cart with rubber tyres and ball-bearings (picture 7). It was slow, but still a bit better than the old bullock cart. The government then called in a foreign expert for advice. He suggested replacing the bullock with thoroughbred horses, which would move faster. [To be continued…]

There is a parallel between this story and the development of a large irrigation scheme. The engine represents the headworks, the dam and/or barrage, the biggest structure, the source of power where the tap is turned on or off. It is built first – because without it there can be no water to divert. It is always completed before the distribution system. So therefore farmers near the dam often in the early years of a scheme can be offered far more water than they are supposed to have when the scheme is completed. When it is completed they might be expected to grow a low water-demanding crop like wheat. But for the moment they can grow valuable and water-demanding rice. This is represented by the farmer hijacking the engine. The head farmers become rich, and buy off officials and politicians, so that water allocations are changed. Finally, the chassis with a bullock represents the case when later some of the tail reaches are finished. The farmers get neither the amount of water they were supposed to get, nor does it come with reliable frequency. But they do get something.

There was, of course, a finished project in the mind of the expert: and it would all have worked according to ahistorical reductionist principles – and indeed the engine did. But big irrigation projects take decades to bring to completion, and in that time the projects and society interact with each other. The farmers' perception and understanding of what was going on was different, and they saw different utility in the component parts. The complete story then clearly becomes historical – and it clearly has no end.

The dimensions of understanding

The dimensions of understanding and explanation which span the different perspectives of the different actors – from engineers, to farmers, to politicians, to bureaucrats, to social scientists – are shown in Table 4.1. The universal and ahistorical principles of physics are relevant to the performance of the scheme: but so are the teleological and reflexive attributes of consciousness. An irrigation scheme involves political struggles over resources as much as it does engineering.

Over the last 150 years the explanation and understanding in South Asia of both what large-scale irrigation systems actually are, and how they are built and operated, have moved from the left-hand columns in Table 4.1 to the right-hand. This is mostly a reflection of intellectual development, itself reflected in political institutions. The idea of biological complexity (column three) is entirely dependent on the articulation of evolutionary theory (Plotkin 1997); the 'triumph' of liberal democracy in the twentieth century is more problematical, but at least in part is the result of the acceptance of social reflexivity (column four) at 'the end of history' (Fukuyama 1992), consistent with Beck's views on media and society. The thinking behind the table is given more fully in Chapman (1985, 1997).

The farmer and water in South Asia

The traditional calendar of the farmers of north India uses 24 divisions of the solar year – the 24 *nakshatra* (asterisms) with an average length of approximately two weeks. The farmers who grow rainfed rice have for millennia sown in Rohini – the two weeks when the first pre-monsoon showers are expected – but most of all they wait for the rain in the four parts of Hathiya (the Elephant), at the end of the growing season, when the milk enters the grain and swells it. Hathiya has four legs: if it rains in the first there will be a crop of iron; if in the second as well, a crop of copper; if in the third as well a crop of silver; and if in the fourth as well, then a crop of gold and all the widows can expect to remarry (Chapman 1983b).

This simple saying is testimony to the impact that the timing of the provision of water can have on crop productivity. The quantity of water alone is not a good enough guide, as pointed out forcefully by one of the early

Table 4.1 Analytical possibilities of understanding and explanation

	Complication	*CompliFIcation*	*Complexity*	*ComplexiFIcation*
Understanding	Nil/or cosmological	Social and historical context	Environmental	Consciousness (philosophy, ideology, values) Language
Explanation	Non-teleological Reductionist Universal (statistical)	Non-teleological Reductionist Deterministic Certainty Universal Ahistorical Bounded	Non-teleological Holistic (and reductionist in part) Emergent Historical Evolutionary	Teleological Holistic Emergent Reflexive Uniquist Historical Unbounded Uncertain Language
Future states	Calculable prediction	Replicable performance	Dependent on contested imaginations Experimental trials	Probabilistic correlation
Modes of analysis of two examples				
Irrigation	km of channels No. of outlets No. of regulators Sedimentation Salinity	Integrated designed system	Ecosystem Plant growth Genetic diversity	Political mandate Crop localisation/ protective/ productive Communication Perception Corruption
People	Numbers Required food supply Demographics Malthusia	Robots	Patient of allopathic medicine	Language Personality Ambition Mood Aggression Altruism
Two examples of fields of practitioners				
Government	Mouths to feed	Technical irrigation project	Agricultural research into HYV	Representation and redistributive policies
Disciplines	'Hard' sciences	Engineering	Biological sciences	Arts and humanities, reflexive social science

British advocates of large-scale irrigation who rebuilt the Grand Anicut (barrage) in the Kaveri delta in Tamil Nadu (then Madras Province), General Sir Arthur Cotton:

> It will be well perhaps to remark on some mistakes which are almost universal on this subject. The first is, that if a tract has plenty of rain, there is no necessity for Irrigation.
>
> One plain answer to this is, that the Famine in Orissa occurred after a Monsoon of 60 inches. The question is not how much rain falls, but how it falls. In Orissa 30 inches fell in June and July. There was then a pause of six weeks during which the whole crop perished, and the 30 inches that fell in September could not restore it.
>
> Again, when we say 'irrigation', we always mean the complete regulation of the water, that is including draining; and so there is never a season when there is not at some moment excess of rain, which requires to be carried off by a system of drains.
>
> It is this regulation of the water that is needed, and which so abundantly repays the cost of works. God gives us the rain, but as in everything else, he leaves something for us to do, which if we are too indolent to do, we must suffer for it (1877: 33).

Cotton's schemes used run-of-the-river barrages in south India: they were concerned with the regulation of water during the monsoon, the principal season of cultivation for most of South Asia. But there is another way in which the regulation of water can be important. For most of the year under 'natural' conditions in much of South Asia potential evapotranspiration exceeds precipitation by a wide margin. Simply, this means that although there is warmth enough for plant growth, there is inadequate moisture[2] for productive agriculture. This suggests storing water in large dams so that regulation is effective between seasons, and not just within a season. Cotton thought about this quite a lot, and proposed an absolutely massive dam for the central Deccan. But in Cotton's day the technology for large-scale dams had not yet been developed and tested.

The farmers and rulers of South Asia through the millennia have adopted different means of regulating water. Former imperial rulers like Firoz Shah built large-scale gravity-flow canals on both banks of the Yamuna in the mid-fourteenth century, and in the seventeenth century Shah Jahan extended it with aqueducts and cuttings to Delhi. They fell into disuse, but were restored by the British. Near the major rivers inundation canals have been dug since before the time of Moenjodaro in the third millennium BC on the lower Indus, to lead the flood waters on to adjacent land. In the south Deccan, in Tamil Nadu in particular, smaller streams are dammed by shallow bunds, to create tanks (small reservoirs) which will irrigate a few hundred hectares of land downstream of them. The command area may then drain into the next tank – at the height of the monsoon it seems as if much of the land is under water, but the submergence of land is more than compensated by the gain in productivity in the command. There is often a social match: one tank for one village. Away from streams and rivers, farmers have dug wells – some simply holes in the ground, some lined with brick. The water then has to be lifted, traditionally by animal or human power, more recently by diesel and electric pumps as well. The command area may vary from just one small field to perhaps several farms (for a modern deep tube well). Since Independence in 1947, large-scale dams have been built on most of the Deccan rivers, and along the Himalayan front from Pakistan to Bhutan, and in the major catchments in Sri Lanka. More are being planned and built, but to a chorus of increasingly strident opposition. The remainder of the chapter looks at these developments over time, using the framework developed above.

British canal irrigation schemes in India

Although there were precedents built by older imperial powers, from the mid-nineteenth century the British expended continual effort on developing ever larger and more integrated irrigation systems, particularly in the upper Ganga basin and in Punjab. In appearance many of the attributes of these schemes seem dictated by environmental circumstance. The headwork barrages were located where major rivers left the foothills of the Himalayas and debouched on to the plains. Since the technology for building large dams did not develop until after the First World War (large-scale earth movers and concrete) the schemes were run-of-the-river. The Upper Ganga Canal built by Major Probey Cautley was a specific attempt to bring the Doab north of Delhi under Pax Britannica. The canal ran in an area where there were traditional wells, and some writers (e.g. Whitcombe 1972) have argued that the rising water table brought by the canal damaged such local systems of irrigation. The canal would bring prosperity and increase government revenues, it would mitigate the threat of famine, but it also put power in the hands of British irrigation officers.

A visiting Australian politician in the 1890s expressed very clearly a kind of post-Malthusian view of the situation:

> As the real notion of irrigation in India is to maintain life, and its success lies in minimising famine, it brings those who would sum up the case for and against it fairly face to face with an old problem of history, pertaining in some degree to all races, but especially under Asiatic conditions. Progress in numbers is readily measured, and at each census the totals of the Indian Empire are enlarged.... Among the most potent means of this rapid growth in the population is unquestionably the irrigation, which not only makes agricultural settlement closer wherever it obtains, but provides the vegetable food of the Hindus for countless thousands beyond the schemes. It may be held to have saved the lives of millions who would otherwise have perished, and to have enabled them to beget millions more, whom it now assists to maintain.
>
> Is this a real gain? Does it deserve the name of progress? Does it benefit either the individual or the race? Many will reply without hesitation in the affirmative; but surely in so doing they confuse the size of a nation with its eminence – they mistake quantity for quality? (Deakin 1893: 53).

Of course the British did not just impose and run the schemes. From the beginning society interacted with them, subverted them, and indeed undermined many principles of their application – particularly with regard to enhanced land revenue:

> …the subordinate bureaucracy, Punjabi in composition, obtained substantial though often illegal benefits from the canal colonies. Government salaries were only one source of income for this class: another was the craft and corruption in which it was so intimately involved. The extractive system, as this process may be collectively termed, was a pervasive influence on social relations in the canal colonies, not only in the relations of the state with the colonists but among the settlers as well (Ali 1988: 158).

> The weakening of (the State's) administered effectiveness was reflected in several other aspects of agrarian and tenurial conditions. Indeed, the collapse in efficiency was so advanced as to cast grave doubts on the degree of bureaucratic and social order in the canal colonies (Ali 1988: 182).

The British did not, however, change their design principles in consequence; these remained based on geographical circumstance and engineering response. They did change the duty of water (the area to be irrigated for a given unit of water at the head of the system) as the railways reached Punjab and enabled the commercial production of cotton for export, rather than food crops alone for food security. This story is told in Michel (1967).

Post-Independence canal building and the big dams

Although democratically elected, the Congress government of independent India led by Nehru assumed something of the autocratic manner of the British Raj, merged with the centrist philosophy of a semi-socialist state. Nehru wanted science to be the handmaiden of progress in modern India. In irrigation this meant not just the continuation of large-scale canal building, but also now the construction of large dams, which Nehru thought of as the temples of modern India. The Bhakra Dam on the Sutlej in the Himalayas reached a height of 207 m, ultimately to provide water for the world's then longest irrigation canal leading far south into the desert of Rajasthan.[3] This and other schemes were celebrated also by external observers. In the early 1950s the Ford Foundation, which was closely allied to many of Nehru's rural development policies in India, financed an American professor of political science, Henry Hart, to write a book about the new projects. The forerunner of the Damodar Development Authority in India and the inspiration for many schemes had of course been the Tennessee Valley Authority, in whose area Hart had grown up. Perhaps he was therefore wont to see progress.

> As to irrigation, India's unprecedented growth is even more evident. No country with any substantial area already irrigated approached India's achieved expansion of five per cent per year, 1951–54. This is the more remarkable, considering the fact that India already has one-fourth of the world's irrigated acreage. India need no longer ask, in short, when she will attain the dynamism of the West, but only when the rest of her economy

will catch the dynamism of her river valley projects. Unobtrusively, they have become the pacemakers for the world (Hart 1956: 280).

There is no doubt at all that irrigation was a key component of the success (in terms of gains in both yields and total production) of the Green Revolution (Chapman 2001). Indeed the use of fertilisers and new HYVs without the use of controlled water can even be a retrograde step. But the importance of irrigation does not specify which scale of system to develop. During and since the Green Revolution the biggest increases in irrigated acreage in many parts of South Asia have been through the exploitation of groundwater using tube wells. There is an irony in this too: sometimes the groundwater is provided by leaky canal systems – so that the canals remain important, but not in the way originally intended. Post-hoc the joint use of surface and groundwater has been named – conjunctive use.

The pursuit of large-scale schemes continued, but there were very many growing doubts. During the British period it was already known that quite often there was a discrepancy in the performance of the head and the tail of a system. The discrepancy might be accounted for in terms of siltation and choking of the system, excessive leakage to groundwater, or blockage by weeds – all of these allied to poor maintenance. But it might also be accounted for by increasing corruption and political subversion. In independent India all of these problems got worse. Seckler (1981) reckoned that they were performing half as well as they might, and others put utilisation as low as 20 per cent: 'By this standard the public sector industries such as fertiliser and steel, about whose under-utilised capacity so much is heard, seem relatively efficient' (Wade 1980: 347).

The late 1980s saw the beginnings of a new strand of thinking about the management of canal systems. It is perhaps best represented by the fact that the Consultative Group on International Agricultural Research (CGIAR), a science-dominated organisation linked to the World Bank that had backed the Green Revolution (Chapman 2001), for the first time founded a new centre based on social science and management – the International Irrigation Management Institute (now the International Water Management Institute), in Sri Lanka. Chambers (1988) reckoned that far more could be achieved by making existing capacity work better, than by building new capacity.

But the big project juggernaut has proved hard to stop. Not only is there a very large and well-skilled lobby of engineers, it also seems palpably 'obvious' that in a country where water is in desperately short supply in urban areas, where agricultural production has only been increasing a little faster than population, where some districts are still drought-prone and unprotected, that allowing the monsoon to run to the sea is a waste. The appeal of Cotton's simple argument of 150 years ago still holds.

The costs and benefits of large-scale schemes

There are short-term difficulties when making large-scale alterations to the hydrological cycle, many of which were not predicted when the schemes were planned. The water may take days or weeks to flow from head to tail in large systems covering millions of farmers. Thus the system is too big to adapt to the varying pattern of rainfall that may happen over the command area. Water may be delivered at a time when rainfall is already adequate, or vice versa. There are also long-term consequences which are rarely predicted well. Canals do not just provide water, they cut across and impede natural drainage systems. The canals, like the railways and their embankments and swampy borrow-pits, were thought to be associated with an increase in malaria – although of course in the mid-nineteenth century no one knew how it was transmitted, and many did not believe in correlations.[4] If the groundwater table reached too close to the surface, so that water moved up by capillary action and evaporated, then land became increasingly saline. In the end the land becomes unusable for agricultural purposes. Some estimates suggest that up to one-quarter of the land ever irrigated in Pakistan now suffers from salinisation, some of it sufficiently badly that it is completely lost to cultivation (Hussain 1997). There are means of remediation, but they are difficult and expensive, and only make sense if the problem will not be allowed to recur. One answer to salinisation has been the use of tube wells, which lower the water table, and produce a downward movement of water through the soil horizon, rather than the unfavourable upward one. The Left Bank Outfall Drain project in Sindh in Pakistan's Indus delta is designed to drain and lower the water table over an area of 500,000 ha, was costed at $640 million (1985) and was predicted to need 12 years for completion (Felton 1997).

Systems can also gradually wear out and fail. Gates and regulators wear out, or perhaps they are tampered with. The canal system can be choked with sediment, particularly if there are changes in the season in which water is abstracted from the rivers. In the case of the Ganges Canal, Cautley used clean water in the *rabi* system from the post-monsoon flow of the river. Now water is also abstracted in the high monsoon season when it is turbid, so the gradient of the canals is no longer productive of the mean-stable regime. New silt extractors have to be built into the system. Even something as simple as the cross-section of an unlined canal can change. On the Sone [Son] system in Bihar the canal banks during the colonial time were reserved for canal officers on horseback: I have heard it said[5] that farmers found on the banks were whipped. But now in democratic India farmers take their buffalo to the canal to wallow in the water. They tread the banks down. They can also be used as mobile regulators to back water up and push more into an outlet.

Not only do systems wear out, new methods of assessing their performance are devised. Some of the most difficult figures to estimate are the losses during transmission – the water that is at the head that never reaches fields or crops, but is lost en route. The figures found in textbooks are often repeated citations of each other, but actually sometimes from one source on one branch of one canal. Clearly a system which can be calculated as profitable and productive with 30 per cent losses may well not be if losses are as high as 60 per cent. New systems may be lined to reduce losses and reduce waterlogging and salinisation. Old schemes may be rehabilitated in part by the installation of lining, although this may not always be a suitable and economic response (Murray-Rust *et al.* 1997; Dhillon 1997). In Sri Lanka losses are arbitrarily set at 25 per cent, but probably exceed it. In Andhra Pradesh, 'Losses due to seepage and evaporation are allowed arbitrarily at the rate of 1.83 cumecs [cubic metres per second] per million m^2 of wetted perimeter without taking soil conditions into consideration. No losses are allowed for lined canals' (CBIP 1981: 7 cited in Chambers 1988: 75). In part, therefore, there has been a shift of investment. In states and provinces in India and Pakistan where there are large existing schemes, more and more effort is being put into maintenance and rehabilitation. Partly this is also tied to changes in the administration of the system, from the colonial command and control, to a more participatory structure. Farmers are helped to form water users' associations, and the control of the lower levels of the system is handed over to them. This is now established in the literature as 'turnover'. In some cases farmers are allowed to buy and sell their allocations of water among themselves, to allow for a better efficiency of water use while preserving social equity within the command.

But in cases where major water resources have not been harnessed, and in cases where regional deficiencies in current water supply are highlighted, the pressure is still on to build big new schemes. These are often built as multi-purpose projects. Dams provide power, irrigation water, flood control, urban water and sometimes other resources such as fish. These purposes do not fit together easily. For power generation and irrigation, as much water should be stored as soon as possible. For flood control dams should be left as empty as possible. So for power generation the early monsoon flow should be stored, in case it fails later; while for flood control, the early flows should be released, in case there is heavy rainfall later. For urban water and urban power, demand is usually maximised in the hot season from March to July – the time when agriculture is dormant. So it is not possible to release water for both purposes at the same time.

Next we come to some of the safety fears over large dams. They can induce seismic activity. The reason why they can do this is sometimes associated with the weight of water in the lake, and sometimes because water under pressure seeps into faults and lubricates them. In India the filling of the Koyna Dam in Maharashtra in 1967 triggered an earthquake of 6.5 on the Richter scale, killing 200 people and making thousands homeless, and also caused some damage to the dam itself (Singh 1997: 156). The risk of earthquakes is greatest in the Himalayas. The engineers believe they can build a dam to withstand earthquakes, but the sceptics think that we may yet see a major failure. Other major failures have been known. Rising water in a reservoir can destabilise steep hillsides. Massive rockslides into reservoirs not only send floods overtopping the main structure, but of course reduce reservoir capacity as well.

Finally, but not least, we come to the social impact. Dams result in the eviction of inhabitants from the submergence zone – something once justified in terms of local pain for national gain. But it has become clear that compensation schemes for 'oustees' or PAPs (project-affected persons) have hardly ever worked. Communities and ways of life are destroyed. That development and urbanisation do this in general on a large scale anyway, does not alter the fact that social trauma in the case of dams is highly visible, very localised, and the cause very specific. This problem is highlighted over the UK's current involvement in the Ilisu Dam in Turkey, built by the Turkish state but displacing a Kurdish ethnic minority.

The Narmada project in India

In a period of autocratic rule what kind of projects are built depends upon the abilities and whims of the autocrats and their professional advisors. It is not unusual for central planners to favour large projects – visible statements of power, but also suggestibly more accountable. It is possible to look at a large structure and say where the investment went, far more easily than knowing the quality of education that has resulted from equivalent funding strewn through thousands of villages.

In an era of increasing openness and greater concern for good governance, it might be thought that the large schemes would be abandoned, particularly in the face of orchestrated hostility. That they have not been is testimony both to the appeal of the simple logic of harnessing resources, but also to the social and political structures that proponents and opponents find themselves in. In contested situations, one side will turn to find allies, to form a bigger coalition against its opponents. The allies sign up at a price, so the scheme becomes bigger. The opponents too will seek to find allies, and escalate the dispute to higher levels.

The multi-component project on the Narmada river has its origin in the juxtaposition of a little harnessed resource in one state, Madhya Pradesh, 'near' the drought-prone states of Gujarat and Rajasthan. From 1946 onwards different irrigation and hydroelectric schemes were proposed. The Bharuch Weir Scheme with a full reservoir level (FRL) of 49 m was actually started in 1961, with Nehru himself laying the foundation stone. But subsequently another site slightly higher upstream was proposed with a height of FRL 91 m, to conduct water further into Gujarat. And then, as more power was sought from hydroelectricity and more water was demanded in Gujarat and Rajasthan, the planned height of the new main dam, the Sardar Sarovar, escalated to 97 m and then 140 m – high enough to drown several other hydroelectricity sites, and to create a reservoir stretching 200 km upstream. Put simply, in the political struggle to get the scheme accepted, more benefits were promised to more states and people, causing the scheme to escalate to the size of the largest yet known on earth: a canal 250 m wide at its head, with a capacity of 40,000 cumecs, a world record (Verghese 1994), 450 km long, connecting to 75,000 km of distribution system. The opposition to the dam, chiefly centred around the tribal people who are being evicted, but led by redoubtable external campaigners, has forged ever bigger alliances with northern NGOs like the International Rivers Network opposed to such large-scale development. The latter 'green' group has had a better grip on the English language press (Chapman *et al.* 1997; Chapman 2000a) in India, although there is a vernacular press in Gujarat which champions the rights of the 20–40 million people who expect drinking and irrigation water from the scheme. The latest recruit to the anti-dam movement is the Booker prize-winning novelist, Arundhati Roy, who uses her undoubted literary skill to lambast this and all similar projects (Roy 1999). One of her key points is that the figures are unreliable, that we do not know which of many conflicting estimates is right or wrong. The 'facts' are twisted and used to suit any current stance. Will there be a few tens of thousands of oustees, or millions of project-affected people? It depends whom you read. The opponents also claim that the environmental costs of lost biodiversity in the submergence zone, salinisation in the command area, plus downstream effects on riparian communities, and reduced fisheries in the coastal zone are not properly taken into account. They further believe that experience shows that the full system as originally designed will never be completed. In sum the costs are higher and the benefits less than stated, and the project(s) should be cancelled.

The consequence of this contest is that the World Bank pulled out of funding the dam in 1993 after being pushed into a belated and eventually very damning environmental impact assessment. Then the actual construction of the dam was halted for a number of years by cases taken to the Indian courts. Very recently the court order staying work has been rescinded, and the building has now restarted – the delays obviously contributing to a worsening financial performance. Fisher (1995), the editor of a compendium which includes representatives of all sides in the dispute, concludes that it was the power and manipulation of language, the power to portray what the engineers see as clean and sustainable power in the guise of wanton environmental destruction that halted the project:

> The Narmada case is a clear demonstration that the use of language has real-life consequences. In a positive light new language can encourage productive dialogue and the focusing of a paradigm if it is open to constant scrutiny and revision. The debate over the Narmada contributes to and sheds light upon the movement toward an emerging paradigm of sustainable development (Fisher 1995: 458).

The engineers think that hydroelectricity is green power and sustainable. The opponents think that big projects with major environmental and social impacts are not sustainable. While sustainability is the mantra, it matters who succeeds in owning the word.[6]

I imagine I am not alone in being puzzled by the widespread use of rather grand but often vacuous words, by people whose intelligence I would not normally doubt. During the Cold War 'we' called our side 'The Free World' – despite the fact that it included dictatorships in Africa, Latin America and Asia (and Europe too before 1975). Words such as 'sustainable', 'development', 'environment' have no objective meaning, like impressionist paintings which conjure images in the minds of viewers, but when looked at close up, dissolve into meaningless daubs. I have a simple hypothesis – that the desire for allies which provokes the growth of coalitions creates coalitions which need to fight beneath a symbolic banner. These vague and imprecise words are exactly suited to providing banners for such coalitions. To be sure, it may in part be a reciprocal process – the banner once hoisted may indeed attract new members – but in the first instance I believe the desire for allies provides most of the momentum.

So the Narmada Dam is not a complex set of interlinked projects, it is simply an 'unsustainable' dam.

The contributors to Fisher's book span from the left columns to the right columns of Table 4.1, from the ideas of design to the experience of enactment, from the idea of ends to the idea of means. Fisher (1995: 11) says

> that insofar as sustainable development implies an ongoing process, it is as important to be concerned with the actual means employed in the development process as it is to anticipate appropriate ends. This conviction has led [them] to elevate the debate about the Sardar Sarovar Project from a debate about whether the anticipated benefits of the Sardar Sarovar Project are desirable or even attainable to a concern about whether development as it is currently practiced by national governments and multilateral funders follows an appropriate or inappropriate process.

The shift from ends to participatory means is also remarked upon by Uphoff (1992), who even goes so far in his analysis of irrigation management to suggest the evolution of a post-Newtonian social science, i.e. one in which the observer and the observed are not differentiated in a one-way system: all participants are both.

The Kalabagh Dam in Pakistan

At Independence in 1947 the great integrated irrigation schemes of the Punjab were divided between India and Pakistan. The Indus Waters Treaty of 1960 gave India ultimate complete control of the waters of the Beas, Sutlej and Ravi rivers, and gave to Pakistan control of the Indus, Jhelum and Chenab rivers. India had already started its Bhakra–Nangal project (on the Sutlej), but the new treaty consolidated the position, and was backed by World Bank loans and grants that helped both Pakistan (at Tarbela on the Indus, Mangla on the Jhelum) and India (in addition to Bhakra, at Pong on the Beas, and at Thein on the Ravi) to develop major new dams in the Himalayan foothills.

The Tarbela Dam is 143 m high, the world's biggest rock-filled structure, impounding a lake of 243 km^2 and with a generating capacity of 3,500 MW. It diverts water for irrigation into the western parts of Pakistan's Punjab – but only as an integral part of the overall management of the Indus basin, on which Pakistan's agriculture and population depend. Mangla is almost as impressive (Michel 1967; http://www.osl.cs.uiuc.edu/~jamali/kalabagh).

Like the Indian Himalayan dams, both these dams suffer from high rates of siltation. To some extent it is possible to sluice out some of the silt, but nevertheless even at planning stage the major reservoirs were not expected to function for much more than 100 years. In 1997 the Bhakra's life expectancy had been cut to 49 years (Singh 1997: 139).

In Pakistan the downstream effects of this trapping of silt are said to have reduced the load from 200 to 36 Mt/year – though the source does not say where, and in the 3,200 km journey from the Himalayas to the delta, the river has again the energy and the opportunity to scour and pick up silt. For all that, there are many claims about how the delta is being adversely affected (Meadows and Meadows 1999; see also Bradnock and Saunders, this volume).

The net result of the current position is that the Water and Power Development Authority of Pakistan believes it necessary to replace capacity being lost at Tarbela and Mangla. Its proposal is to build another dam 240 km downstream of Tarbela, near Attock, known as the Kalabagh Dam. This would flood the valley of the tributary Kabul river, probably flooding the town of Nowshera (depending on the final height of the dam).

The issue is contested inside Pakistan, but not with the level of publicity that surrounds the Narmada Dam(s), nor with the same international attention. The reason is, of course, that the Pakistani political system has veered more towards the autocratic model of British colonialism and the Moguls, and has proved less capable of sustaining a contested democracy. The power of the Water and Power Development Authority (WAPDA) within Pakistan far exceeds that of the Ministry of Water within India. But there are opposition groups. The Sustainable Development Policy Institute of Pakistan contests the figures used for water availability in the Indus system, and the rates of sedimentation in the reservoirs. In any case, it concludes that it is possible to raise the height of Mangla to mitigate the storage losses. There are also major inter-provincial rivalries, Sind being deeply suspicious that the dam will augment the resources of the Punjab at its own expense.

Partly as a result of its bruising experiences over the Narmada Dam, the World Bank has been instrumental in establishing the World Commission on Dams, before which both India and Pakistan have given evidence. There is even official admission by Pakistan that there are important difficulties to be faced (WCD 2000, Earthscan Publications, London). Whether or not the dam will be built is yet to be decided. But Pakistan's agricultural economy relies almost entirely on surface and groundwater irrigation. People think they know what debt they owe to Tarbela and Mangla, and it would be as rash of any Pakistani politician or general simply to turn his back on the project as it would be for the Chief Minister of Gujarat to say he had lost interest in completing the Narmada Dam.

Dams in Nepal

Many of the major Ganges tributaries flow from Nepal. Approximately 15 per cent of Nepal is covered by permanent snow, and there are 80 glaciers on the High Himalayas. These two sources of water make the larger rivers perennial. Even so, 72 per cent of the river runoff comes between July and September, a severe limitation on many river uses. But Nepal has a potential 83,000 MW of hydroelectric power – 2.7 per cent of world potential – yet so far only 0.05 per cent has been developed. But with huge opportunity comes huge cost and huge risk. Nepal's GNP is about $2bn and the Chisapani project alone would cost $3.7bn, at 1983 prices (Still and Kirkby 1991; Rasheed 1995). But Nepal overlooks the impoverished plains of Uttar Pradesh and Bihar. The possible generating capacities and the staggering heights of the proposed dams are shown in Table 4.2. As with the dispute in India over the Narmada, so with these projects the different sides are

Table 4.2 Large-scale projects for Nepal

Project river and dam	Capacity (MW)	Height (m)	Reservoir capacity (million m³)	Submerged area (ha)
Karnali (Chisapani)	10,800	268	16,210	33,900
Mahakali Pancheswar	1,335	262	4,800	4,750
Mahakali Poornagiri	1,065	156	1,240	6,500
Kosi High Dam	3,000	269	9,370	19,500

Source: Rasheed (1995).

ranged against each other. In the early 1990s, the International Rivers Network (IRN), prominent in helping Medha Patkar's anti-Narmada movement, also worked to stop the controversial $1bn Arun III hydroelectric project in eastern Nepal. The IRN claims that together with NGOs and local activists, it helped file the first ever claim to the World Bank Inspection Panel, which exposed the Bank's faulty project planning and appraisal process, thus causing the Bank to cancel its proposed $175 million loan in 1995. But some observers such as B.G. Verghese (1990, 1992) believe the export of electricity to India could be Nepal's biggest revenue earner, and that in the end the dams will be built. This is a debate which may well intensify as not only power, but water itself, becomes more and more of an issue for the growing urban populace of the plains.

India has some Himalayan Gangetic districts in the new state of Uttaranchal (formerly part of Uttar Pradesh (UP)) and has started constructing a dam on the Ganga at Tehri, which will provide power, and urban and agricultural water for the plains below. It has become hotly contested, and a project taken on board by the erstwhile frontman of the Chipko Movement (the famous tree-hugging campaign in the mountains of Uttar Pradesh), Sunderlal Bahuguna. He and the leader of the Narmada Bachao Andolan, Medha Patkar, have helped each other in publicity stunts. The reason that it is being built and will be built despite all opposition and despite fears that it will not survive earthquakes, is because over 500 million people, nearly 10 per cent of the human race, live in the Ganges/Brahmaputra valley. Currently the urbanisation level of the plains is not much more than about 25 per cent – but the area is urbanising rapidly, and by 2050 the figure will have gone beyond 50 per cent, of a population twice as big as now. The cities will increase their populations four times over. Other dams seem equally desirable from Delhi's point of view.

East of the Tehri Dam, the next major Ganges tributary, the Mahakali, marks the border between India and Nepal for much of its upper course. Since the river shifts its course, there has on occasion been doubt about the exact delineation of the border. In 1920 a treaty between (British) India and Nepal allowed the development of the moderately sized Sarda barrage for irrigation in UP. To secure the east bank within Indian territory, a transfer of land was agreed, which was compensated in part by money and in part by the counter-transfer to Nepal of 1,600 ha of forest – but there has since been dispute about where this land was. The treaty, which seemed to give no benefit to Nepal, has been resented inside Nepal ever since. In the last quarter of the twentieth century there have been several momentous developments. Nepal has changed from an absolute monarchy to a representative democracy, in which new political parties and their alliances have proved fragile. In India the pressure to develop the resources of the hills is immense – *vide* the Tehri Dam. But despite this, India signed a treaty with Nepal in 1996, which in essence provides for the development of the Pancheswar Dam on the Mahakali, the world's highest rock-filled dam. The arguments within Nepal about the constitutionality of the treaty, and the terms which Nepal negotiated (poor ones), have fractured parties, brought down governments, and spawned a Maoist terrorist insurgency. Gyawali and Dixit (1999) characterise the Nepalese elite as rushing to sell their assets to the Indian developmental state, without adequate analysis of the risks, rates of return, environmental costs etc. As an alternative to the seductive allure of a mega-project, they suggest: 'The alternative approach of national capacity building, local government participation, use of cheap reliable electricity to give national industries a competitive edge, etc., would obviate much of the ills in Nepali body politic' (p. 563).

Conclusions

Human capacity to intervene in the hydrological cycle has escalated enormously in the last 150 years. Technology allows us to conceive of ever greater projects. The needs of growing populations and growing economies also seem to point to the need to develop water resources on these newer greater scales. But we have been forced to learn, even if slowly, that as the scales escalate, so do the uncertainties. These refer both to the data about things we think we do or should know about, like discharges, but which of course are subjected to endless variability, and also to the truly unsuspected impacts, like the pollution of land around the Hirakud reservoir with molybdenum at dangerous levels, and also to the experiences of the people who should not be obstacles to a project, but its beneficiaries. To understand how these dimensions interact, it is necessary first to put them in the context of the explanatory structures, the intellectual world views, that envelop different times and different professions. These considerations apply to experiences in most countries. They are shown in strong relief in South Asia because of the extreme seasonality of the climate, and the demands of a very large population not only for food, but also for development. Secondly, in the case of the different country examples offered here, it is also clear that the limitations put on the risk takers vary according to the kind of political institutions which have evolved so far. The biggest challenge to big engineering and the greatest hope for a new sustainable engineering based on participatory approaches is to be found in the most plural and democratic country – India. In Pakistan the survival of the autocratic idea of government (see Chapman 2000b) and the threats to the press make such a challenge less likely. In Nepal, the promise of big engineering has been partly responsible for blocking progress towards a stable democracy.

Notes

1. These ideas are worked out much more fully in Chapman *et al.* (1997).
2. Clearly this has to be qualified by reference to plant types, root systems and soil types.
3. Now known as Indira Gandhi Nahar.
4. Continuing the quote from Cotton cited above: 'It will be well perhaps to remark on some mistakes which are almost universal on this subject. The first is.... A third is that irrigation as a rule produces fever' (1877: 33). He dismissed the argument about fever, peremptorily observing that he never suffered from any, and he certainly lived to a vigorous and active old age, hardly stopping until he died at the age of 96.
5. In conversation with canal officers during fieldwork in the Sone command in 1986.
6. Scale is a key issue here: there is no necessary link between hydroelectricity schemes and *large* dams (see Simon 2000).

References

Ali, Imran (1988) *The Punjab under Imperialism.* Princeton University Press, Princeton, NJ.

Allan, S., Adam, B. and **Carter, C.** (eds) (2000) *Environmental Risks and the Media.* Routledge, London.

Beck, Ulrich (1995) *Ecological Politics in an Age of Risk,* translated by Amos Weisz. Polity Press, Cambridge.

Beck, U. (2000) 'Foreword', in Allan *et al., op. cit.*: xii–xiv.

Buckley, R. (1880) *The Irrigation Works of India and their Financial Results.* W.H. Allen, London.

Buckley, R.B. (1908) *Facts, Figures and Formulae for Irrigation Engineers.* E and F N Spon, London (also Calcutta, Simla and Bombay).

Chambers, R. (1988) *Managing Canal Irrigation: Practical analysis from South Asia.* Oxford and IBH, New Delhi.

Chapman, G.P. (1983a) 'Underperformance in Indian irrigation systems: problems of diagnosis and prescription', *Geoforum*: 267–75 (Special Issue on Geography and Development).

Chapman, G.P. (1983b) 'The folklore of the perceived environment in Bihar', *Environment and Planning A,* **15**(7): 945–68.

Chapman, G.P. (1985) 'The epistemology of complexity' in Aida *et al., op. cit.*

Chapman, G.P. (1997) 'Thinking about canal irrigation systems', in Wooldridge, *op. cit.,* Chapter 2 'Invited Lectures': 37–51.

Chapman, G.P. (2000a) '"Other" cultures, "other" environments, and the mass media' in J. Smith (ed.), *The Daily Globe: Environmental change, the public and the media.* Earthscan, London, Chapter 10: 127–50.

Chapman, G.P. (2000b) *The Geopolitics of South Asia: From early empires to India, Pakistan and Bangladesh.* Ashgate, Aldershot.

Chapman, G.P. (2001) 'The Green Revolution', in V. Desai and R. Potter (eds), *The Development Handbook*. Arnold, London.

Chapman, G.P., Kumar, K., Fraser, C. and **Gaber, I.** (1997) *Environmentalism and the Mass Media: The North–South divide*. Routledge, London.

Chapman, G.P and **Thompson, M.** (1995) *Water: And the quest for sustainable development in the Ganges Valley*. Mansell Publishing, London.

Cotton, A. (1866) *The Famine in India*. Lecture at the Social Science Congress, Manchester.

Cotton, A. (1877) *The Madras Famine (with appendix containing a letter from Miss Florence Nightingale)*. Cambridge University Library Pamphlet: no reference details.

Cotton, A. and **Hope, E.R.** (1900) *General Sir Arthur Cotton: His life and work*. Hodder and Stoughton, London.

Deakin, A. (1893) *Irrigated India*. Thacker and Co., London.

Dhillon, G.S. (1997) 'Techno-economic evaluation of Punjab's irrigation development efforts', in Wooldridge (ed.), *op. cit.*: 393–401.

Felton, M. (1997) 'Placing agriculture and development issues on the agenda of irrigation projects: a case study of engineering bias and institutional complexity on the Left Bank Outfall Drain Project, Pakistan', in Wooldridge (ed.), *op. cit.*: 165–75.

Fisher, W.F. (ed.) (1995) *Toward Sustainable Development? Struggling over India's Narmada River*. Columbia University Seminar Series.

Forrest, D.W. (1974) *The Life and Work of a Victorian Genius*. Elek, London.

Fukuyama, F. (1992) *The End of History and the Last Man*. Penguin, London.

Government of India (1972) *Report of The Irrigation Commission* (4 vols and Atlas). GOI, New Delhi.

Govardhan, V. (1993) *Environmental Impact Assessment of Tehri Dam*. Asian Publishing House, New Delhi.

Gyawali, D. and **Dixit, A.** (1999) 'Mahakali impasse and Indo-Nepal water conflict', *Economic and Political Weekly*, **34**(9), 27 February–5 March: 553–64.

Hart, H.C. (1956) *New India's Rivers*. Orient Longman, Bombay.

Hussain, M.A. (1997) 'Management of impermeable saline-sodic soils', in Wooldridge (ed.), *op. cit.*: 303–11.

Maloney, Clarence and **Raju, K.V.** (1994) *Managing Irrigation Together*. Sage, New Delhi.

Meadows, A. and **Meadows, P.S.** (eds) (1999) *The Indus River: Biodiversity, resources, mankind*. Oxford University Press, Karachi.

Michel, A.A. (1967) *The Indus Rivers: A study in the effects of partition*. Yale University Press, New Haven and London.

Murray-Rust, D.H., van der Velde, E.J. and **Merrey, D.J.** (1997) 'Impacts of lining, maintenance and operations on water delivery performance in Pakistan', in Wooldridge (ed.), *op. cit.*: 403–12.

Pearson, K. (1914–30) *The Life, Letters and Labours of Francis Galton* (3 vols). Cambridge University Press, Cambridge.

Plotkin, H. (1997) *Darwin, Machines and the Nature of Knowledge*. Harvard University Press, Cambridge, MA.

Rasheed, K.B.S. (1995) 'Nepal's water resources: the potential for exploitation in the upper Ganges catchment', in Chapman and Thompson (eds), *op. cit.*, Chapter 5: 90–103.

Roy, A. (1999) *The Cost of Living: The greater common good and the end of imagination.* Flamingo, London.

Seckler, D. (1981) 'The new era of irrigation management in India'. Mimeo, Ford Foundation, New Delhi.

Simon, D. (2000) 'Damm(n)ed development?', *Third World Planning Reviews,* **22**(2), May: iii–ix.

Singh, S. (1997) *Taming the Waters: The political economy of large dams in India.* Oxford University Press, Delhi.

Still, M. and **Kirkby, J.** (1991) *The Atlas of Nepal in the Modern World.* Earthscan (in association with the ETC Federation), London.

Stone, Ian (1984) *Canal Irrigation in British India: Perspectives on technological change in a peasant economy.* Cambridge University Press, Cambridge.

Uphoff, Norman T. (1992, 1996) *Learning from Gal Oya: Possibilities for participatory development and post-Newtonian social science.* Cornell University Press, Intermediate Technology Publications, London.

Verghese, B.G. (1990) *Waters of Hope: Integrated water resource development and regional cooperation within the Himalayan–Ganga–Brahmaputra–Barak Basin.* Oxford–IBH, New Delhi.

Verghese, B.G. (1992) 'Wealth, welfare, water – developing the eastern Himalayan rivers', *Contemporary South Asia,* **1**(2): 193–202.

Verghese, B.G. (1994) *Winning the Future: From Bhakra to Narmada, Tehri, Rajasthan Canal.* Konark Publishers, Delhi.

Wade, R. (1980) 'India's changing strategy of Irrigation development' in E.W. Coward, (ed.), *Irrigation and Agricultural Development in Asia.* Cornell University Press, Ithaca, Chapter 16: 345–64.

Wade, R. (1982) 'The system of administrative and political corruption: canal irrigation in South India', *Journal of Development Studies,* **19**(3): 287–328.

WCD (World Commission on Dams) (2000) Earthscan Publications, London.

Whitcombe, E. (1972) *Agrarian Conditions in Northern India,* Vol. 1: *The U.P. under British Rule 1860–1900.* University of California Press.

Wooldridge, R. (ed.) (1997) *Asian Regional Symposium on Maintenance and Operation of Irrigation/Drainage Schemes for Improved Performance Hydraulics Research.* Wallingford, and Ministry of Water Resources, People's Republic of China.

Internet and other electronic sources

http://irn.org/programs/india
http://osl.cs.uiuc.edu/~jamali/kalabagh
http://www.teriin.org/
Tata Energy Research Institute TERI CD Data Base Teddy Online+ 1998 Energy-Environment

Part Two The politics of development

CHAPTER 5

The shock of reform
The political economy of liberalisation in India

Stuart Corbridge and John Harriss

Introduction

The economy of India has been restructured since 1991 in a manner that has won the support of its middle classes.[1] In place of the Licence and Permit Raj which Jagdish Bhagwati once described as inefficient and perverse to the point of being 'Kafkaesque' (Bhagwati 1993: 50), the reforms that were introduced by Finance Minister Manmohan Singh in the summer of 1991, and which have been largely continued by his successors, including the Finance Ministers of the Bharatiya Janata Party (BJP)-led governments which came to power in 1998 and 1999, have opened India's borders to private capital from overseas as never before, and have secured changes in the country's trading and financial economies that few would have predicted. In just seven years the government of India, aided by many state governments, dismantled the major part of the system of industrial licensing that had taken shape over the previous 40 years; substantially relaxed the provisions of the Monopolies and Restrictive Trade Practices Act of 1969; provided incentives for foreign equity investment in high-priority industries; encouraged joint ventures in these and other industries; reduced the import weighted average tariff from 87 per cent in 1990 to 20.3 per cent in 1997–8; and reduced the cash reserve requirement which links commercial banks to the Reserve Bank of India from 25 per cent in 1990 to around 10 per cent in 1997–8.[2] More recently, the government has restated its intention to divest some state assets, and it has discussed proposals for the abolition of the Food Corporation of India, greater 'targeting' in the operations of the public distribution system, and the gradual phasing out of the subsidies which reduce the costs of water, fertiliser and power to some of India's farmers.

According to Stephen Radelet and Jeffrey Sachs, the reforms have dismantled 'a considerable portion of the Licence Raj', and 'with Nehruvian socialism succumbing to history in the same year as the Soviet Union' the actual and potential growth rate of the Indian economy may have 'increased to more than six per cent per year' (Radelet and Sachs 1997: 57). But the triumphalism that is expressed in this unpleasant equation between Soviet and Nehruvian socialism does not mean that the reforms have been as successful as some neo-liberals would like to believe: India continued to face balance of payments crises in the late 1990s, the incidence of rural poverty was probably increasing at this time, and the year-on-year growth rate of the economy in the 1980s was also close to 6 per cent. Moreover, the reforms have not borne out the view of politics that informs most neo-liberal accounts of structural adjustment in India. This chapter is not mainly concerned with the political obstacles that might hinder the future development of the reform agenda in India, which is the usual manner in which politics and economics are combined in discussions of liberalisation. Our concern instead is to read the history of the reforms with a view to exploring three competing accounts of the political process in India. The chapter begins, therefore, with a review of neo-liberal writings on the politics of economic management in India. It then moves on to a pluralist account of the reforms, and the work of Rob Jenkins. Jenkins considers how and why some members of India's traditional 'rent-seeking' elites have benefited from the reforms (despite neo-liberal predictions that they would inhibit change), and he pays attention to the ways in which the reforms have been pushed through at national and State levels by social elites who are well versed in the 'dirty politics' that animate India's democracy. If there is a weakness in Jenkins's work it is that it assumes that the reforms have been driven only by the state, and then by stealth, and not in accordance with particular class interests. The argument of the last major part of the chapter is that the partiality of the reforms in India (both in terms of a continuing bias against the poor, and in terms of their urban-industrial focus) should be understood as the outcome of a political settlement which favours the very different interests of India's rural and urban middle classes, and which can reasonably be described as an 'elite revolt'. The implications of this argument are discussed in a brief conclusion to the chapter.

Rent-seeking and the impossibility of reform

Although the Nehru–Mahalanobis model for the structural transformation of the Indian economy met with some resistance from private capitalists in the 1950s, most economists were active supporters of Indian planning until at least the mid-1960s, including those who later became celebrated as critics of dirigiste development. Indeed, it was not until 1970, when Jagdish Bhagwati and Padma Desai published their account of India's experiments with industrial and trade policies since 1951, that a sustained critique was made of planned development in the subcontinent.[3] The Bhagwati and Desai study was one of six country studies that were produced as part of the OECD Development Centre's ambitious research project on industrialisation in transitional (post-colonial) economies. As such it took on some of the colours of the more general volume on economic development written by Ian Little, Tibor Scitovsky and Maurice Scott, which John Toye describes 'as the basis of the neo-classical approach to development policy' (Toye 1987: 119). Little and his co-workers argued that many developing countries had lost their way by allowing a concern for national self-sufficiency to dominate the more urgent agenda of maximising the rate of economic growth (Little *et al.* 1970). This led to the protection of inefficient local industries behind high tariff walls and overvalued exchange rates. Bhagwati and Desai tested the validity of this argument with reference to India, where, they argued, high rates of protection applied across a wide range of domestic industries, in the process giving rise to an indiscriminate rather than selective policy of import substitution.[4] Bhagwati and Desai further argued that India's industrial development was being paid for, in part, by an overvalued currency (unlike in South Korea), and was helping to swell an already bloated bureaucracy as various restrictions were imposed on those economic agents who were obliged to approach the government to support their ventures.

The critique of Indian planning that began with Bhagwati and Desai was later generalised into a more biting critique of the 'rent-seeking society' that India supposedly had become. Following the initial and more careful critique of rent-seeking behaviour that was offered by Anne Krueger in respect of foreign trade – Krueger suggested that rents from import licences could be valued at 5 per cent of India's national income in 1964 (Krueger 1974)[5] – scholars including Bhagwati and Srinivasan, and Roy, suggested that India's political and bureaucratic elites had grown fat on the rents which they were able to exact from business leaders at home and abroad. Two consequences of this rent-seeking behaviour were the endemic corruption of India's public life (as politicians encouraged a growing competition in rent-seeking activities) and the misallocation of economic resources. Licences were awarded to companies with economic and political muscle, or in proportion to existing levels of economic capacity (regardless of levels of utilisation), and time was frittered away by business leaders and some farmers in the search for contacts in India's corridors of power.

In Roy's analysis, quite logically, the upshot of India's development as *the* rent-seeking society, was that economic reform was all but impossible. 'Even a limited liberal agenda [he wrote] would appear to be still-born. Incumbent politicians, government officials and the public-sector unions in general would vigorously oppose any reduction in government intervention in the economy for fear of losing the rents and sinecures of the status quo' (Roy 1984: 67).

Dirty politics and the management of reform

Revisiting the 1980s

Roy's analysis may not be as sophisticated as that of some other neo-liberals, and certainly it lacks the subtlety that is evident in Bardhan's neo-Marxian account of the 'strangulating embrace' which locked both rural and urban bourgeoisies into a political economy of slow development in the 1970s (Bardhan 1984), but the pessimism that informs Roy's work is consistent with his account of the predatory state in India. This pessimism also informs the faith which many neo-liberals continue to place in various exogenous or endogenous 'shocks' that must beset the Indian economy from time to time, and which provoke a reform agenda by default. The reforms that began in India in 1991 are explained by neo-liberals with reference to the fiscal crisis which deepened in the late 1980s and early 1990s, under the governments of Rajiv Gandhi, V.P. Singh and Chandra Shekhar, and which emerged from the unwillingness of successive administrations to match increased expenditures on various 'demand groups' with revenues raised from taxation or from the sale of goods and services. In early 1991, the story goes, the poor performance of the 'real' Indian economy, which suffered from what Raj Krishna called a Hindu rate of growth, was mirrored by such a rapid deterioration in India's public finances that the country's fiscal deficit ballooned to nearly 9 per cent of GDP and its reserves of foreign currency dwindled to the point where they could finance only two weeks' worth of imports. It was against this background, with Moody's and Standard and Poor downgrading India's credit ratings, that economic reform became necessary in India and was enacted notwithstanding the vested interests of the country's rent-seeking elites. Moreover, it was the threat of a return to high budget deficits in the late 1990s that encouraged Deepak Lal, a strong proponent of what has been called a 'second round of reforms in India' (including privatisation and labour market reforms), to maintain that the very scale of the central government deficit – 'about 6 per cent of GDP, one of the world's largest', according to the World Bank (1998: xiii) – would compel any responsible government to make fresh reforms in recognition of the fact that 'globalization is now an irreversible process' (Lal 1999: 46). As Bhaduri and Nayyar once put it, writing from a perspective that is critical of the reforms, the government of India's recognition of various and recurring economic 'crises' allowed it to turn 'the necessity of going to the IMF [which brokered a first round of reforms] into a virtue' (Bhaduri and Nayyar 1996: 50) – just as most neo-liberals would have expected and desired.

It is obvious that various 'shocks' did affect the Indian economy in the years leading up to 1991. The promise made by V.P. Singh in 1989, during his election campaign, to write off the debts of small farmers, did not help India's public finances, any more than did the Gulf crisis in 1990–1, when the price of crude oil more than doubled and remitted incomes from the region fell sharply.[6] But if these events left India facing a crisis of liquidity, it is not clear that the country was approaching bankruptcy or that its earlier models of 'urban-biased' development had 'failed' in the 1970s and 1980s (or had remained unchanged through this period). Jayati Ghosh has noted that a collapse in India's balance of payments in 1991 did not 'reflect any major adversities in the real economy: agricultural growth and industrial output remained normal, and inflation was not abnormally high for the period' (Ghosh 1998: 324). The 1980s, indeed, had been judged by many commentators to be a period of economic success and even 'take-off' in India. In its report *India: An industrialising economy in transition*, the World Bank had declared that: 'The 1980s...are showing that growth of at least 5 per cent a year is achievable. Investment is being sustained at nearly 35 per cent of GDP [and] a consistent 92–94 per cent of this investment is being financed by domestic savings, which have been growing' (World Bank 1989: 1).[7] The World Bank might also have noted that India's growth performance in the 1980s was such that the economy 'grew almost twice as fast as the average rich industrial country and about three times as fast as the typical poor developing country' (Adams 1990: 79–80). This development, moreover, paved the way for a significant reduction in the percentage incidence of rural poverty (from around 57 per cent in 1970–1 to about 33.7 per cent in 1989–91 according to the National Sample Survey Office method of poverty reduction pioneered by Bagicha Singh Minas), and for an associated improvement in the productivity of many of India's manufacturing sectors, including chemicals, consumer durables and high-technology services. All this happened, it should be noted, precisely when levels of rent-seeking in the Indian political economy were supposedly at their peak, and with very little support from the pro-market reforms that were introduced by Rajiv Gandhi in his government's first budget of 1985. Mr Gandhi may have spoken eloquently, if not always grammatically, about India 'doing a Korea' under his leadership, but Deepak Lal was being only a little unkind when he wrote, in an essay first published in 1988, that 'the translation of [Rajiv's] good intentions was at best hesitant, and largely window dressing' (Lal 1999: 35). John Adams rightly surmised that 'the scant degree of reform that was achieved [in the late 1980s] can hardly have been sufficient to have caused a doubling of the rate of per capita economic growth in an economy as large and diverse as India's' (Adams 1990: 81).

Politicians as patrons of reform

This testimony to the strength of some parts of the Indian economy in the 1980s, even allowing for what Partha Chatterjee called 'the mindless spending spree' which was indulged in by Rajiv Gandhi and his colleagues (Chatterjee 1997: 201), points up some aspects of what is missing from neo-liberal accounts of the reform process in India. It is not necessary to turn a blind eye to the failings of the Indian economy in this period – low productivity growth in some sectors; massive subsidies to the rural and urban middle classes; an often unhelpful attitude to foreign capital – to acknowledge that planned development had brought a measure of spatial and sectoral balance to the Indian economy, and that agriculture, in particular, had prospered in the wake of the Green Revolution. Although stabilisation policies were required in 1991, it is not clear that these policies should have led on to structural adjustment. The fact that they did so 'is an indication', Ghosh reminds us, 'of the much greater significance by this time of finance capital, both domestic and international, in changing the course of the Indian economy and economic policies' (Ghosh 1998: 324). The reforms, in short, were engineered by and for certain interest groups in India, with the help, of course, of the recently elected (minority) Congress government of Narasimha Rao and the IMF; they were not an inevitable or 'natural' response to the fiscal and payments 'shocks' of 1991.

But if the reforms were 'engineered' in a more active fashion than neo-liberalism implies, the question remains as to who acted as the designer(s) and to what ends. Ashutosh Varshney has suggested that pressures for reform were building in the 1980s from among the urban middle classes, but that the Congress government of Rajiv Gandhi was unable or unwilling to push through an effective agenda for liberalisation (Varshney 1996). In the 1990s, in contrast, the government of Narasimha Rao was able to neutralise the opposition of the Communists and the Janata Dal to its reforms by offering to support these parties in their stuggles with the BJP, not least in regard to the question of reservations.[8] Liberalisation became an option in the 1990s in part because of the three-cornered nature of parliamentary politics, and because of a perceived need to counter the rise of Hindu nationalism.

It is in the work of Rob Jenkins, however, that we find a more insistent account of the decisions that have been taken by New Delhi and various State governments to continue the reform agenda, even where these decisions would seem to threaten the interests of established elites, and not least those 'elites' (such as organised labour) that dominate what Varshney calls the arena of 'mass politics'. Jenkins notes that the reforms have been more radical and continuous than some commentators allow, in part because of the rise of 'competition States' within India's federal democracy. The competition waged by the governments of Tamil Nadu and Maharashtra to host a Fiesta (Ford Motor Company) assembly plant is just one instance of the changing context for economic policy in India (the battle, incidentally, was won by Tamil Nadu in 1996). Political bosses from the more prosperous States have been keen to extend the reforms to stake out their independence from New Delhi and to improve the financial situation of their provinces.

Jenkins has further argued that complaints about the powers of organised labour are often misplaced, and that:

> despite laws forbidding firms from dismissing workers, many chronically loss-making companies have simply locked their factory gates as elected state governments looked the other way, preferring to let this practice proceed quietly.... Union leaders argue that while they have shown flexibility by toning down labour stridency, 'management militancy' has been on the rise, abetted by state governments. This is borne out by the statistics: while the number of person-days lost because of strikes decreased by almost half, from 12.43 million in 1991 to 6.6 million in 1994, the number lost due to management lockouts has actually increased over the same period (Jenkins 1997: 162).

The willingness of the government of Rajasthan to develop a new mining policy in 1994, or of the communist government of West Bengal to court foreign multinationals (even as it denounced Manmohan Singh's surrender to the IMF)[9], or of the government of Orissa to break up the State electricity board, are also treated by Jenkins as evidence that the reforms in India, which began with a big bang in New Delhi, are being 'increasingly implemented in the form of successive micro-reforms in different states, at different times, and under different political circumstances' (*ibid.*: 160).[10]

Perhaps more significantly, it is Jenkins's wider argument that the reforms in India have been pushed through as quickly and consistently as they have been because they have been guided by politicians who are experienced in building coalitions, and who have worked behind the scenes in

India's democratic polity 'to blunt the edge of opposition [to reform]'(*ibid.*). Jenkins thus has no truck with those proponents of 'democracy-in-general' or 'good governance' who seem to think that democracy is only (or mainly) about open and competitive politics, accountability and transparency, and who constantly bemoan the meddling of politicians in matters 'economic'.[11] He notes instead that 'governing elites at many levels of the Indian polity were attracted by the potential of liberalisation to provide new sources of patronage to substitute for some of those forfeited by the shrinkage of the state's regulatory role' (*ibid.*: 6), and he suggests that the more skilled and accomplished of these politicians – a group that ranges from Narasimha Rao and Deve Gowda to Sharad Pawar and Biju Patnaik[12] – have sought to capture many of the benefits of liberalisation for themselves and their supporters by means of 'obfuscatory and manipulative tactics' which seek to neutralise opponents or less deserving followers. In addition, then, to 'outright pilfering', these power brokers make use of the following tactics: 'shifting unpleasant responsibilities and blame to political opponents, surreptitiously compensating selected interests, concealing intentions, reassuring and then abusing the trust of long-time political allies, and obscuring policy change by emphasising essential continuity' (*ibid.*: 9). In Jenkins's view, the reform-initiating state in India is at once a democratic state and a dirty state, and this is why it has been so effective.

An elite revolt? The partiality of the reforms

The pluralist view of politics in India that informs Jenkins's work stands in contrast to the more anodyne accounts of the state and politics that are to be found in some neo-liberal writings on liberalisation. Like Varshney, Jenkins credits India's politicians with playing a more active role in the construction of the reforms than is generally acknowledged in the 'shock and response' model. But Jenkins also contends, more controversially, that the reforms in India have been pushed through by government in abeyance of particular class interests and without regard for an ideology of economic change. Jenkins maintains that, 'Contrary to the conventional wisdom, the projection of an economic "vision" for India played virtually no part in [the] process [of] sustaining adjustment or liberal reforms' (Jenkins 1997: 8), and he takes strong exception to the views of those 'structuralists' (including Pranab Bardhan) who fail to understand that 'nothing so grand as a "social contract" [a vision of reform put to the people openly by the government] is necessarily required, or even desirable, if government proclaims intentions that are more modest than "a project of national economic transformation"' (*ibid.*: 40). In Jenkins's view, this sort of 'wooden Marxist political economy'[13] is unable to grasp the essential messiness and government-directed nature of the reforms, or the fact that: 'The Indian reform programme has succeeded largely because it proceeded by stealth' (*ibid.*: 40).[14]

This interpretation of the reforms has already been challenged by Bardhan on the grounds that it underestimates the scale and nature of opposition to economic reform, and it would seem to be at odds with at least some parts of Jenkins's characterisation of the reforms in western India.

Jenkins suggests that management lockouts in Mumbai, reforms to the public distribution system which favour richer farmers, and new policies which legitimise land thefts or land consolidation, have all been legitimised with reference to an 'ethos of liberalisation – and its attendant rhetoric of necessity – [which] has created the climate in which [certain] actions have become politically justifiable, even if their subversion for private profit remains at least nominally obscured' (*ibid.*: 125). But there are three further reasons for being cautious about some aspects of Jenkins's account, and together these suggest that a project of directed social change has taken place in India, and has consistently favoured some social groups and excluded others, even if the design for that 'project' has lacked a single originating point. To put it another way, a reading of the reforms that works backwards from the end of the 1990s suggests that there is more to the 'Marxian' interpretation than Jenkins allows, and that this is so because the reforms have proven themselves to be unstable, partial and even violent.

Instability

The instability of the reforms has been apparent in the return of high central and State government deficits in the second half of the 1990s (as Lal noted), and of a trading imbalance which worsened from −$2.8 bn in 1991–2 to −$8.9 bn in 1995–6 as manufactured imports flooded the country. This trading imbalance was provoked by an unwarranted appreciation of the rupee in the second half of the 1990s, but it was caused rather more so by government policies which pandered to the consumption demands of the urban middle classes, and which refused to raise taxes on the sizeable rural middle class which had emerged as a heavily subsidised participant in the circuits of agrarian capitalism. The reforms also engendered a measure of resistance from some of India's business houses, not all of whom welcomed the new regime of foreign competition to which they had been exposed. As Bardhan points out, 'patriotism is...the first refuge of laggards in competition' (Bardhan 1998: 124). But if patriotism caused all the 'major political parties (the Left parties, BJP, the Janata Party and even the Congress) [to] play the "swadeshi" tune on this issue at election time' (*ibid.*), it also caused foreign capital to look with concern at their reformist credentials. And this mattered, of course: to finance India's large trade deficits in the 1990s the country needed to attract not only the remittances of Indians working or settled abroad, but also the direct and portfolio investments of foreigners. By pandering to domestic interests (or domestic elites), the government ran the risk of alienating foreign capital.

Partiality

It remains to be seen whether a 'second round of reforms' will bring government into conflict with some of those social groups (and most especially the rural elites) who have blocked its more radical proposals for change, and who have condemned India to run a large fiscal deficit. It is possible that the BJP-dominated government which came to power in 1999 will use the growth of India's budget and trade deficits to argue for a broadened programme of privatisation, and perhaps also for labour market 'reforms'[15] and even a reduction of the power and fertiliser subsidies. In the longer run, however, the reforms will do little to ensure sustainable economic growth or development if funds are not also committed to the educational, health and social capital needs of the country's social majorities. Many neo-liberals would agree with this proposition, but would argue that social spending must first be paid for from the expanded revenue receipts that come from rapid economic growth. But there are reasons to be sceptical of the claim that this will happen soon in India, or, indeed, that it was proper to defer expenditure switching through the 1990s. Jean Dreze and Amartya Sen (1995) suggest that funds could have been found for such projects from increased taxation – not least from the agricultural sector – or from cuts in the many transfers which are still made over to India's middle classes.[16] The fact that this did not happen tells us something about the biased nature of the reforms.

In terms of revenue spending, it is fair to point out, the percentage of funds allocated by central or State governments to social services (including education, health, social security and the welfare of scheduled communities) remained more or less constant at 29.5 per cent from 1990–1 to 1996–7, even as defence spending (revenue and capital) fell from 14.85 per cent of total government revenue receipts to 12.17 per cent over the same period. It needs to be borne in mind, however, that central government spending on social services only just exceeds that on the fertiliser subsidy in rupee terms, and that State expenditures on social services (which far exceed those of central government) fell as a percentage of total State revenue expenditures from 38.98 per cent in 1990–1 to 36.79 per cent in 1996–7).[17] The budget for primary health care was also cut in the early 1990s in favour of other areas of health expenditure.

When it comes to the capital account, moreover, and more importantly, it is clear that spending has been switched away from the social sector since 1990, a point conceded by the World Bank in its *India: 1998 Macroeconomic Update*:

> As part of the fiscal adjustment, direct capital spending by the Central Government (excluding defence capital) fell from about 1.4 per cent of GDP in 1990–91 to 0.7 per cent of GDP in 1997–98, representing 40 per cent of the cut in its deficit. In the States, direct capital spending also fell.... Private infrastructure provision may eventually offset some of these investment cutbacks and raise efficiency. However, in many other areas, such as roads, power transmission, urban infrastructure, primary school buildings, and health infrastructure, a strong case exists for more public investment (World Bank 1998: 11).

Capital expenditure on 'development (social and economic services)' also fell as a percentage of total government revenue receipts: from 15.28 per cent in 1990–1 to 9.50 per cent in 1996–7.

Nor is there much evidence to suggest that improvements will soon be forthcoming. A director of Lazard Brothers, writing in the *The Economic Times* on 16 October 1999, noted that 'it is an accepted fact that growth in GDP is catalysed by investment in the infrastructure sector. The problem remains that the government does not have resources to spend on infrastructure.' Notwithstanding eight years of reforms, the scale of the government's 'internal debt' is such that: 'Almost 50 per cent of the revenue receipt is eaten away by interest payments alone. India's internal debt to GDP ratio stands at a staggering 55 per cent.' It also needs to be borne in mind, or restated, that while government resources are limited, successive governments have refused to make the reform of primary education, for example, a key issue in India, so that, in the words of the editor of the *India Development Report, 1999–2000*, 'roughly 50 million children in the age group of 6–11 are not attending school today. This directly translates to 50 million uneducated citizens of India for decades to come' (Parikh 1999: 6). It further needs to be borne in mind that the estimated 'total annual expenditure required for universal primary education is of the order of Rs.281 billion, i.e. about 2.8 per cent of GDP annually to provide adequate quantity and quality of primary schooling compared to current expenditure of 1.5 per cent on primary education' (*ibid.*: 7). This is a substantial sum of money, but in terms of current spending forgone it is equivalent to no more than two-thirds of the sum of money which the government forgoes in the form of uncollected taxes from the black economy (see Jha 1999: 175).

Violence

The continuing failure of the state to invest in universal primary education could reasonably be described as an act of violence, both against the government's own proclaimed intentions and against those girls and boys who are refused the funds and institutions needed to improve their life chances or choices. This failure, of course, did not begin with the age of economic reforms, but the very visible returns (social, economic, political) which have accrued to many supporters of the reform process – notably those among the urban middle classes – have revealed and accentuated the gaps which are now growing between rich and poor in India. It is possible, too, that the anti-state rhetorics of the reform process have been seized upon by some of India's elites, more or less consciously, to weaken those laws and public institutions which are supposed to protect India's weaker communities against the brutalising conditions that can obtain in some labour markets. The sheer supply of labour in India makes exploitation possible on the basis of absolute surplus value – that is, on the basis of sweated labour rather than on the basis of labour that is combined with new technologies (which is what Marx called exploitation on the basis of relative surplus value).

Jan Breman has long argued along these lines, but the recent work of Barbara Harriss-White offers the most thorough probing of the relationships that might obtain between state power and the patterns of 'accumulative cruelty' which have come increasingly to define relations of capitalist development in India's informal economies (Harriss-White 1997; see also Breman 1996; Balagopal 1992). Harriss-White insists that the recent liberalisation of parts of India's high or formal sector economy must be placed in this wider context. In her view: 'Calls for the radical privatization of the state ignore the effective radical privatization, informalization and now mafianization that South Asian states have been undergoing for much longer than the era of liberalisation in the 1990s' (Harriss-White 1997: 19).

The focus of Harriss-White's work is the vast and diverse informal economy in India, or that sector of the economy which provides at least 80 per cent of non-agricultural jobs. She notes that this economy has evolved its own forms of regulation over many years, with contracts typically being enforced on the basis of the spoken word, reputation or the threat of *goondaism* [thuggery]. She also notes that the growth of the informal economy in Tamil Nadu has been related to patterns of 'state failure'. The inability of the local state to secure funds for its developmental or revenue functions creates a space for private security services or benefit providers. The result, says Harriss-White, is a blurring of the boundary between the state and civil society in rural and urban Tamil Nadu. India's intermediate classes – its rich peasantry and 'lower middle class' – seek to protect their interests against those of labouring people and the formal sector by colonising the state with kith and kin, and by expanding the scope of a shadow state. This shadow state

> comes into being because of the formal state and co-exists with it. It is therefore part of the state. Some elements of the shadow state are played simultaneously by real state players – e.g. corrupt lines of tribute, patronage/clientelage. Other shadow state livelihoods are independent of direct state players – e.g. private armies enforcing black or corrupt contracts, intermediaries, technical fixers, gatekeepers and adjudicators of disputes. Hence the real state is bigger than the formal state (Harriss-White 1997: 15).

Rob Jenkins would surely recognise this shadow state. After all, his own work is sharply focused on the dirty or backside of democratic politics in India. But Jenkins's refusal to entertain the idea of class interests in a broad liberalisation project blinds him to the conclusions which Harriss-White draws from her studies of the local state(s) and the informal economy in Tamil Nadu. Harriss-White insists that: 'As a result of the shadow state and the informal economy, the formal state loses legitimacy. Then state authority will come to reside in the private social status of state agents' (*ibid.*). She also notes that: 'The norms of the shadow state may become society's norms', and that in a world where '"an honest man is he who does not know how to live" [as some maintain in the area in which she has done her research]…there are…extreme penalties in not playing according to the shadow state's rules' (*ibid.*).

What we are beginning to see now in parts of India is a more sustained assault on the idea of the 'servant state', or the importation into the formal sector of many of the norms that have governed (if this is not an oxymoron) the unregulated or informal sector. The idea of a 'servant state' may always have been fragile in India, as it is in some advanced industrial countries, but it can reasonably be argued that respect for state institutions and functionaries (including the elite civil servants, the Indian Administration Service officers) has weakened over the past 20 years, and that it has weakened at an alarming rate during the period of economic reform. The idea that liberalisation must occur 'of necessity' is being used in some parts of India to weaken the protective and developmental capacities of the very state that is meant to ensure that the poor (including some groups within organised labour) can access the benefits of sustained economic growth. It seems plausible, then, to agree with Jenkins that the state in India has proved to be more adept in sustaining a process of reform than some critics of rent-seeking will acknowledge, while also arguing, with Harriss-White, that the contraction of the 'formal' state at the local level has allowed private capitals to renegotiate their relationships with labour in the 'informal economy' to their obvious advantage. The reforms in India need to be understood in this dialectical context. The reforms are not simply about the renegotiation of India's relationships with the global marketplace, nor even are they about the relationships of private capital with the Indian state in the formal economy; the reforms are also about the reworking of the idea of the state itself and of the state's capacity to work on behalf of those who stand outside India's (expanding) social and economic elites. It is in regard to these aspects of the reforms that the charge of 'partiality' is most apposite, and where it is not unreasonable to describe the changes as being induced (or blocked) by or on behalf of India's contending social and economic elites.[18]

Conclusions

It would be idle to pretend that a fully fledged process of economic reform in India could not raise the trend rate of growth in the country, or contribute significantly to the reduction of 'mass structural poverty'.[19] By the same token, it would be misleading to suggest that the reforms in India have been enacted entirely at the bidding of the country's proprietary elites, or that they have worked to their exclusive or uncontested advantage. The reforms, as Jenkins reminds us, have sometimes taken on a life of their own, and the dynamic behind them has sometimes come from officials in the central and State governments reacting to changing conditions in the world economy or to new opportunities in the domestic arena. They cannot be understood, simply, as a set of provoked or even passive responses to various shocks to the Indian economy. To acknowledge these points, however, is not to concede the two further arguments that have been made in this chapter. These arguments concern the scope and sustainability of the reforms, and the possibility that they have attended consistently to certain class interests.

It is unlikely that the process of economic reform in India will be halted in the next few years. The reforms have changed India's relations with its trading partners and erstwhile creditors, and there remain pressures from within the country (including from within the ranks of the BJP), and without, to press ahead with labour reforms, further trade and banking reforms, a privatisation programme, and a more 'responsible' fiscal policy (which might yet entail an attack upon water, power and fertiliser subsidies). It is even possible that the government which came to power in 1999 will make education a priority in fact as well as in rhetoric. The BJP has pretensions to being a government of 'national unity', and it must have some regard for the upsurge in 'plebeian assertiveness' that marked politics in the 1990s as much as did its own agenda of *Hindutva*. The BJP will not survive in power if it poses only as a party of the high or Forward castes, or if it disregards completely the claims upon the state that are being made by India's Backward or even Scheduled Classes.[20] But the pace of these reforms remains open to doubt, and not least because the proposed fiscal changes threaten the livelihoods of many of those upon whom governments are forced to rely for more immediate political support (and not least, now, in the countryside). In the absence of these budgetary changes the government will be hard pressed to raise the extra funds it considers necessary for social or developmental spending, and it will find that monies drain from the exchequer to pay the interest charges that are mounting on borrowed funds. It is likely, too, as many neo-liberals insist, that continuing high government deficits will produce a measure of inflation at home, and that an overvalued rupee will put further pressures on India's trading and current accounts (much as we saw in the late 1990s).

The process of economic reform is never easy, of course, and it might be argued that India has avoided many of the pitfalls that have dogged attempts at reform in eastern Europe or parts of Latin America. After all, inflation in India was still below 10 per cent per annum at the end of the 1990s (after nearly ten years of reform), and the growth rate of the economy remained respectable at 5–6 per cent per annum. In part, though, these figures reflect the continuing strong performance of the agricultural sector, notwithstanding the fact that this sector has been sidelined in many discussions of reform and may yet be threatened by an assault on the subsidies which have helped to reshape it – in some parts of India – as a vigorous capitalist agriculture.[21] If there is a sense in which the reforms in India have steadfastly refused to acknowledge the strengths of India's past development experience, there is also irony in the fact that a process of economic reform that has been targeted on the 'high' or 'formal' (and 'external') economy in India has been underpinned by the performance of an economic sector which has thus far refused to bend itself significantly to the imperatives of the world market (although this is changing). Here, then, is one aspect of the partiality of the reform process in India, and one reason why the reforms are – as yet – most evident in India's towns and cities.

Another aspect of the partiality of India's reforms concerns class. Not all members of India's proprietary elites have benefited – or benefited equally – from the reforms, and voices continue to be raised by some industrialists (for example, through the federations of small-scale industry) about the need for continuing protection against foreign goods or competitors. It is significant, however, that these voices are attended to more quickly than the voices of poorer households who worry about rising prices (as witness the three-percentage-point hike in all non-oil tariffs in September 1997), and it cannot be denied that the broad thrust of the reforms has been to increase the powers of private capital. In the long run, of course, the promise is that the benefits of a higher trend rate of economic growth will trickle down to the poor and poorest, and that government spending will increase (as revenues rise) or will be switched from those activities in which the state should not be engaged (the production of goods and services) to those in which it should play a central role (the regulation of the economy and the building up of social or human capital). In the meantime, however, it is not clear that the long-term trend rate of growth is moving ahead of that established in the mid-1980s, nor is it clear that government will find the funds needed to support those without assets, or to promote their 'access' to the market. The partiality of the reforms is confirmed not so much by the controls which India's proprietary elites may or may not exercise over economic policy (*pace* Jenkins), but by the consistency with which the reforms have failed to promote the economic or political interests of those who are excluded from India's 'new' regime of accumulation. In practice, too, as Harriss-White reminds us, the reforms have eroded further those institutions of state which might once have been turned to by the less powerful as a possible source of redress for the 'hidden injuries of class'. Attempts to solve India's looming fiscal crisis by cuts in state spending – and not least in the States, where financial 'indiscipline' is a particular irritant to the World Bank (and other multilateral agencies) – will only promote a hollowing out of the state's capacity at the local level, and its replacement by a 'shadow state' which is run in many places by mafia groups or members of India's intermediate classes. In certain districts of Bihar it is already the case that 30–40 per cent of development spending (as, for example, on the Employment Assurance Scheme) is creamed off by contractors, and the idea that labourers might take their employers to court for failing to pay the statutory minimum wage is laughable in all but a few Blocks where the labouring poor have been organised by Naxalite groups.[22]

The collapse of the 'idea of the state' in parts of India cannot be laid at the door of the reforms alone – it has been going on for much longer – but the 'idea of reform' has contributed to a new calculus of political (ir)responsibility which threatens to end any semblance of a moral or community order in India's political economy. It is here, perhaps, that the reforms – or the rhetorics of reform – have been most potent. There is little evidence to suggest that the Indian economy is on a new or sustainable growth trajectory in the 1990s: the best that can be said is that conditions in India might have been worse if some measure of stabilisation was not attempted in 1991–3. Nor is there much evidence to suggest that the reform of some parts of India's formal or 'external' economies will be accompanied by reforms in the domestic economy that will threaten the living standards of India's growing middle class. And yet the idea that India is an emerging market, or an economy that is now facing up to a new challenge in the global marketplace, is firmly established in the rhetoric of Indian policymakers and their boosters abroad. Here, perhaps, is Manmohan Singh's major achievement, and here too – in the gap between 'the fantasy and the reality of a globalizing India' (as Bagchi puts it: 1994) – we can sense fresh dangers for Indian and possibly even South Asian politics.

Notes

1. Although this chapter is concerned with the political economy of liberalisation in India, a process that has had dramatic and far-reaching effects on the country since 1991. Liberalisation has perhaps been *the* dominant theme within world development over the last two decades, and within South Asia, India has not been alone in attempting to pursue policies of liberal economic reform. In Pakistan and

Sri Lanka, however, a rhetoric of reform has rarely been translated into effective reform policies, notwithstanding public displays of support for such initiatives from the Bretton Woods institutions. The civil war in Sri Lanka has eroded confidence in the country's economic prospects, and in Pakistan the economy has been weakened by high and increasing levels of defence and debt-service expenditures. The decision of India to increase defence spending by 28 per cent in the fiscal year 2000–1 put more pressure on Pakistan to commit resources to the military – all this at a time when the suspension of American aid (since 1990) has forced the country to borrow at high interest rates from the international capital markets. It can reasonably be argued, too, that the political constituencies which might support liberalisation in Pakistan (notably, a civilian middle class and the international trading community) are relatively less well developed there than in India. In Bangladesh, a small but visible class of entrepreneurs has emerged alongside some foreign capitalists in the wake of huge external aid flows and continuing state support for the private sector. There is little evidence, even so, that Bangladesh is committed to a process of economic reform that will significantly affect the lives of its overwhelmingly poor and rural social majorities. The agenda of liberalisation has thus progressed further in India than elsewhere within South Asia, and, as this chapter argues, this 'progress' raises a number of serious concerns for development across the region.

2. This summary is indebted to Khatkhate (1992: 57–9). The argument here and in the final two sections of this chapter derives very substantially from Chapter 7 of *Reinventing India* (Corbridge and Harriss 2000).

3. Hanson (1966), of course, offered a sustained and sometimes critical commentary on planning in India, but it did not strike the same tone as the Bhagwati and Desai study. Bhagwati has generously reminded his readers, however, that Manmohan Singh's doctoral thesis at Oxford, which was submitted in 1961 and published in 1964, 'argued precisely that India's export pessimism was unjustified' (Bhagwati 1993: 57). The dissertation was supervised by Ian Little.

4. On this, see Joshi (1998). Some neo-liberals oppose import substitution in principle and in all of its guises (see Radelet and Sachs 1997), but others will concede that a focused and temporary strategy of infant industry support (as described by Amsden [1989] and Wade [1990]) did help to lay the foundations of sustained and 'market-friendly' economic development in South Korea and Taiwan. For a fine review of the debate on the 'East Asian miracle' – and for a refinement of the Wade–Amsden thesis which takes account of the changing global environment for trade and industrial policies – see Akyuz *et al.* (1998).

5. Krueger estimated the total value of all rents in 1964 at Rs14,645m., or 7.3 per cent of India's national income (Krueger 1974: 294).

6. In their comprehensive overview of India's political economy from 1964 to 1991, Vijay Joshi and Ian Little write that: 'By the of the decade [the 1980s] the macroeconomic fundamentals were out of joint. Even a strictly temporary shock like the Gulf War was enough to trigger a full-scale crisis' (Joshi and Little 1994: 190–1).

7. For a contemporary critique of the World Bank's (1989) assessment, see Corbridge (1991).

8. In 1990, the National Front government headed by Prime Minister V.P. Singh (as the leader of the Janata Dal Party) moved to adopt some of the recommendations that had been presented to Mrs Gandhi's Congress government ten years previously in the report of the Second Backward Classes Commission (or the Mandal Commission as it is more widely known). The government's decision to extend a system of reservations to include almost half of all central government jobs enraged India's higher or Forward Castes, many of whom turned to the BJP to protect their interests. Although the BJP has never publicly renounced India's systems of employment reservation, it has been less than effusive in its

support for policies of affirmative action or 'compensatory discrimination', and its attempts to build a single and undivided Hindu constituency (in line with its ideology of *Hindutva*) should be seen in part as a programme of nation-building that is at once exclusionary (because of its anti-Muslim or anti-Christian bias) and assimilationist (because of its deliberate lack of attention to caste conflicts). The major Communist party in India, which is especially active in West Bengal, Kerala and Tripura, is the Communist Party of India (Marxist), or CPM.

9. The tendency of some on the Left to blame the IMF for imposing liberal economic reforms on India is inattentive to the domestic constituencies which pressed for liberalisation and which quickly offered their support to Manmohan Singh in the summer of 1991. Manmohan Singh also rejected the charge of a 'surrender to the IMF', but in rather different terms. In an interview with *The Economic Times* (of India) in March 1992, Singh maintained that in the old days of aid dependence, 'the economic affairs secretary of the government of India [had been forced] to go to Paris' to beg for soft money. Now, by contrast, or so Singh maintained, the economic affairs secretary could do business with Western capitalists on equal terms and could look 'every investor straight in the eyes'. It was thus 'utter hypocrisy [he concluded] to say that all these years we were self-reliant and suddenly Dr Manmohan Singh and Mr Narasimha Rao have surrendered the country's sovereignty.'

10. See Weiner (1999) for a less optimistic account of the reforming instincts of some State-level politicians.

11. A particular target for Jenkins is the World Bank and its overdrawn contrasts between the 'new democracies and their authoritarian predecessors' (Jenkins 1997: 5; and compare with World Bank 1992 and 1994). For more general writings on democratic regimes and the politics of reform, see Haggard and Webb (1994), Haggard and Kaufmann (1995) and Przeworski (1991).

12. Narasimha Rao and H.D. Deve Gowda were Prime Ministers of India (successively) in the early–mid-1990s; Sharad Pawar is often described as the strongman of Maharashtra, where his links to various sugar barons have helped to secure his power base inside (and more recently outside) the (official) Congress Party; Biju Patnaik was the Chief Minister of Orissa in the early–mid-1990s, and the leader of the State's Janata Dal Party.

13. The reference is to a comment on p. 34 of Jenkins's doctoral dissertation; a version of this dissertation was published as a monograph in 1999 (Jenkins 1999).

14. Varshney also refers to the possibility that the reforms in India might be expanded by government attempts to create 'larger political spaces for big moves by stealth' (Varshney 1996: 35), but he has in mind a government that would exploit further the politics of identity, or which would convince ordinary men and women that 'deeper and quicker reforms [might be] positively linked to mass welfare' (*ibid.*).

15. Whether such reforms are necessary is a moot point: see Jenkins's earlier remarks on management lockouts in Mumbai, and see our discussion of some recent work by Barbara Harriss-White later in this section.

16. In the public sector (central government), the power of some members of the middle class was evident in their ability to secure pay awards in 1998–9 that were substantially in excess of those recommended by the Pay Commission: the World Bank refers to raises worth '0.6% of GDP more than budgeted, including arrears' (World Bank 1998: 6). In addition, the pay award was 'not accompanied by any measures to increase efficiency and the Pay Commission's proposals to reduce staff by 30 per cent over 10 years and eliminate unfilled positions were rejected as part of the final wage settlement' (*ibid.*).

17. All these figures are calculated from World Bank data (World Bank 1998: Tables A4.2–A4.4). It should also be pointed out that considerable funds have been

spent, since 1993, on the District Primary Education Programme (DPEP), with results that have not as yet been properly assessed. Funding for the DPEP comes mainly from the World Bank (IDA) and the European Union. Large-scale external funding for primary education in India was a development in the 1990s – see Varghese (1999: 131) and World Bank (1997).

18. In other words, as what Sudipta Kaviraj once decribed as an 'elite revolt', or a 'revolt of the elite' (Kaviraj 1997: 19).
19. Deepak Lal distinguishes between mass structural poverty, destitution and conjunctural poverty. While 'income transfers are the only way to tackle private destitution and conjunctural poverty' (Lal 1999: 149 – he has in mind private transfers), 'mass structural poverty can only be alleviated within one generation by sustained and rapid economic growth' (*ibid.*: 148).
20. Yogendra Yadav has detected a second 'democratic upsurge' in India in the 1990s, and has argued that: 'The subaltern acceptance of the democratic invitation inevitably means a reshaping of the political agenda in accordance with their tastes, convictions and expectations' (Yadav 1996: 103). For more on the tensions which can be observed between India's 'elite revolts' (those of economic liberalism and Hindu nationalism) and the reshaping of India's democracy, see Corbridge and Harriss (2000: part III).
21. Although a certain scepticism about the willingness or capacity of central government to tackle the food and fertiliser subsidies seems proper, it should be noted that Chandrababu Naidu was returned to power in Andhra Pradesh in 1999 in spite of increasing electricity charges there. In Andhra, it seems, farmers can be persuaded to pay for electricity if they are assured of receiving a good service in return. In Punjab, by contrast, the State Electricity Board was facing bankruptcy in late 1999 after the State government promised free electricity for farmers.
22. These figures drawn from Corbridge *et al.* (2000).

References

Adams, J. (1990) 'Breaking away: India's economy vaults into the 1990s', in M. Bouton and P. Oldenburg (eds), *India Briefing, 1990*. Westview, Boulder, CO, and The Asia Society: 77–100.

Akyuz, Y., Chang, H-J. and **Kozul-Wright, R.** (1998) 'New perspectives on East Asian development', *Journal of Development Studies*, 33(3): 4–36.

Amsden, A. (1989) *Asia's Next Giant: South Korea and late industrialisation.* Oxford University Press, New York.

Bagchi, A. (1994) 'Globalising India: the fantasy and the reality', *Social Scientist*, **22**: 18–27.

Balagopal, K. (1992) 'Economic liberalism and decline of democracy: case of Andhra Pradesh', *Economic and Political Weekly*, **XXVII**: 1958–62.

Bardhan, P. (1984) *The Political Economy of Development in India.* Blackwell, Oxford.

Bardhan, P. (1998) *The Political Economy of Development in India*, 2nd edn. Oxford University Press, Delhi.

Bhaduri, A. and **Nayyar, D.** (1996) *The Intelligent Person's Guide to Liberalization.* Penguin, New Delhi.

Bhagwati, J. (1993) *India in Transition: Freeing the economy.* Clarendon, Oxford.

Bhagwati, J. and **Desai, P.** (1970) *India: Planning for industrialization.* Oxford University Press, London.

Bhagwati, J. and **Srinivasan, T.** (1975) *Foreign trade regimes and economic development.* Columbia University Press, New York.

Breman, J. (1996) *Footloose Labour: Working in India's informal economy.* Cambridge University Press, Cambridge.

Chatterjee, P. (1997) *A Possible World: Essays in political criticism.* Oxford University Press, Delhi.

Corbridge, S. (1991) 'The poverty of planning or planning for poverty? An eye to economic liberalisation in India', *Progress in Human Geography,* **15**: 467–76.

Corbridge, S. and **Harriss, J.** (2000) *Reinventing India: Liberalization, Hindu nationalism and popular democracy.* Polity, Cambridge and Oxford University Press, Delhi.

Corbridge, S., Williams, G., Srivastava, M. and **Veron, R.** (2000) 'State performance and spaces of empowerment in Bihar and West Bengal, India'. Working Paper 1, ESRC Research Project.

Dreze, J. and **Sen, A.K.** (1995) *India: Economic development and social opportunity.* Clarendon Press, Oxford.

Ghosh, J. (1998) 'Liberalisation debates', in T. Byres (ed.), *The Indian Economy: Major debates since Independence.* Oxford University Press, Delhi: 295–334.

Haggard, S. and **Kaufmann, R.** (1995) *Political Economy of Democratic Transitions.* Princeton University Press, Princeton, NJ.

Haggard, S. and **Webb, S.** (eds) (1994) *Voting for Reform: Democracy, political liberalisation and economic adjustment.* Oxford University Press, London and World Bank.

Hanson, A. (1966) *The Process of Planning: A study of India's five year plans, 1950–1964.* Oxford University Press, London.

Harriss-White, B. (1997) 'The state and informal economic order in South Asia'. Mimeo, Queen Elizabeth House, Oxford.

Jenkins, R. (1997) 'Democratic adjustment: explaining the political sustainability of economic reform in India'. Unpublished D.Phil. dissertation, University of Sussex.

Jenkins, R. (1999) *Democratic Politics and Economic Reform in India.* Cambridge University Press, Cambridge.

Jha, S. (1999) 'Tax evasion, amnesty schemes, and black income: theory, evidence, and issues', in K. Parikh (ed.), *India Development Report 1999–2000.* Oxford University Press, Delhi, 165–76.

Joshi, V. (1998) 'Fiscal stabilization and economic reform in India', in I.J. Ahluwalia and I.M.D. Little (eds), *India's Economic Reforms and Development: Essays for Manmohan Singh.* Oxford University Press, Delhi: 147–68.

Joshi, V. and **Little, I.M.D.** (1994) *India: Macroeconomics and political economy, 1964–1991.* World Bank, Washington, D.C.

Kaviraj, S. (1997) 'The general elections in India', *Government and Opposition,* **32**: 3–24.

Khatkhate, D. (1992) 'India on an economic reform trajectory', in L. Gordon and P. Oldenburg (eds), *India Briefing, 1992,* Westview, Boulder, CO, and The Asia Society, 47–70.

Krueger, A. (1974) 'The political economy of the rent-seeking society', *American Economic Review*, **64**: 291–303.

Lal, D. (1999) *Unfinished Business: The Indian economy in the 1990s*. Oxford University Press, Delhi.

Little, I.M.D., Scitovsky, T. and **Scott, M.** (1970) *Industry and Trade in Some Developing Countries*. Oxford University Press (for OECD Development Centre, Paris), Oxford.

Parikh, K. (1999) 'Overview', in K. Parikh (ed.), *India Development Report 1999–2000*. Oxford University Press, Delhi: 1–24.

Przeworski, A. (1991) *Democracy and the Market*. Cambridge University Press, Cambridge.

Radelet, S. and **Sachs, J.** (1997) 'Asia's reemergence', *Foreign Affairs*, **76**(6): 44–59.

Roy, S. (1984) *Pricing, Planning and Politics: A study of economic distortions in India*. Institute of Economic Affairs, London.

Singh, M. (1964) *India's Export Trends*. Oxford University Press, London.

Toye, J. (1987) *Dilemmas of Development: Reflections on the counter-revolution in development theory and policy*. Blackwell, Oxford.

Varghese, N.V. (1999) 'Mass education', in the Alternative Survey Group's *Alternative Economic Survey, 1991–98*. Rainbow Publishers, Delhi: 130–2.

Varshney, A. (1996) 'Mass politics or elite politics? India's economic reforms in comparative perspective'. Mimeo, Department of Government, Harvard University.

Wade, R. (1990) *Governing the Market: Economic theory and the role of government in East Asian industrialisation*. Princeton University Press, Princeton, NJ.

Weiner, M. (1999) 'The regionalization of Indian politics and its implications for economic reforms', in J. Sachs, A. Varshney and N. Bajpai (eds), *India in the Era of Economic Reforms*. Oxford University Press, London: 261–95.

World Bank (1989) *India: An industrializing economy in transition*. World Bank, Washington, D.C.

World Bank (1992) *Governance and Development*. World Bank, Washington, D.C.

World Bank (1994) *Governance: The World Bank's experience*. World Bank, Washington, D.C.

World Bank (1997) *Primary Education in India*. World Bank, Washington, D.C. and Allied Publishers, New Delhi.

World Bank (1998) *India: 1998 macroeconomic update*. World Bank, Washington, D.C.

Yadav, Y. (1996) 'Reconfiguration in India politics: State Assembly elections 1993–95', *Economic and Political Weekly*, **XXXI**: 2–3.

CHAPTER 6

Regionalism, Decentralisation and Politics
State reorganisation in contemporary India

Emma Mawdsley

Introduction

An enduring issue for the post-colonial nations of South Asia are the tensions between the demands for greater regional autonomy and the centralising urges of national governments. The independent governments that followed colonial rule generally tried to concentrate effective power within a strong political core, often seeing such centralisation as a means of strengthening weak and potentially fragmenting states. Examples include the highly centralised federal system in India, and the strong civilian and military regimes of Pakistan (Jalal 1995). But over the last 50 years these governments have faced a wide variety of pressures and challenges from communities and political parties seeking greater autonomy and devolution of power to the regions. The most 'extreme' expression of such regional demands are the struggles for outright secession. In 1971, for example, East Pakistan fought a war against West Pakistan to become the independent country of Bangladesh. Ongoing secessionist movements in South Asia today include the conflicts in Kashmir, the north-east of India, and in north and north-east Sri Lanka. More frequent, if less spectacular, are the other forms of regional struggles which do not challenge sovereignty of existing countries by trying to separate from them, but rather seek to change the balance of various territorial powers within them. These could include demands for greater political representation for a particular region (for example, in the Baluchistan and Sind provinces of Pakistan); increased economic and fiscal control (as demanded by many State[1] governments in India); or the promotion and protection of particular cultural identities (such as in certain *adivasi* or tribal movements in India). Actors in these regional struggles include social movements, the media, formal political parties, government institutions and personnel, the police, activists, military forces, and freedom fighters/guerrilla insurgents (depending on your point of view).

This chapter examines how different regional demands and mobilisations have challenged various centralising trends over the last 50 years or so in South Asia, focusing on India. In particular, it will consider some of the democratic and developmental possibilities of one form of regional change – the creation of new federal States in India. A word of warning though – the 'balance' between the centre and the regions should not be seen as a simple linear relationship or a zero-sum equation (where one 'side' must always win or lose at the expense of the other). The connections between centralising urges and decentralising pressures are complex and multidimensional. They are influenced by formal and informal politics, long-term historical trends and contemporary events, and by unforeseen and unintended outcomes as well as political strategies and planning. An exploration of some of the interweaving aspects of federalism, the creation of new States, and the 'regionalisation of politics' in India over the last 50 years should provide an introductory case study of some of these complexities and changes. The chapter will start with a historical context to regionalism, and why South Asian governments tend to feel threatened by regional and decentralising demands. This helps explain some of the choices made at the time of Independence about federalism in India, and how the map of the new States was drawn up and later contested. The chapter then places these shifts within post-colonial political changes, and concludes by considering some of the benefits and problems of such regional devolution.

The context

South Asia is highly diverse in cultural, ethnic, linguistic and religious terms. In different places and at different times it has witnessed a vast and changing panorama of competing kingdoms, states, republics and empires, as well as looser territorial arrangements. It was not simply that borders, rulers and systems of government changed as various political entities rose and fell in different parts of South Asia, but even notions of what it meant to control territory. In what is now Nepal, for example, the very concept of a fixed boundary demarcating the extent of 'sovereignty' was alien to indigenous understandings of 'nation' or 'state' (Burghart 1984). The British attempted to fix boundaries more securely, but even during the colonial period the subcontinent was made up of a complex array of directly ruled British provinces, 'Princely States', independent kingdoms and more marginal areas of uneasy truces and treaties, especially around the borders of what is now north-east India, and the 'North-West Frontier' in present-day Pakistan and Afghanistan (Box 6.1). Thus, when India and Pakistan won independence in 1947, they each inherited a confused patchwork of territories which had never previously been united as whole 'nation-states', and which had diverse histories and experiences of different forms of political and territorial power (Schwartzberg 1992).

Another colonial legacy was a pattern of profound regional economic disparity. The cities of Madras, Calcutta and Bombay, for example, became colonial India's great industrial centres (as did Karachi for what is now Pakistan), while other areas were under- or even deindustrialised (Kumar 1983).[2] Other areas were developed for different purposes, such as the imperial administrative centre of New Delhi, the military towns of Dehra Dun and Ranchi, and the hill resorts of Nainital, Shimla and Ootacamund. But, overall, the imperatives of the colonial political economy led to a significant geographical and sectoral concentration within the economy. This resulted

---

Box 6.1   Territory and rule in colonial India

By the end of the colonial period, the British ruled directly only about one-half to two-thirds of India, the rest constituting more than 550 'Princely States'. These had local rulers but they were often effectively under the thumb of Britain. They ranged from States like Hyderabad, with an area of 207,000 km² and a population of 18 million, to 'States' like Varnoli Nana in western India which, rather endearingly, had an area of 2.5 km² and a population of 96 at Independence (Spate *et al.* 1971). Moreover, many of the Princely States were extremely fragmented. Baroda, one of the bigger States, was divided between 4 or 5 large non-contiguous areas and about 30 smaller parts. As Spate *et al.* (1971: 164) put it:

[T]he boundaries of the old regime were often arbitrary, the old Princely States in particular for the most part having neither rhyme nor reason but owing their fantastic assortment of sizes and shapes to historical accidents not always of an edifying kind.

The territorial map that independent India inherited in 1947 had not been the outcome of planned decisions, but was the result of:

[A] process of annexation, and on the basis of strategic and political considerations rather than on any rational basis.... [T]he infra-structure of the polity that we inherited in 1947 was a confused mosaic created by a foreign imperial power unmindful of the valid basis for the territorial organisation of the sub-continent (Khan 1992: 39).

---

in the pre-eminence of certain regions and cities and the neglect or under-development of others, as well as the relatively poor development of services and industry compared to forestry and aspects of agriculture. Fifty years later the effects of this unevenness can still be observed. Even the colonial organisation of labour continues to have contemporary impacts, including regional tensions. In the nineteenth century, for example, Indian Tamils were brought over as indentured labourers to work on the British tea estates of Ceylon (now Sri Lanka). The complexity of the colonial inheritance, however, is indicated by the fact that their descendants have played virtually no part in the bloody secessionist struggle for a Tamil Eelam (homeland) against Sri Lanka's Sinhala and Buddhist majority, being waged by the Tamils of the north and east of the island, who can claim over 2,000 years of residence (Bose 1994; Stokke and Ryntveit 2000).

Culturally, colonial rule often resulted in the widening of social divisions, most notably and devastatingly between Muslims and Hindus, but also between castes and ethnic groups (Breckenridge and van der Veer 1994; Corbridge 2000). Deliberate strategies of 'divide and rule'; the desire to categorise people into 'natural' groups, based on nineteenth-century racist anthropology; and the institutionalisation of these real and constructed differences in laws, political constituencies and policies – all encouraged social and cultural divisions. In some respects indirect rule such as the British exercised over the Princely States, rigidified and strengthened social power structures which had until then been fluid and dynamic. Some of these social divisions had a territorial dimension because of the concentration of particular groups – like the Sikhs of Punjab and the Muslims of present-day Pakistan and Bangladesh. There were existing differences between particular groups, and sometimes there was animosity. But the colonial policy of 'divide and rule', and the practices of rigidly classifying people and ascribing to them all sorts of fixed characteristics, did much to widen and foster divisions within society. Some of these have fed into post-colonial regional identities and struggles.

For India and Pakistan, Independence was both a time of great rejoicing and great anxiety – how would these fragile new countries hold together given their incredible diversity and complex history? The very moment of their birth had been the moment of their separation, apparently under-lining the inability of different peoples (here religiously divided between Hindu and Muslim) to exist together in one country.[3] What if other ethnic, religious, linguistic and regional groups followed by demanding their own sovereign states? The last thing that the leaders of the anti-colonial strug-gles wanted was for their newly won countries to fragment into tens or hundreds of different (and perhaps warring) regions and nationalities. One outcome of this anxiety was a strong 'nationalist discourse'. In India, for example, Jawaharlal Nehru[4] took every opportunity to stress the idea of India's 'natural' geographical, historical, cultural and political unity (e.g. Nehru 1960). He and other leaders depicted the historic fight for Independence as a unified 'national' resistance to British rule, rather than the more varied set of struggles it actually had been (Chatterjee 1986; Guha and Spivak 1988; Amin 1995; Bose and Jalal 1997). The nationalist leaders were trying to forge a 'new-old' sense of what it was to be 'Indian' among the many different peoples of this vast geographic area, and thus, build, create and consolidate the nation of India (Khilnani 1997).

Allied to these nation-building efforts were the processes of creating the sovereign Indian state – its laws, boundaries and institutions. A major concern of the architects of the new Indian state was to ensure that it would hold together. The dominant nationalist leaders therefore drew up a Constitution which, although federal in name, gave significant power to the central government over the State governments. They also reorganised the new map of India in a way which they hoped would undermine regional sentiments and loyalties. These two attempts to grant more power to the centre and less to the regions are the subject of the next two sections of this chapter.

The federal balance

In 1947, Jalal (1995) argues, the decision about India's political form was in little doubt. Federalism was less of a choice than a necessity given the coun-try's vast size and diversity, and its history of division, conquest and colonisation. Federalism implies a commitment to the ideal of 'unity in diversity'. In theory it allows for dual national and regional identities, and the sharing of decision-making between the centre and the regions (King 1982; Smith 1995). But in practice, as Graham Smith (1996) notes, 'federalism' has a wide range of meanings and possible outcomes, both as a political ideology and as an institutional arrangement. If we take the example of three other federal countries – Canada, Nigeria and Germany – we can see that the historical context for federalism differs for each one, as do their federal ideologies, and their contemporary experiences of it. Although they are all federal states, each is unique. In the case of India, although there are a number of important institutional and social factors that work against centralisation, its Constitution is generally recognised to be significantly biased (compared to other federal countries) in favour of an economically and politically strong union or central government vis-à-vis the States (Dandekar 1987; Brass 1994). For example, although the States' Constitutional autonomy is real, it is limited (Lewis 1995), and the central government has relatively more freedom to dismiss State governments, raise and distribute revenue, and intervene in a wide range of economic and political spheres. Although the balance of centre–State responsibilities have been subject to review (notably by the Sarkaria Commission appointed in 1983: Saez 1999), there has not been any significant shift in Constitutional power towards the States.[5]

Paul Brass (1982) has famously argued that the centralising drives of the state under Mrs Gandhi worsened regional and other societal tensions rather than containing or managing them. He suggests that the centralisation of power, decision-making and control of resources in one of the world's most culturally diverse and socially fragmented countries had unintended results. These include the erosion of the effectiveness of some political organisations; the declining ability of the central government to implement development plans in the States and localities; and the heightening of ethnic, religious, caste and other regional and cultural conflicts. In other words, holding on to power and decision-making too tightly at the centre may have caused problems rather than solved them. This 'over-centralisation thesis' has led many commentators to suggest that administrative, territorial and political decentralisation would strengthen the nation-state of India, rather than, as many at the centre fear, weaken it (Kothari 1989). The debates on how to achieve this have focused on strengthening and supporting local democratic institutions (such as the village councils or *panchayats*), Constitutional reform (such as devolving more power to the States), and encouraging greater political transparency.[6] Another possibility is through the creation of more, smaller States, and it is this aspect of the debate which is taken up in more detail and discussed in the next section.

Creating and contesting the 'internal' map of India

We have seen how India's national leaders produced a relatively centralised Constitution – at least partly in response to fears about fragmentation and division – and in this section we will turn to a geographically parallel process by looking at the debates over the internal map of India.

As far back as 1905, the Indian National Congress[7] had supported the idea of reorganising the map of India around the distribution of the major languages (King 1997). It was believed that 'linguistic States' would encourage greater administrative efficiency and economic development than the existing mosaic of multilingual States and provinces. But in the immediate aftermath of Independence, none of the main leaders of the national movement was keen to make immediate changes to redraw State boundaries in line with the major language groups (Vanhanen 1992), and together with the amalgamation of the Princely States, pre-existing territorial divisions were broadly retained (Figure 6.1).

When he was reminded about his earlier commitment to reorganisation, Nehru admitted that he was 'never very enthusiastic about linguistic provinces' (quoted in Das Gupta 1988: 148). He was worried that linguistic

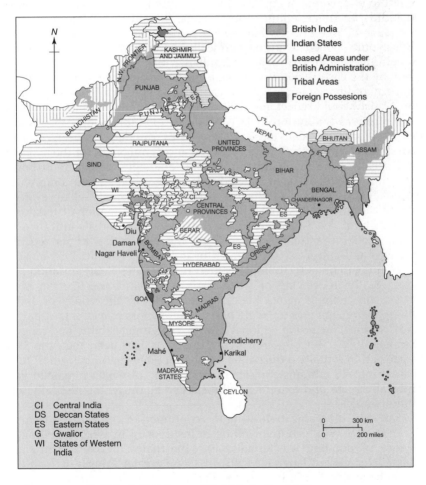

Figure 6.1 Political divisions of British India, 1946

127

States would promote regional loyalties and identities at the expense of a national loyalty and identity. Given the bloody and traumatic experience of the Partition between India and Pakistan in 1947, and the fact of India's great cultural diversity, these concerns were understandable. Moreover, for Nehru in particular, Independence signalled not just freedom, but the possibility of modernity. Europeans had both romanticised and despised India as part of the mysterious, superstitious and backward Orient (Said 1978; Inden 1990). Indian nationalists argued that colonial rule had held them back, condemning millions to live in ignorance and poverty. Freedom would allow the economic growth, technological development and national pride which would mean that India could join the ranks of 'advanced' nations. In the meantime, Nehru and other nationalists felt, the traditional suffocating and backward identities of region, caste, religion and ethnicity held Indians back from becoming modern, achieving citizens. The map of India that emerged after Independence was, at least in part, an attempt to undermine or smother these 'traditional' identities and divides – including language/ethnic groups. Thus, the old British provinces and the small and large Princely States were rather clumsily amalgamated into political units which 'were economically, administratively, linguistically and culturally illogical' (quoted in Vanhanen 1992: 70).

But, by the early 1950s, it was clear that State politicians, regional elites and many ordinary people wanted change and specifically the creation of States which would reflect linguistic and cultural patterns and differences. The centre was finally forced to concede in 1953 following protests and riots in Madras (Spear 1965). Madras was divided between Andhra Pradesh for the majority Telugu speakers, and Tamil Nadu for the majority Tamil speakers (although other minority languages remained in both). It was clear that the issue of State boundaries required a more comprehensive policy study, and in the same year the government set up a States Reorganisation Commission (SRC) to look into the matter of territorial reorganisation. The SRC came to the conclusion that India's unity would be enhanced rather than weakened by the recognition of its regional languages and cultures (Government of India 1955: 45). In 1956 most of its recommendations were implemented in the States Reorganisation Act, leading to the redrawing of some territorial boundaries to form more linguistically homogeneous States, particularly in the south of India. Although language was the main basis, other factors played a tacit part in these decisions, including regional culture, economic viability and religion (Oommen 1990). But although many demands had been met, the 1956 reorganisation actually reduced the number of States in India from 27 to 14 (plus 6 centrally administered territories), many of which remained geographically huge and extremely ethnically, religiously, linguistically and culturally mixed.

Despite the reorganisation, struggles continued in and across a number of States, some of which involved fierce insurgency. In 1960, the bilingual province of Bombay was divided into the States of Maharashtra and Gujarat after violent language riots. In 1966, 'greater Punjab' (which had already been divided in two by the Partition of India and Pakistan) was divided further between Punjab, Haryana and Himachal Pradesh. In the

1970s several States were created in the north-east, with ethnicity clearly one of the guiding factors. The former Portuguese colony of Goa, which remained in Portuguese hands until 1961 when it was incorporated into India as a Union Territory, became a State in 1987.

Each of these changes represented very different struggles and circumstances, and each needs to be contextualised within its own specific history. But one relatively constant factor up to the 1990s was that successive central governments of India tended to be opposed to territorial changes, and new States were usually conceded only after considerable social and political pressure and sometimes violence. But the situation from the 1990s to the present appears to have shifted quite significantly. There has been an increase

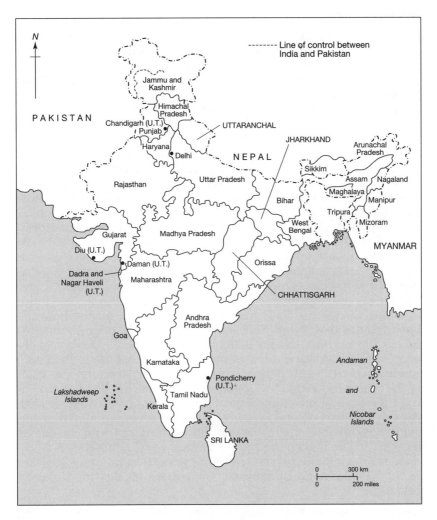

Figure 6.2 The 'new' States of India, 2000

in the number and intensity of regional movements, but the real difference is that various mainstream national political parties, and even different central governments, have started to support the creation of new federal states. Most significantly, in November 2000, the Government of India created the three new states of Uttaranchal (from the hill region of Uttar Pradesh, see Box 6.2), Chhattisgarh (from eastern Madhya Pradesh, see Box 6.3), and Jharkhand from South Bihar.[8] Why should some of the major national political parties have become more supportive of changing the internal boundaries of India through the creation of new States?

Box 6.2   Uttaranchal

Uttaranchal ('Uttarakhand') is the new Himalayan state formerly part of Uttar Pradesh. Before Independence the western half was a Princely State, and the eastern half was ruled directly by the British. At Independence it was joined with the massive plains State of Uttar Pradesh (UP), despite the arguments of some of the hill people that it should form a separate State because of its very different mountain geography.

In 1994 a massive popular regional movement exploded in the hills. The final straw, according to the activists, was yet another piece of legislation from the government of Uttar Pradesh, which failed totally to take into account the geographical and cultural differences of the mountains and the mountain people. Moreover, they argued, ever since Independence, the State of UP had exploited the area economically (especially its forests and water resources), while giving it little in return. Protesters felt that a separate State would mean they could have a more sensitive, knowledgeable, transparent and accountable government. Opponents argued that a State of Uttaranchal would have to depend heavily on central funds because of its limited resources, and that the regional agitation was either the result of cultural chauvinism or a way for the regional elites to secure jobs and money from the opportunities that a new State government would offer. Of interest here is the fact that various national and State parties have actually supported the demand for a separate Uttaranchal State, including when they were in government in the State and at the centre. The result is that after various political shenanigans and hitches, the Indian central government ratified the State of Uttaranchal on 9 November 2000. Although there was some violence during the Uttaranchal movement, and it was certainly very intense, the government was in no way forced into this action. Rather, the various parties were finding new political reasons (and not always very creditable ones) to support the division of Uttar Pradesh and Uttaranchal. The reasons for this are discussed below.

References: Aryal (1994); Rangan (1996); Mawdsley (1997), (1998); Guha (1989).

The present government offered the public surprisingly little detail on the reasons for its decision to create three new States, broadly claiming that their creation would improve administrative efficiency and give greater democratic transparency. In March 1998, for example, the Prime Minister, Atal Behari Vajpayee, was quoted as being 'of the view that the formation of these three States was necessary for the proper development of these areas' (*Deccan Herald*, 31 March 1998). The Home Minister stated that in general he favoured smaller States in the interests of growth and development: 'The rationale behind this decision was the administrative problems created because of the very large size of the States of Bihar, Uttar Pradesh and Madhya Pradesh' (*Times of India*, 11 June 1998). But despite the political rhetoric, the evidence suggests that most parties are not motivated only or even primarily by the merits or demerits of creating more federal units, but by considerations of short-term political gains. Indeed, sometimes these seemed to be more important than considerations of the financial, social or political viability of the proposed States. With regard to Chhattisgarh, for example, it has been suggested that:

> Nothing except electoral arithmetic seems to have prompted the entire spectrum of political parties relevant to Madhya Pradesh to support the formation of a separate Chhattisgarh State...privately the leaders of almost all parties admit that the move is devoid of any logic and could prove detrimental to both the States (*Deccan Herald*, 10 September 1998).

To look at this change in political logic, we need to turn to the last of the three interconnected themes in this exploration of regional challenges to the power of the centre – the shifting terrain of party politics.

The changing political economy and the 'regionalisation of politics'

It has been suggested that there has been a noticeable shift in the attitude in many of the larger political parties in India, including some with a 'national' presence, towards the issue of the creation of new States. It must be admitted that over the last few years support has often been qualified, inconsistent over time, and there have been discrepancies between different parts and levels of the political parties and individuals within them. Nevertheless, however qualified it is apparent that there is more general support for the creation of certain new States than ever before. Moreover, this has been true even when some parties are in power, rather than just in opposition (it is a political truism the world over that parties will claim one thing when they are in opposition and do another when they are in government). I will argue that one explanation for this is a growing political opportunism, which can be situated within some of the larger changes taking place in the Indian political economy over the last decade or so, and which has much to do with the balance of party political power between the centre and the States.

It is widely recognised that in the mid-1960s there was a critical shift in India's political economy, marking a change from what Rudolph and

Rudolph (1987) have termed 'command' to 'demand politics'. Very roughly, they mean that the central government has moved from a rather dominant position, where through consensus and coercion it could effectively direct policy – the 'command' phase – to a position in which central governments are increasingly battered by, and unable to moderate 'demands' from, a wide variety of social and political actors. One of the best studies of this widely identified transition is Atul Kohli's (1990) exploration of India's 'growing crisis of governability' (see also Kohli 1994). Kohli analyses the increasing 'strain' that emerged in the Indian polity after the mid-1960s, demonstrated in the absence of enduring coalitions, a growing political ineffectiveness in dealing with important problems, and an inability to accommodate growing political conflict (including regional demands) without resorting to force and violence. Kohli blames this on an uncontrolled politicisation within both the state and civil society, which resulted in the incapacity of the state to 'simultaneously promote development and to accommodate diverse interests' (Kohli 1990: 14). In other words, he suggested that the state was being pulled too many ways by various powerful interest groups, and was increasingly unable to make long-term, best-interest, autonomous decisions.

Closely related to this process was the decline of the Congress Party, and the sometimes vibrant, sometimes ugly scramble for power by a host of smaller parties, many of which are highly region- or State-specific (such as the AIADMK in Tamil Nadu; or the Communist Party of India – Marxist – in Bengal). These regional parties have little or no chance of taking direct power at the centre as they are only represented in one or two, or at best three States. But, given the current inability of a single party to win power decisively at the centre, their support in multi-party coalitions has become critical. Thus the BJP-led coalition government formed in 1999, for example, was made up of no less than 23 parties. And, over the 1990s, central politics entered a new phase of coalition governments in which smaller and regionally based parties had an increasingly influential voice in federal government. For example, the AIADMK, under the charismatic and manipulative leadership of Jayalalitha, practically held the first BJP-led government of India to ransom by threatening to withdraw support from the national coalition if it did not concede her demands with respect to Tamil Nadu. One result has been that, in contrast to the relative stability of previous decades, the 1990s witnessed a rapid 'turnover' of central governments, with four general elections between 1996 and 1999 alone, only the last producing a government which seemed likely to last its term.

How does this relate to the support for the creation of new States? In these circumstances, politics has become competitive to an even finer degree. A few seats either way can now topple State and even central governments. The issue of new States can win critical votes where they are popular, or they can be used as a way of breaking up bigger States in which political opponents are entrenched, and so reduce their power, a consideration which undoubtedly influenced the decision to create Jharkhand State.

Another very important factor influencing regional movements and politics over the last decade has been the shift in economic regime towards a more open liberal economy which has taken place since 1991. The shifts, opportunities and threats of liberalisation have inflected in multiple and complex ways with regional trends and tensions. For example, the fact that firms and State governments can now bypass New Delhi and negotiate directly with TNCs provided a considerable incentive for certain groups in Jharkhand (also known in the 1990s as Vananchal), which has very rich mineral resources, to support separation from North Bihar and to enter directly into foreign investment deals. See Corbridge and Harriss's chapter in this volume for a further discussion of these issues.

Political opportunism is hardly new in India, in relation to the creation of new States as much as anything else, and short-term strategic decision-making on this subject is not an entirely novel situation. What is perhaps different is the degree and depth to which various 'national' parties have been drawn in (not just their State units, and not just when they are in opposition), as well as various important smaller parties. The centre has not suddenly become powerless, and will no doubt continue to resist those regional movements which it thinks are inappropriate or divisive.[9] But at present, territorial reorganisation is very much back on the political agenda for India, revitalising debates on federalism, regionalism and decentralisation. This brings us to a key question – even if, at the party level, the issue is being driven by cynical political self-interest, will greater regional devolution improve governance and administration, economic growth and the lives of ordinary people? The next section reflects briefly on the impact that the creation of new States might have for the people living in them, and considers some of the potential and pitfalls of this path to greater devolution. Again it should be stressed that these generalisations should not mask the unique nature of each regional struggle and the circumstances in which it is set.

For and against smaller States in India

How do people feel about the creation of new States? Some, many perhaps, are fearful that it represents a lurch towards the 'Balkanisation' of the country, as it will open a Pandora's box of regional demands which may ultimately result in the disintegration of India. As we have seen, this was a fear of the early nationalists, and some commentators warned of the threat of disintegration (Harrison 1960, 1963; Dikshit 1975). The global resurgence of ethno-nationalist conflict in the 1980s and 1990s (Hobsbawm 1996; Castells 1997) and the tragic example of Yugoslavia have served to rekindle these fears. A former Chief Minister of Bihar, for example, stated that 'divisions of Bihar will jeopardise national and regional interests and affect national unity and integrity besides encouraging separatist and extremist forces' (quoted in the *Times of India*, 16 September 1998).

But we can take issue with this gloomy projection. First, to take the example of the three regions at the centre of current debates, although the struggle in Jharkhand was often violent, and the demand for statehood in Uttaranchal was intense, the decision to make the Chhattisgarh region into a State was certainly not the act of a reluctant government succumbing to overwhelming popular protest (Box 6.3). Indeed, a grassroots movement

---

Box 6.3 Chhattisgarh

Chhattisgarh covers 135,000 sq km in the eastern part of former Madhya Pradesh, parts of which are very isolated. The language, Chhattisgarhi, has about 7 million speakers (there are also other languages spoken in the region), and it is an area of rich *adivasi* history and culture. Currently about one-third of the population is designated as 'tribal'. Relative to other contemporary and past regional movements, the demand for a Chhattisgarh State was rather subdued and low-key. The Chhattisgarh Rajya Sangarsh Morcha was formed in 1999 by a previous Chief Minister of Madhya Pradesh, and led some rallies, *bandhs* (strikes) and courting of arrest (a classic Gandhian tactic). Thus, although the demand had an element of popular support, this was not accompanied by an intense struggle or violent insurrection. Why then was the demand for a new State supported by political parties at the centre? The major political parties in Madhya Pradesh – Congress and the Bharatiya Janata Party – have both claimed to support the idea of a new State since 1993, even to the extent of incorporating it in their election manifestos. But after an especially poor electoral showing in the region in the early 1990s, the BJP decided to play the issue more strongly. Courting popularity, and making much of the regional issue helped them win six seats in the 1996 national elections. In the campaigning before the 1998 national elections, A.B. Vajpayee (who was later to become the BJP Prime Minister) promised the people of Chhattisgarh a new State if they voted for them – and won all 11 seats. Most political analysts felt that the new State was principally the outcome of these electoral calculations – that it had become an issue that national parties were willing to use as a way of winning support, and was no longer beyond the pale as something divisive and dangerous. But the parties which support separation claim that they are motivated by the need to divide this enormous State, to facilitate a more efficient and sensitive development process, and to recognise the cultural – and especially the *adivasi* – identity of the region.

References: Singh (1982); Flueckiger (1991); Shukla (1999).

---

for a Chhattisgarh State was barely evident, and the central government obviously did not envisage the creation of this or the other two States as representing a threat to India's unity. The experience of 50 years of independence as a sovereign 'nation-state', during which crises in Punjab, Kashmir and the north-east have been weathered (if not always finally resolved), has perhaps engendered more security and confidence in India's continued unity among policy-makers.

Second, with the exception of certain struggles in the north-east (such as Bodoland, which has a moderate group seeking statehood and a more extreme group demanding secession), none of the regional movements display any serious separatist intention. In the future, of course, all or part of a particular movement could start to make secessionist demands, but this seems unlikely. These mobilisations are directed for the most part against their respective State governments, which are usually identified as the primary site of neglect, oppression and 'internal colonialism', rather than the central government. In Jharkhand, for example, it was the government of Bihar that was blamed for much of the region's exploitation (Corbridge 1987) and, as noted in Box 6.2, the same is true of the Uttaranchal movement and the State of Uttar Pradesh. As Brass (1994) suggests, regional groups are pushing for greater access to political power and control over the government purse, not cultural or linguistic separation. This is not to say that the movements for greater territorial autonomy in other regions will not be encouraged by these developments. Following the positive noises made about Uttaranchal, Chhattisgarh and Jharkhand, Kodava speakers recently demanded their own State of Kodagu, currently a part of Karnataka (*Deccan Herald*, 31 August 1997); as have Karbis, Kukis and other ethnic groups in the north-east (*Deccan Herald*, 2 April 1998). Two new political parties in Andhra Pradesh, both established in 1997, have made statehood for the Telengana region their main electoral plank (*Deccan Herald*, 16 December 1997), while there have been increased stirrings of regional demands in Bundelkhand, Vidharba, eastern and western Uttar Pradesh, the Vindhya region and elsewhere. But although they will require careful handling, these regional demands need not necessarily be damaging to the unity of India – as we saw above, Paul Brass has persuasively argued that India needs to be more decentralised, and one way might be through the creation of smaller States. A number of senior political commentators are sympathetic to the demands for greater regional autonomy. Rajni Kothari, for example, suggests that 'A large part of such assertions [of ethnic identities demanding more autonomous spaces for themselves] need to be considered as natural concomitants of the democratic struggle for achieving a more participant and decentralised polity and economy' (Kothari 1989: 13).

In his 'manifesto for change from the present centralised, dysfunctional, anachronistic union system' to an 'equipoised, co-operative and contemporary federalism', Khan (1992) argues that India needs to return to 'socio-cultural ecology' as the basis for political–administrative organisation. Territorial reorganisation is one of the measures required in order to transform the large, administratively unwieldy, politically troublesome and economically uneven States into a more 'rational' map of States, based upon economic viability, socio-cultural homogeneity and political and administrative manageability (see Figure 6.3). From this perspective, the regional movements of post-colonial India can be seen as one expression of the increasing political engagement of different, and often marginal, social groups, who are demanding a more participatory and decentralised polity.[10] However, both Rasheeduddin Khan and Rajni Kothari recognise, as Graham Smith puts it, that 'there is no basis in political theory for claiming that smaller territorial units are necessarily more hospitable to democratic politics' (Smith 1996: 398). Neither are smaller territorial units necessarily going to be more administratively efficient or 'developmentally' effective. But there is much that is positive and plausible in the arguments for smaller States, both for the specific regions under review,

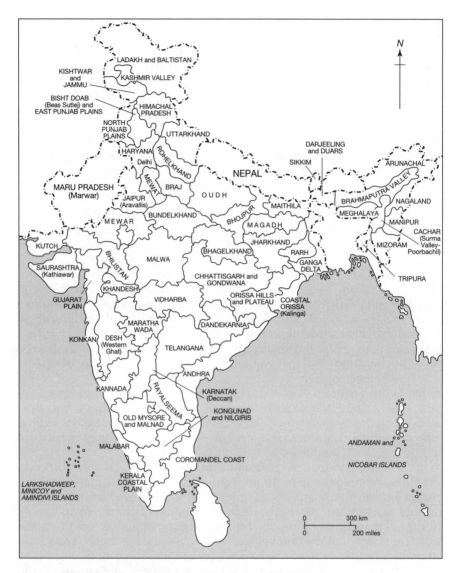

Figure 6.3 The socio-cultural subregions of India
Source: after Khan, R. (1992).

and even, arguably, as an agenda for a second round of States reorganisation (the States Reorganisation Act of 1956 was the first). Many of the current States of India are administrative leviathans, and the sheer physical distance ordinary people, politicians and bureaucrats have to cover can act to alienate groups and regions, and hinder sensitive or well-managed development planning and initiatives. Smaller States may well allow more manageable administrative loads, greater understanding of and commitment to the local region, and greater proximity between people and political and institutional centres of power.[11]

However, there are also problems with this agenda for a more 'rationally' organised map of India. First, under present administrative and government structures, the new States would require an expensive multiplication of capitals, parliaments, ministries, courts and other institutions of government. Ideally, reorganising the States would lead to faster economic growth through more effective development measures – but this is by no means a definite outcome. Second, in terms of democratic functioning, these regional mobilisations, like other social movements, can cover up partial and elite interests and manipulations, especially in marginal areas. The people and groups who are likely to get the most benefits of money and power from the new States are those who are already (relatively) the rich and powerful. There can be no automatic assumption that a new State would lead to greater social or political justice, or more rapid economic growth. Finally, India has a long history of 'sons of the soil' movements, which although diverse in form and nature, usually aim at the exclusion of 'outsiders' from the State/region, either through expulsion or by giving the 'native' population preference and privileges (Weiner 1978; Das Gupta 1988). The creation of new States could go hand in hand with new or enhanced chauvinism and violence against 'outsiders'. None of these problems are inevitable, and the various outcomes of territorial reorganisation would be highly place-specific and context-dependent, but these general concerns suggest caution must be exercised.

Conclusions

As various colonies achieved independence, many of the national elites who inherited power felt that it was in their interest to maintain the centrally oriented institutions, Constitutions, bureaucracies, armies, fiscal arrangements and perspectives that the colonial rulers had left behind. This was and is a common practice across the post-colonial world, and there is a substantial literature on the many legacies of colonialism today (see for example Mamdani (1996), Shankar (1999), Murray (2000) and Philpott (2000)). In the case of India, a strong central state was seen as critical to hold together this vast, diverse and newly forged country, and also to direct its path towards modernity, industrialisation and economic growth. But this model has been subject to a range of challenges,[12] of which the demand for new States, and the redrawing of India's internal map, is just one. Other debates have centred on the devolution of more power and decision-making further down the bureaucratic chain and to directly elected bodies (like State governments and, right at the grass roots, to the town and village councils). In a few more extreme, and often tragically bloody cases, the demand for more regional autonomy and recognition has taken the form of outright demands for secession. These continuing tensions between the centralising urges of the national governments and bureaucratic institutions on the one hand, and the vast array of different groups, citizens, political institutions and regions on the other, owe much to South Asia's colonial past, and its impact on shaping the nature of its post-colonial regimes. Nevertheless, as we have seen in the more detailed discussion of the creation of new States, political contexts change over time, as do people's perceptions and the circumstances in which struggles evolve and take shape.

The ways in which the arrangements of power, and the 'visions' of the states of South Asia (and indeed the world) are being challenged and reshaped by demands from below offer one of the most vibrant areas of study today. The movements for a greater decentralisation of power are often captured by partial and elite interests, but there is also some evidence of increased grassroots political empowerment – Uttaranchal is one example where elements of both processes can be identified. For better or worse, it seems likely that the debates and tensions between centralising and decentralising pressures are likely to continue and increase in the post-colonial states of South Asia, and that they will prove to be one of the most important issues that face them in the twenty-first century.

Notes

1. To distinguish between them, I shall use the upper case 'State' to refer to the federal units of India (e.g. the State of Karnataka), and the lower case 'state' to refer to the idea in the abstract (e.g. state–society relations, the Indian state).
2. This remains a controversial subject. Other commentators are more positive about Indian industrialisation during the colonial period – although they tend to focus on its growth rate and relative global size rather than its unevenness. See Lidman and Domerese (1970) and Lal (1988).
3. Note the use of 'apparently' here. There are Hindu citizens of Pakistan, and nearly as many Muslim citizens of India than there are Muslims living in Pakistan or Bangladesh! The politics of Partition are too long and contested to go into here, but many commentators would argue vigorously against the idea that Hindus and Muslims form different 'nations' and are inevitably or always divided (Jalal 1994).
4. Nehru was a great leader of the freedom fight and independent India's first Prime Minister (from 1947 to 1964).
5. Critics note that we must be wary of depending too much on a 'Constitutionalist' account of centre–State relations and the balance of power. Even when the Constitution has not changed, changing political realities and contingent circumstances have meant that the actual balance of power between different States and the centre has fluctuated over time. This is briefly discussed in the section on the 'Regionalisation of Politics' below, which explores the way in which India has moved from being dominated by a single strong party at the centre to a more coalitional model, whereby a number of smaller parties have a share in power.
6. On all of these issues, see for example, Rajashekara (1997), Williams (1997), Babulal and Menaria (1990), Dhavan and Pant (1997), Diwedi (1999), Desai and Imrie (1998) and Bajpai (1997).
7. The Indian National Congress (usually known as the Congress Party, or just 'Congress') was the key vehicle of opposition to British rule, and became the first party elected to government in independent India. It towered over the political scene until the early to mid-1990s, and although it has since been reduced in political stature, it remains a powerful player.
8. The region has long been known as 'Jharkhand' (literally meaning, 'land of the forests'). For some, the 'true' 'Jharkhand' would also include contiguous areas in

West Bengal, Orissa and Madhya Pradesh. However, the Government (and the relevant State Governments) were not prepared to countenance this.
9. The sometimes brutal oppression and suspension of human rights in parts of the north-east, for example, demonstrate just how far the central government is prepared to go if it does not wish to concede greater regional autonomy (Verghese 1997; Pakem 1997).
10. Others include women's rights organisations (Kabeer 1994; Calman 1995), forest struggles (Pathak 1994; Jewitt 1995), anti-firing range/land activism (Routledge 1992), fishermen's movements (Shah 1988), farmers' movements (Bentall and Corbridge 1995; Brass 1995), Dalit movements (Omvedt 1995; Mendelsohn and Vicziany 1998, Pushpendra 1999), anti-dam movements (Roy 1999; Baviskar 1995), slum dwellers' organisations (Standing 1991) and a nascent gay movement (Joseph 1996).
11. For discussions of the links (or lack of them) between decentralisation and more effective and/or transparent government, see World Bank (1999–2000) and Hemming *et al.* (1997).
12. This model has been challenged far beyond the issue of how decision-making and political power should be shared out across the state and among its citizens. It has also been questioned in terms of the desirability of its goals and directions. Radical movements and commentators have, in diverse ways, contested the whole vision of industrialisation and 'modernity' – at least as it is currently being pursued. Many of the movements and struggles alluded to above are engaged upon this sort of more fundamental critique of state power.

References

Amin, S. (1995) *Event, Metaphor, Memory: Chauri Chaura, 1922–1992.* University of California Press, Berkeley, CA.

Aryal, M. (1994) 'Axing Chipko', *Himal,* **7**(1): 8–23.

Babulal, F. and **Menaria, R.K.** (1990) *Sarkaria Commission Report and Centre–State Relations.* Sahitya Bhawan, Agra.

Bajpai, K. (1997) 'Diversity, democracy and devolution in India', in M.E. Brown and S. Ganguly (eds), *Government Policies and Ethnic Relations in Asia and the Pacific.* MIT Press, Cambridge, MA.

Baviskar, A. (1995) *In the Belly of the River: Tribal conflicts over development in the Narmada Valley.* Oxford University Press, New Delhi.

Bentall, J. and **Corbridge, S.E.** (1995) 'Urban–rural relations, demand politics and the new agrarianism in northwest India: the Bharatiya Kisan Union', *Transactions of the Institute of British Geographers,* **21**(1): 27–48.

Bose, S. (1994) *States, Nations, Sovereignty: Sri Lanka, India and the Tamil Eelam movement.* Sage, New Delhi.

Bose, S. and **Jalal, A.** (eds) (1997) *Nationalism, Democracy and Development: State and politics in India.* Oxford University Press, Oxford.

Brass, P. (1982) 'Pluralism, regionalism and decentralising tendencies in contemporary Indian politics', in A.J. Wilson and D. Dalton (eds), *The States of South Asia: Problems of national integration.* Hurst, London: 223–64.

Brass, P. (1994) *The Politics of India since Independence,* 2nd edn. Cambridge University Press, Cambridge.

Brass, T. (ed.) (1995) *New Farmers' Movements in India*. Frank Cass, Ilford.

Breckenridge, C.A. and van der Veer, P. (eds) (1994) *Orientalism and the Post-Colonial Predicament*. Oxford University Press, New Delhi.

Burghart, R. (1984) 'The formation of the concept of nation-state in Nepal', *Journal of Asian Studies*, **44**(1): 101–25.

Calman, L.J. (1995) *Toward Empowerment: Women and movement politics in India*. Westview Press, Boulder, CO.

Castells, M. (1997) *The Power of Identity*. Blackwell, Oxford.

Chatterjee, P. (1986) *Nationalist Thought and the Colonial World: A derivative discourse?* Zed Books, London.

Corbridge, S.E. (1987) 'Industrialisation, internal colonialism and ethnoregionalism: the Jharkhand, India, 1880–1980', *Journal of Historical Geography*, **13**(3): 249–66.

Corbridge, S. (2000) 'Competing inequalities: the Scheduled Tribes and the reservations system in India's Jharkhand', *Journal of Asian Studies*, **59**(1): 62–85.

Dandekar, V. (1987) 'Unitary elements in a federal constitution', *Economic and Political Weekly*, **32:** 1865–70.

Das Gupta, J. (1988) 'Ethnicity, democracy and development in India: Assam in a general perspective', in A. Kohli (ed.), *India's Democracy: An analysis of changing state–society relations*. Princeton University Press, Princeton, NJ.

Desai, V. and Imrie, R. (1998) 'The new managerialism in local governance: North–South dimensions', *Third World Quarterly*, **19**(4): 635–50.

Dhavan, R. and Pant, R. (1997) 'Back to the future: *panchayats* and governance in India', in G. Hassall and C. Saunders (eds), *The People's Representatives: Electoral systems in the Asia–Pacific Region*. Allen & Unwin, St Leonards, NSW, Australia.

Dikshit, R.D. (1975) *The Political Geography of Federalism: An inquiry into origins and stability*. Macmillan India, Delhi.

Diwedi, O.P. (1999) 'Governance and administration in South Asia', in K.M. Henderson and O.P. Diwedi (eds), *Bureaucracy and the Alternatives in World Perspective*. St Martin's Press, New York.

Flueckiger, J.B. (1991) 'Genre and community in the folklore system of Chhattisgarh', in A. Appadurai, F.J. Karom and M.A. Mills (eds), *Gender, Genre and Power in South Asian Expressive Traditions*. University of Pennsylvania Press, Philadelphia, PA.

Government of India (1955) *Lok Sabha Debates on the Report of the States Reorganisation Commission*, Vols. I–III. Lok Sabha Secretariat, New Delhi.

Guha, R. (1989) *The Unquiet Woods: Ecological change and peasant resistance*. Oxford University Press, New Delhi.

Guha, R. and Spivak, G. (1988) *Selected Subaltern Studies*. Oxford University Press, New Delhi.

Harrison, S. (1960) *India: The most dangerous decades*. Princeton University Press, Princeton, New Jersey.

Harrison, S. (1963) 'Hindu society and the state: the Indian Union', in K.H. Silvert (ed.), *Expectant Peoples: Nationalism and development*, Random House, New York.

Hemming, R., Mates, N. and **Potter, B.** (1997) 'India', in T. Ter-Minassian (ed.), *Fiscal Federalism in Theory and Practice*. IMF, Washington, D.C.

Hobsbawm, E. (1996) 'Identity politics and the Left', *New Left Review*, **217**: 38–47.

Inden, R. (1990) *Imagining India*. Blackwell, Oxford.

Jalal, A. (1994) *The Sole Spokesman: Jinnah, the Muslim League and the demand for Pakistan*. Cambridge University Press, Cambridge.

Jalal, A. (1995) *Democracy and Authoritarianism in South Asia: A comparative and historical perspective*. Cambridge University Press, Cambridge.

Jewitt, S. (1995) 'Europe's "Others"?: forestry policy and practices in colonial and post-colonial India', *Society and Space*, **13**(1): 67–90.

Joseph, S. (1996) 'Gay and lesbian movements in India', *Economic and Political Weekly*, 17 August: 2228–33.

Kabeer, N. (1994) *Reversed Realities: Gender hierarchies in development thought*. Verso, London and New York.

Khan, R. (1992) *Federal India: A design for change*. Vikas Publishing House, Delhi.

Khilnani, S. (1997) *The Idea of India*. Penguin, London.

King, P. (1982) *Federalism and Federation*. Croom Helm, London.

King, R.D. (1997) *Nehru and the Language Politics of India*. Oxford University Press, New Delhi.

Kohli, A. (1990) *Democracy and Discontent: India's growing crisis of governability*. Cambridge University Press, Cambridge.

Kohli, A. (1994) 'Centralization and powerlessness: India's democracy in a comparative perspective', in J.S. Migdal, A. Kohli and V. Shue (eds), *State Power and Social Forces: Domination and transformation in the Third World*. Cambridge University Press, Cambridge.

Kothari, R. (1989) 'Federalism', *Seminar*, May.

Kumar, D. (ed.) (1983) *The Cambridge Economic History of India*, Vol. II. *c.1757–1970*. Cambridge University Press, Cambridge.

Lal, D. (1988) 'Ideology and industrialisation in India and East Asia', in H. Hughes (ed.), *Achieving Industrialisation in East Asia*. Cambridge University Press, Melbourne.

Lewis, J.P. (1995) *India's Political Economy: Governance and reform*. Oxford University Press, New Delhi.

Lidman, R. and **Domorese, R.J.** (1970) 'India', in W.A. Lewis (ed.), *Tropical Development, 1880–1913*. Allen & Unwin, London.

Mamdani, M. (1996) *Citizen and Subject: Contemporary Africa and the legacy of late colonialism*. Princeton University Press, Princeton, NJ.

Mawdsley, E.E. (1997) 'Non-secessionist regionalism in India: the Uttarakhand separate state movement', *Environment and Planning A*, **29**: 2217–35.

Mawdsley, E.E. (1998) 'After Chipko: from environment to region in Uttaranchal', *Journal of Peasant Studies*, **25**(4): 36–54.

Mendelsohn, O. and **Vicziany, M.** (1998) *The Untouchables: Subordination, poverty and the state in Modern India*. Cambridge University Press, Cambridge.

Murray, M. (2000) 'Configuring the trajectory of African political history', *The Canadian Journal of African Studies*, **34**(2): 376–86.

Nehru, J. (1960) *The Discovery of India*. Meridian, London.

Omvedt, G. (1995) *Dalit Visions: The anti-caste movement and the construction of an Indian identity*. Orient Longman, Hyderabad.

Oommen, T.K. (1990) *State and Society in India: Studies in nation-building*. Sage, London.

Pakem, B. (ed.) (1997) *Insurgency in North East India*. Omsons, New Delhi.

Pathak, A. (1994) *Contested Domains: The state, peasants and forests in contemporary India*. Sage, New Delhi and London.

Philpott, S. (2000) *Rethinking Indonesia: Postcolonial theory, authoritarianism and identity*. Macmillan, Basingstoke.

Pushpendra (1999) 'Dalit assertion through electoral politics', *Economic and Political Weekly*, **34**(36): 2609–18.

Rajashekara, H.M. (1997) 'The nature of Indian federalism: a critique', *Asian Survey*, **37**(3): 245–53.

Rangan, H. (1996) 'From Chipko to Uttaranchal: development, environment and social protest in the Garhwal Himalayas, India', in R. Peet and M.J. Watts (eds), *Liberation Ecologies: Environment, development, social movements*. Routledge, London and New York: 205–26.

Routledge, P. (1992) 'Putting politics in its place: Baliapal, India, as a terrain of resistance', *Political Geography*, **11**(6): 588–611.

Roy, A. (1999) *The Greater Common Good*. Indian Book Distributors, Bombay.

Rudolph, L. and **Rudolph, S.H.** (1987) *In Pursuit of Lakshmi: The political economy of the Indian state*. University of Chicago Press, Chicago and London.

Saez, L. (1999) 'The Sarkaria Commission and India's struggle for federalism', *Contemporary South Asia*, **8**(1): 41–64.

Said, E.W. (1978) *Orientalism*. Routledge and Kegan Paul, London.

Schwarztberg, J.E. (1992) *A Historical Atlas of South Asia*, 2nd Impression. Oxford University Press, New York and Oxford.

Shah, G. (1988) 'Grassroots mobilisation in Indian politics', in A. Kohli (ed.), *India's Democracy: An analysis of changing state–society relations*. Princeton University Press, Princeton, NJ.

Shankar, G. (1999) 'Where the present is haunted by the past': disarticulating colonialism's legacy in the Caribbean', *Cultural Dynamics*, **11**(1): 57–88.

Shukla, H.L. (1999) *Chhattisgarh in Making*. B. R. Publishing, Delhi.

Singh, K.S. (ed.) (1982) *Tribal Movements in India*. Manohar, New Delhi.

Smith, G.E. (ed.) (1995) *Federalism: The multiethnic challenge*. Longman, London and New York.

Smith, G.E. (1996) 'Russia, ethnoregionalism and the politics of federation', *Ethnic and Racial Studies*, **19**(2): 391–410.

Spate, O.H.K., East, G.W. and **Fisher, C.A.** (eds) (1971) *The Changing Map of India: A political geography*. Methuen, London.

Spear, P. (1965) *A History of India: From the sixteenth century to the twentieth century.* Penguin, Harmondsworth.

Standing, H. (1991) *Dependence and Autonomy: Women's employment and family in Calcutta.* Routledge, London.

Stokke, K. and **Ryntveit, A.K.** (2000) 'The struggle for Tamil Eelam in Sri Lanka', *Growth and Change,* **31**(2): 285–304.

Vanhanen, T. (1992) *Politics of Ethnic Nepotism.* Sterling Press, New Delhi.

Verghese, B.G. (1997) *India's Northeast Resurgent: Ethnicity, insurgency, governance, development,* 2nd edn. New Delhi.

Weiner, M. (1978) *Sons of the Soil: Migration and ethnic conflict in India.* Princeton University Press, Princeton, NJ.

Williams, G. (1997) 'State, discourse and development in India: the case of West Bengal's Panchayati Raj', *Environment and Planning A,* **29**(12): 2099–112.

World Bank (1999–2000) *World Development Report, 1999–2000. Entering the Twenty-First Century.* Oxford University Press, New York.

CHAPTER 7

NGO–Government Relationships in South Asia
The case of India and Bangladesh

Vandana Desai

Introduction

Närman has argued that for many years non-governmental organisations (NGOs) have been 'something of a sacred cow' in the development debate (1999: 168), among aid donor countries as much if not more than among 'recipients'. In recent years, and especially after the collapse of the Soviet Union and the emergence of a powerful neo-liberalist agenda, NGOs have often been seen as providing a bridge between capital and a poverty-focused development agenda. South Asia has been one of the richest areas of NGO experiment, and every country of South Asia has a tradition of NGO involvement in areas ranging from the environment to the meeting of basic health needs. From the outset relationships between NGOs and the formal sector were often distant, sometimes deliberately so. Närman has argued that in Sweden, where in the 1970s and 1980s NGOs were widely seen as a major channel for development, there has been an increasingly strong link between the government and NGO sectors, and a growing debate about the effectiveness of the role of NGOs in the development process (Närman 1999). Pushpa Pathak has argued in the context of her study of urban poverty reduction in India that 'in India co-operation between governments and NGOs appears to be a necessary and effective strategy for economic empowerment of urban poor men and women' (Pathak 1999: 261), although Main (1999) has suggested that Calcutta's experience suggests that NGOs have ranged widely both in their focus and in the strength of their relationship with government.

This chapter reviews the growing and changing NGO–government relationships, in the context of two South Asian countries where NGOs have played a particularly active role, India and Bangladesh. Several debates have emerged on the state–NGO relationship, calling for analysts to examine state officials and their ideologies, state policies and institutions, and state definitions of the parameters of politics. While recognising that there are, indeed, areas of common experience and interest between the state and NGOs in South Asia, this chapter questions why certain NGOs tend to collaborate with the governments, while others do not. Why do the governments and

144

NGOs collaborate reluctantly at times? This emerging relationship requires NGOs to rethink their positions, strategies and relationship with government. Finally, this chapter asks how to facilitate an enabling environment for NGOs and governments.

The chapter begins by analysing the emerging role of NGOs in South Asia under structural adjustment programmes. It then provides a critical review of the changing NGO–state relationship by comparing the social, political and economic contexts in India and Bangladesh, through particular themes such as service provision, accountability and co-ordination. It also highlights the changing nature of the state and how that influences the relationship. Finally, some challenges and issues facing the relationship between the NGO sector and government in South Asia are elaborated.

Emerging role and importance of NGOs in South Asia

In South Asia, NGOs are organisations that are generally formed by professionals or quasi-professionals from the middle or lower middle class, either to serve or work with the poor, or to channel financial support to special focus NGOs such as community-based organisations (CBOs) or grassroots organisations (GROs) of the poor. NGOs generally are non-membership organisations and have salaried employees. Young professionals have started joining NGOs and are motivated by job prospects in the sector. As the NGO sector has matured, job prospects rather than the commitment to social change motivate young professionals who run the sector today. They have also formed intermediary NGOs or public service contractors that provide research, consultancy and training services to the state, smaller NGOs and CBOs. While GROs and CBOs can be defined as organisations formed by members of the low-income community, most of whom offer their services voluntarily, it is also becoming common for CBOs to have salaried employees.

South Asia has a very diverse 'non-profit sector' which includes other types of organisations such as caste associations, business associations, promotion of culture and so on. Similarly voluntarism[1] is an integral part of South Asian society, and dates back to ancient times when it operated in the fields of education, medicine and cultural promotion, and in crises such as droughts and famines.

According to Korten (1990) there are three types of NGOs: those with a welfare orientation which provide services such as famine, flood relief or child sponsorship; those with a modernisation or developmental orientation, which support development projects that enable the poor to provide for their own basic needs; and those with an empowerment or conscientisation orientation, which see poverty as the result of political process and are committed to train communities to enter these processes in order to bring about social change. In South Asia, as elsewhere, it is not uncommon for more than one orientation to exist within the same NGO.

Corbridge and Harriss (in Chapter 5 of this book) have explored the background to the neo-liberalism which has characterised South Asian development since the early 1990s. They recognise that the economic libe-ralisation which South Asian countries have now adopted came about not only under the pressure of 'conditionality' from the World Bank and International Monetary Fund but also from domestic political interests to which governments have been responsive, even where they have not been the chief architects of programmes. It is clear that South Asian govern-ments have not fully understood their implications; nor have the programmes gained an electoral mandate.[2] Developments in the West con-tinue to be a powerful influence in decisions about the appropriate role of the state in developing countries such as those in South Asia. Liberalisation has required reduced expenditure by the state on public services. Initially the World Bank argued that any hardships in the post-liberalisation period were temporary blips. Later the Bank did agree to monitor the grassroots effects of adjustment, and to provide packages of short-term alleviation policies designed to target the worst hit 'vulnerable' sectors of populations and to relieve their most immediate problems, by encouraging the role of NGOs in targeting the poor. With the onset of these programmes, NGOs are generally promoted to deliver public services through grants from donors or governments as states withdraw from such activities (Robinson 1997). As a result governments have been prepared to channel increasing amounts of official aid to and through NGOs. Although comprehensive and accurate data are hard to obtain, there is some evidence to suggest that the rise and growth of NGOs relate to the increasing availability of official funding under this new policy agenda (Edwards and Hulme 1992).[3] Therefore, NGOs can advance their gentle anarchism based on the common objectives of both sides – reversing the marginalisation of the poor, alongside the selective government privatisa-tion of development efforts.

There is no question that NGOs have emerged as important agents in South Asia since the 1980s. With their meteoric rise as a new development force, most donors and some governments now agree that development would benefit from increased collaboration between the government and NGOs. For some they embody the development of a 'civil society' which is seen as a third force set against both state and market. NGOs have vital roles to play in development to maintain a thriving civil society and democratisation; hence governments and development agencies such as the World Bank or the Asian Development Bank are increasingly working with NGOs and promoting growth of the sector.

NGO–government relationships: collaboration or co-option?

Failures in structural adjustment have contributed to a renewed concern with the capacity of states and their support of the institutional conditions within which markets and citizens can flourish. These conditions include active civil society, 'good government' and effective public administration. The term 'governance' is often used to describe this broader view of the way that government should function (Kooiman 1999; UNDP 1997; World Bank 1992). Similarly, the term 'public management' is used to differentiate a new more open and entrepreneurial style from bureaucratic public administration (Minogue 1998; Dunleavy and Hood 1994). How far particular states have the capacity to move beyond a bureaucratic mode to the more interventionist and managerial role of partnering, promoting and regulating other actors is a key question (World Bank 1997; Larbi 1999; Batley 1999).

Should NGOs work with the state or against the state? Närman (1999: 168) has suggested that from one donor perspective, that of Sweden, 'trends in the NGO sector seem to be moving towards closer co-ordination with the rest of Swedish aid'. As part of that greater co-ordination the role and achievement of NGOs have come under closer scrutiny. In South Asia few states had a positive track record on social development/grassroots development, and many in South Asia judged states as doomed to perpetuate NGO subordination. Power and authority emanated from the state, and their inhabitants had limited agency. There are, however, observable differences in state–NGO relationships in various parts of South Asia, according to the social and institutional histories of NGOs, specific local conditions, institutional behaviour, local politics and actions of local agents. Governments' stance towards NGOs can be non-interventionist, encouraging, offering partnership, seeking co-option or controlling.

In India NGOs had a low priority in the state's policy dialogue with donors when addressing issues of structural adjustment, macroeconomic policy and structural reform. Consequently, state elites retained a strong role in decisions about development. Meanwhile, in Bangladesh, the 1990 World Bank review of poverty and public expenditure repeatedly compared NGO activities favourably with those of the state, and recommended the expansion of NGOs to supplement government efforts and provide for improvement in delivery of services. The review saw NGOs not as competitors to the state, but rather as potential new partners in a mutually advantageous collaborative project.

Social programmes financed by the World Bank and a host of state-sponsored social funds have co-opted NGOs into implementing state development plans in order to mitigate the harshest effects of structural adjustment programmes, one example being the Sri Lanka Poverty Alleviation Project (Hodson 1997). On the other hand, many governments have themselves been keen to harness NGOs for their own ends. The impact of the 'lost decade' associated with the 1980s debt crisis and declining international terms of trade for the majority of developing countries has been well documented, and governments have had little choice but to concede the important role of NGOs in providing social services. From their perspective, drafting NGOs into government-sponsored social funds poses the least threat to state legitimacy. In fact, these linkages can actually serve to boost state legitimacy, both among international donors (because civil society involvement, in the form of NGOs, complies with donor demands for good governance) and among the electorate (due to the provision of better services, which may be associated with the state development programme even if delivered by an NGO). At the same time, it has been noted that some governments direct social funds towards social welfare provision for politically sensitive groups, in order to minimise opposition to the state (for example, to the Muslim community in India, particularly after the communal riots of 1992 – see Sharma 1995).

NGOs are considered to be useful partners of governments and of development agencies as they are able to deliver certain types of programmes more effectively than others – because of their capacities for reaching the poor, or for mobilising people 'at the grassroots' and for encouraging participation, or, more simply, because they are thought to be able to do things cheaply. NGOs can offer essential 'local knowledge' about local conditions and the poor, delivering services to vulnerable and difficult-to-reach population groups. NGOs can point out how programmes can better serve different communities, and be tailored differently for different ethnic groups. By sensitising authorities to the ideas and preferences of the poor, they can help attune projects to their real needs, and through social mobilisation they can help the poor organise themselves, either to form co-operatives, demand changes, or ensure the accountability and probity of the project officers (Clark 1995).

These are vitally important roles, best served by NGOs, community-based organisations and others in civil society who are close to the poor and enjoy their trust. Some argue that NGOs are *better* at development than government or official donors. This is a sterile line of argument. NGOs will never be able to do many things or acquire the scale of government programmes. Their value is that they do *different* things. There are many reasons to promote close state–NGO relations. NGOs may be cost-effective, work in remote areas or be innovative. But their most important potential is for engaging citizens, particularly the poor, in shaping the decisions that affect them and in allocating associated resources. Popular participation is the main argument for improving state–NGO relations and fostering an enabling environment for NGOs, because ultimately 'development' is what is done *by* people, not *to* people.

Governments, therefore, need NGOs to help ensure that their programmes are effective, well targeted, socially responsible and well understood. The public need NGOs because of their services and their mobilising capacity, helping citizens express their voice, or challenge authority. Parliamentarians need them for policy guidance, for feedback on what people want, and as watchdogs in monitoring public programmes and enhancing the accountability of officials. But in general, civil society helps attune governments to their populations and strengthens mechanisms of democracy and accountability.

The creativity and impact of NGOs are understudied and imprecisely ascertained. The scale of NGO operations is still small in many South Asian countries, yet it is approaching significant impact in others. A recent account by Mahbubul Karim (1996: 132) estimates that NGOs now operate in 50 per cent of all villages in Bangladesh. What remains uncontested is his assertion that issues such as land tenure policies are the most difficult for

NGOs to influence, while NGOs such as Grameen Bank, Proshika and Bangladesh Rural Advancement Committee (BRAC), or Sarvodaya in Sri Lanka, are widely praised as models to be replicated.[4]

The impact of states upon NGOs is absolutely central in defining the role that NGOs can play in national development, for it is governments which give NGOs the space and the autonomy to organise, network and campaign (Fowler 1991). Of course, it is difficult to generalise about state–NGO relations as local political networks are always diverse. Government relations with NGOs have tended to range from tolerance to antagonism; at the very least, tensions have risen and fallen over time. Of course this differs between individual NGOs. NGOs are not homogeneous and it is perfectly clear that some NGOs will seek accommodation with the state while others will prefer to oppose it.

The attitudes and nature of the state also clearly influence its relationships with NGOs. NGOs which demand greater participation in setting and implementing the development agenda, for instance, are unlikely to be looked on favourably by a state that is resistant to democratic reform. On the other hand, even authoritarian states may tolerate NGOs that are politically opposed to them, if the state benefits from the subsidies gained by NGOs implementing development programmes as found in Bangladesh and Pakistan (Sen 1999: 328).

NGO–government relationships can generally be characterised by the hostility of politicians, party workers, local elites, lower-level bureaucrats and lower-level employees of the state towards NGO activity. Local power structures pose problems even for NGOs that are implementing the state's development programmes. Proper implementation of these programmes may alter material conditions, access to resources and information, status and positions of power, thereby threatening existing power structures at the local level. It is not surprising that those who are adversely affected will attempt to resist the loss of power. Conflicts arise between lower levels of bureaucracy and NGO officials in the form of laws pertaining to registration, income and taxation of NGOs. In my own field research in Mumbai (Bombay), many NGOs reported that in recent years they had frequently endured harassment by local bureaucrats regarding these issues (Desai 1999). Most NGOs are interested in what will work within their chosen geographical areas: what becomes irksome for NGOs working with the state is the slow procedural progress and the constant changes in policies, though progressive (at times), that punctuate each and every step in programme implementation. A further difficulty has been that of ensuring adequate co-ordination among various government departments. Continuing restrictions by government on access to details of future policies, programmes and projects cause difficulties for NGOs, while the government agencies feel that NGOs choose simply to disregard what is on offer domestically before approaching foreign donors for funds (Desai 1999). Most of the large NGOs or internationally reputed NGOs get called for government funding and programmes, while small and medium size NGOs get excluded. It is a small group of privileged NGOs that are invited on boards, to meetings or briefings and in programmes.

The nature of NGO–government relationships in India and Bangladesh

Here, I look at this question of collaboration or co-option with particular reference to India and Bangladesh. This issue has grown in importance during the 1990s: with the so-called transitions to democracy beginning to emerge in Bangladesh, greater attention is being given to the kinds of institutions that permit democratic openings, giving rise to opportunities for NGO collaboration with state organisations rather than opposition and conflict. In contrast, a majority of the states and municipal governments in India have long been open to such collaboration, seeking links with NGO activities, in order to overcome the financial crises existing at the local level. Both the national contexts and the particular activities undertaken by NGOs have been important in determining these differing relationships between the state and NGOs in India and Bangladesh: below, I consider the impact of each of these factors in turn.

The socio-political and economic context under which NGOs emerge tends to influence state–NGO relationships. Individuals who form NGOs under authoritarian regimes to bring about social change are likely to be distrustful of the state, while individuals who form NGOs for more pragmatic reasons, such as alternative sources of employment, are less likely to be hostile towards the state. For example, an NGO formed at a time of political repression develops an identity and culture that may make it difficult to trust the state, even if a transition to democracy takes place.

In India, the political context within which NGOs operate has meant that their relations with the state have not always been 'cordial' (Robinson 1993: 95). An era of co-operation between NGOs and the state in the early post-independence period (1947 to the late 1950s) was followed by the emergence of antagonism between the NGOs and the state in the 1960s and 1970s; this has been followed by increased control by the state in the 1980s and 1990s (Sen 1999).

Since the independence period the state has viewed NGOs as potential allies in reaching down to the electorate. Successive five-year plans have given NGOs a role in rural development (usually in rural social welfare provision), channelling substantial government funds to them in an attempt to subvert the entrenched elite interests (upon which the Congress Party's support base had largely been built) and to reach down directly to the grassroots, giving rise to what Clarke has called 'one of the most significant examples of co-operation between state and NGOs in the developing world' (1998b: 15). However, the increasing professionalism of the sector, combined with a growing tendency among many NGOs to criticise rather than support state development policy, has given rise to a more antagonistic relationship.

In the post-independence era significant elements of social development were left to the NGO sector; the state elite that inherited power enjoyed prestige that no individuals or organisations could oppose. The Congress Party dominated India's political space, but could also forge some semblance of authority at all levels of society through patronage linking the national elite with the regional and local elites. A democratic structure within the Congress which extended to the grassroots, along with its organisational ability to knit together the state and society, also resulted in a relatively monolithic state. A healthy relationship existed because actors in both institutions had common objectives and shared beliefs about development (Sen 1999).

In the second period between the 1960s and 1970s, there was growing dissatisfaction with state-led development in India. The leftist parties, government officials, academics and Indian Christians influenced by liberation theology believed that state-led development reached only the elites, that it had little effect in alleviating poverty and that the development model widened the gap between the rich and the poor. Clearly, such beliefs were not conducive to collaboration with the state. The Congress lost its close network of local-level political connections which, along with the apathy of other political parties, created a political vacuum at the grassroots (Sen 1999).

On re-election in 1980, the Congress Party organised a major inquiry into the activities of NGOs (the Kudal Commission), which was noted in particular for its harassment of NGOs associated with the political opposition during the election campaign (Sen 1999). At the same time, the emergence of issue-based networks campaigning for a variety of causes such as housing, women's rights and the environment began to place greater pressures on the state. This was also a period during which NGOs became involved in long-standing religious, social and cultural cleavages which were thought to be threatening national cohesion (Sen 1999). With the Seventh Five-Year Plan in the late 1980s the government launched an initiative which was decried by NGOs as an attempt to steer them towards service provision and away from political activities. The foundation in 1986 of the Council for Advancement of People's Action and Rural Technology (CAPART), ostensibly to promote the role of NGOs in agricultural technology and rural development, but which in practice became a vehicle for state control over NGOs, was heavily condemned by the NGO sector (Clarke 1998b; Farrington and Lewis 1993). Administered by the Ministry of Rural Development, and largely distrusted by many rural NGOs, CAPART was an attempt by the state to release funding to NGOs to support its own development strategy and mitigate the effects of structural adjustment in rural areas (Sen 1999). State–NGO relations were further strained in 1986 when the government established the Council of Voluntary Agencies with an express mandate to develop a Code of Conduct for NGOs which would prevent them from participating in commercial or political activities and which would call for greater financial accountability (Clarke 1998b). Opposition to the imposition of such a code was fierce and ultimately the government was unable to implement it (Clarke 1998b). However, Sen (1999: 342) suggests that the opposition was voiced mainly by certain officials from established NGOs and that the idea for the Code of Conduct originated from a particular NGO official who had worked closely with the state for several years. This suggests, as Sen points out, that not all NGOs are united in their opposition to the state; and moreover, that particular NGOs seek accommodation with it.

In recent years the state in India has increased and diversified government funding for development programmes aimed at poverty reduction by NGOs. The Seventh Five-Year Plan 1987–92 stated that serious efforts would be made 'to involve voluntary agencies in various development programmes' and set aside Rs2.5bn for NGOs (Sethi 1988). Similarly the Eighth-Five Year Plan (1992–7), called for increased participation of NGOs in improving the delivery of social services and in ensuring people's participation for micro-level planning (Government of India 1992). This is well illustrated in the National Housing Policy (NHP), for privatisation of housing provision, and NGOs are expected to deliver housing services to low-income communities.

Meanwhile, in Bangladesh continuing dependence on aid from ex-colonial powers has ensured ongoing tensions regarding national sovereignty, and, until the 1991 elections, the military basis of the state raised fundamental questions about its internal legitimacy. In this context, the development of the NGO sector is inherently controversial. NGOs are Bangladeshi but foreign-funded, and they claim to speak for the poor majority in a way that potentially threatens the scope and legitimacy of state aspirations to embody the nation (White 1999: 311). Some donors see the promotion of better contracts between NGOs and government departments as a way of increasing the efficiency of the latter (Bebbington 1991). Foreign aid has undoubtedly played an important role in stabilising the state in Bangladesh and has influenced the relationship between the state and NGOs. Suspicions in the Planning Commission after liberation that aid was likely to compromise Bangladesh's independence quickly gave way to the recognition that the state could not function without it (Faaland 1981, quoted in White 1999: 311). NGOs in Bangladesh have had a particularly high profile since the disastrous floods of 1988, when they were at the forefront of relief and rehabilitation. These floods were experienced as a national crisis, and parallels were frequently drawn with the liberation struggle, when in a similar way differences and self-interest were forgotten in the enthusiasm to reconstruct the country after the devastation of war. This pattern was reconfirmed in 1991 when the NGOs again were among the first on the scene after the devastating cyclone hit the south-east of the country (White 1999: 310).

In Bangladesh government employees do not like to have their departments' programmes compared, almost always unfavourably, with those of NGOs. There is also without doubt both envy and distrust among higher government officials of the economic resources and increasing political influence wielded by senior NGO leaders. In both Bangladesh and India in the 1980s and 1990s, new types of NGOs began to emerge, which were formed by retired bureaucrats, business people and political parties. These types of NGOs seem to have a good relationship with the state because of their founder-members' connections with state officials. Funding for these NGOs is also secured through these connections: the links can often take the form of patron–client relationships. These sort of intermediary NGOs and public service contractors maintain a good relationship with the state, as they rely upon it as their principal source of projects.

The nature of NGO activity and programmes thus largely determines state–NGO relationships. Most states welcome NGO activity which is oriented towards welfare and modernisation. This is especially true if the services provided by the NGOs substitute or supplement the state's efforts. NGOs which provide services are non-profit organisations that sell their services to aid donors and governments to implement development programmes. Such NGOs[5] are often driven by markets rather than by the values and motives that are generally associated with NGOs. These NGOs obtain projects from governments or donors either by competitive tendering or through subcontracting on a non-competitive basis. A distinction should be made between government policies that promote or facilitate NGO initiative and policies for NGO collaboration/co-operation.

NGO interaction in the context of housing and land policies for the urban poor, or activities that are empowerment or conscientisation oriented often generates an antagonistic reaction from the state. There seems to be a realisation among the NGO sector that empowerment-oriented projects cannot be sustained without providing material benefits to their beneficiaries. The state is not willing to accept alternative development models introduced by empowerment-oriented NGOs,[6] or feels threatened by such NGOs. It can be argued that these policies have reduced the political space for NGOs (Sen 1999). Similarly in Bangladesh, in the political upheavals surrounding the forced resignation of President Ershad in December 1990, the NGOs were the subject of bitter criticism because they supported the pro-democracy movement, and took part in the monitoring of the parliamentary elections.

Eviction of the poor and their relocation to the outskirts of the city by the city authorities were common in Mumbai until the mid-1980s, despite the stated policy since the early 1970s of upgrading existing slums. As Hasan has shown (Chapter 11 in this book), the dramatic growth in 'slum' populations was common across South Asia, and represented a challenge to governments in Pakistan as well as India, though the planning response was not always the same. The state of Maharashtra was also unwilling to provide land for the poor, despite legal provision under the Urban Land Ceiling and Regulation Act of 1976. NGOs oriented towards the empowerment processes became heavily involved in demanding low-income housing for the poor and insisting that these poor people had equal rights to live in the city. During these times NGOs were seen as honest brokers between state and society, which fostered broader-based social and political participation. This is a key part of the appeal of NGOs for the poor, as this relationship with NGOs guarantees a strong committed partner, which might serve as protection in times of crisis for the poor. These NGOs had a very conflictual relationship with the state, while other NGOs involved in welfare activities in the delivery of services (e.g. health, education) had a collaborative relationship with the state (Desai 1999). An innovative approach to urban health management, funded by the World Bank entitled India Population Project V, was implemented by the Bombay Municipal Corporation (BMC). By 1991 15 health posts (centres) had been established in collaboration with NGOs. Another collaboration in the health sector is that of the BMC collaborating with NGOs for leprosy control and eradication through geographical division of various municipal wards between different NGOs for total leprosy control.[7]

This means that there is a powerful incentive for the NGOs to shy away from more controversial approaches and settle for the safe aspects of welfare provision. This process may lead to a widening rift between well-resourced service providers and poorly funded social mobilisation organisations. In South Asia the dominant trend has been towards co-optive relations between governments and NGOs, where the former have sought to place various limits and restrictions on NGO functions and activities (see Heyzer *et al.* 1995).

The NGO–government relationship: issues and challenges

Controlling NGOs

The kernel of the NGO–state relationship is usually the framework of laws and regulations which govern the formation and operation of NGOs. Ideally this framework is fully enabling while instilling some discipline. Laws which hamper the formation of independent NGOs, which deny citizens' rights to join or support NGOs, or which subject NGO operations to strict government control and unpredictable intervention, clearly fetter the NGO sector. Conversely, where laws are so lax that anyone can form an NGO, register it for tax and fundraising advantages, and where there are no rules for ensuring a modicum of transparency and accountability, the public are unprotected from unscrupulous NGO operators.

Many states sought to gain a measure of control over NGOs through an overhaul of legislation governing NGO sector activities. Debates over NGO legislation in many countries have often been highly politicised (at times acrimonious) in which both governments' and NGOs' fears and suspicions about each other have become condensed into wrangling over civil liberties, the influence of foreign donors in sovereign countries, and state attempts to harness NGOs. In Bangladesh for example, Tvedt suggests that NGOs and the government were involved in a 'tug of war' (1998: 183) in 1992 over increasing state efforts to control what were considered to be subversive and divisive activities of many NGOs, fuelled by Western ('imperialist') funding.

In India, the state increased its control on foreign NGOs and foreign funding for NGOs, as it believed that it should nurture its own developmental capabilities and also alleged that NGOs were involved in subversive activities (see Sen 1999). The Foreign Contribution Regulation Act (FCRA), enacted in 1976, aimed at maintaining surveillance of NGOs which received foreign funds. NGOs receiving funds had to get clearance from the ministries of Home Affairs, Finance and External Affairs, and the ministry which had jurisdiction over the programme to be funded. Many empowerment-oriented NGOs were denied FCRA registration, which meant they could not access foreign funding. Many NGO leaders even today have argued for liberalisation of the FCRA on the grounds that liberalisation has been taking place in all other sectors of the Indian economy. The Act empowers the state to ban any organisation from receiving foreign contributions, should the state consider that organisation to be a political rather than 'neutral' NGO (Sen 1999).

Similarly in Bangladesh, NGOs have to register with the Government NGO Affairs Bureau to receive funds from outside the country. Registrations increased from 494 in 1990–1 to 986 in 1994–5 and by far the greatest increase within this was due to local NGOs, the registrations of which rose from 395 to 848 over the same period (World Bank 1996: 5, quoted in White 1999). Concentration within the NGO sector means that it now contains a number of very large organisations, some of which can rival government departments in size. The feeling that NGOs are favoured by the foreign donors is obviously one that brings its own tensions.

Tensions between the state and the NGOs are most clearly evident in government moves to tighten control of NGO activities, and particularly their access to foreign funding. By 1989 the system for approval of projects had virtually broken down in Bangladesh. Structurally, state suspicion of NGOs was expressed in a highly complex and inaccessible bureaucratic procedure; less formally, state–NGO tensions found expression in the demand among government officials for extra payment if applications were to be processed (White 1999: 312).

Staffing issues in NGOs

Many NGO staff have been recruited into government office, strengthening ties between the state and the NGO sector. For example, representatives of the NGO sector were involved in establishing a Co-ordinating Committee for Vulnerable Children (CCVC) in Mumbai which encouraged the government to work with NGOs and which recognised the right to organise and to participate in structures outside of the state. Further provision for their participation followed; NGOs (following years of campaigning) assisted in the drafting of the Institutional-Plus Programme which provides for NGO representation in planning and implementing government programmes for street children in the city of Mumbai.

Many NGOs are staffed by urban, educated, middle-class elites with no substantive roots in underprivileged groups. Those in the best position to take advantage of the donor penchant for NGOs are often urban-based educated elites, professionals or ex-government officers or civil servants who have access to information, other civil society organisations, and donors. This is particularly true in the large metropolitan cities and capitals where donor organisations and foreign embassies tend to congregate. As a result, the urban middle classes are over-represented within this growing NGO sector and often lack mass-based poor constituencies. Under these circumstances, the extent to which the agendas of such NGOs represent the real concerns of marginalised groups is a real concern, for they may simply reflect the types of activities which donors are willing to fund or which urban-based elites deem to be important (Farrington and Bebbington 1993). Accordingly, urban NGOs in India have yet to make their mark where entrenched power networks pose a formidable obstacle to urban NGOs' attempts to lobby for issues such as housing rights of the slum dwellers in metropolitan cities of Mumbai, Kolkata (Calcutta) and Delhi.

NGOs are often internally undemocratic, characterised by authoritarian or charismatic personalised leaderships, competitive, riven along class, gender, religious and ethnic faultlines, and steered by either the state, or donors, or both. Internal undemocratic workings of NGOs and NGO sectors pose problems for their supposed role in democratising work. In extreme cases, NGOs are set up to access government or donor funding, whereby members, beneficiaries and entire projects are fabricated in order to win funds which are promptly embezzled (Clark 1995). More generally, NGOs must be seen as a constituent part of the culture in which they work. In Bangladesh, Wood (1997) argues that hierarchical and authoritarian social structures which encourage dependence and deference will often be replicated in NGOs. In particular, the tendency for NGOs to be headed by one charismatic (often founding) leader can stifle individual staff autonomy and discourage wider staff participation in decision-making processes, which is ironic given the role NGOs are supposed to play in empowering their beneficiaries. It is not a problem that is restricted to Bangladesh or India, but is a concern common to many countries in the South.

Impact of increased donor funding to NGOs

The increased availability of funding via the social programmes has also given rise to confrontation, competition and sometimes convergence between NGOs as they scramble for a piece of the funding pie, and ultimately for their organisational survival. As NGOs become co-opted into government programmes, other, more radical NGOs have been marginalised and find themselves unable to operate or to attract funding.

The increased availability of large-scale funding has been one of the primary factors driving NGO growth in the 1980s, encouraging the proliferation of social welfare organisations, which often had little or no political agenda. Edwards (1999) has highlighted a general trend towards donor-funded service provision at the expense of political activities in India: such a trend has also been noted in Bangladesh (Wood 1997; White 1999).

The impact of increased direct donor funding on NGOs' internal bureaucratisation, management structures and relationships with beneficiaries is well documented (Edwards and Hulme 1992; Hulme and Edwards 1997). Reflecting on the experience of NGOs in Bangladesh, White (1999) suggests that NGOs are increasingly adopting formal procedures required by donors and developing the conservatism and self-protection usually associated with state agencies. This has been compounded by the donor-funded growth of some of the country's largest NGOs which have become 'formidable institutions.... Increased size has inevitably meant increased distance from the grassroots, and the early pioneering vision has been replaced by an ethic of efficiency and professionalism' (White 1999: 321). White associates the large amount of funding received by a number of Bangladesh's largest NGOs with the fact that they have become relatively autonomous both from the grassroots and from the donors themselves. BRAC (Bangladesh Rural Advancement Committee), for example, generates approximately one-third of its own income and is large enough to be independent from any one single donor. This may be positive in terms of reducing dependency, but there are deeper issues of accountability, legitimacy and representation, as White (1999: 321–2) asks:

> Who owns BRAC? To whom is its executive of the 'great and the good' accountable? Nominally the NGOs are accountable to their members – the Grameen Bank, for example, states that its members are its 'stake-holders' – but it is not clear that there are any institutional mechanisms to ensure that this nominal 'stakeholding' is transformed into an effect-ive 'shareholding' which guarantees members a real say.

Service provision

Another issue is that increasing dependence on one-off contracts could increase NGO financial insecurity. Some may find that service delivery contracts may reduce an NGO's own space for reflection and action (see experiences from South America in Aguirre and Namdar-Irani (1992)). Undoubtedly, many NGOs see the provision of certain services as properly within the mandate of government and are unwilling to take the role of substituting for services. They would argue that NGO provision is poten-tially unsustainable and places a key role of the government in the hands of NGOs which are unelected and, therefore, potentially unaccountable for their actions. By contrast, Western neo-liberal ideology traces the achieve-ments of NGOs and offers prescriptions for how NGOs could contribute further to democratic development given more favourable conditions or funding, which has tended to push NGOs towards becoming public service contractors in India and Bangladesh.

The influence of the neo-liberal agenda has been far-reaching. Edwards and Hulme (1995) drew attention to the fact that NGOs could not be expected simultaneously to fulfil their roles as social welfare providers and as democratic actors in the political process as donors and academics had imagined. The key issue here is that if NGOs are taking up greater roles in social welfare activities as more donor funding becomes available, then the legitimacy of their claims to work with and represent the interests of the poor and disenfranchised on a political level come under threat.

NGOs find themselves steered towards social provision and away from grassroots activism. The resulting 'crisis of legitimacy' which many NGOs are facing entails the loss of linkages with grassroots constituencies (and with it the close relationships required in order to represent their interests in policy-making fora). Accountability now flows upwards to the donor rather than downwards to poor beneficiaries, and grassroots organisations are excluded from policy dialogue (Bebbington 1997; Farrington and Lewis 1993).

NGOs are wary of increased links with government for fear of losing their identity and autonomy, and of becoming public service contractors for the state. On the other hand, some government officials question the repre-sentativeness of NGO leaders. In some cases, the role of the NGO seems to be to support and legitimise government policy rather than to question it. In other words, NGO participation serves to legitimise the status quo, not to challenge it.

NGOs' role in civil society

Civil society is a more problematic sphere of competing interests across both state and society, in that it is not an inherently democratic space. It is also far too simplistic to suggest that the mere existence of an NGO sector can be expected to boost both civil society and the democratic process. Sheer numbers of NGOs in India and Bangladesh cannot be used as an indicator of the strength of civil society, especially given the diversity that exists within the sector.

Few attempts have been made to take critical stock of the wider picture of NGOs' political role. At best, it is evident that NGOs' role in democratising development has been overstated; at worst, it is possible that NGOs are becoming key actors within the wider neo-liberal project. Clarke (1998a) argues that NGOs have grown in India to fill the institutional vacuum caused by the weakness of political parties and trade unions.

NGOs in India or Bangladesh have not been active participants during election campaigns, particularly at the local level, either through directly participating in electoral contests, supporting particular parties and personalities or by influencing the agendas of other candidates. However, through their wider actions, NGOs have sometimes undermined the interests and patron-clientelism of the elite which had become entrenched within the political realm over several decades since Independence.

Both in India and in Bangladesh, the state in the post-structural adjustment period has not been able to meet the demands of the poor and the grassroots organisations. At the same time, the NGOs demonstrated a notable lack of enthusiasm for participating in the political mobilisations of poor people that occurred during the 1980s and 1990s. For instance, NGOs took no role in mobilising poor people against structural adjustment programmes that would adversely affect their living standards. The weakening of state agencies and grassroots organisations, coupled with the strengthening of NGOs, which are heavily dependent on international donor funds, undermines the institutional capacity of South Asian countries to define and defend alternatives to the development agenda articulated by international financial institutions and development agencies. This mutes voices of opposition and fundamentally weakens democratic political processes. It is ironic that NGOs, which generally see themselves and are often seen by others as agents of democracy, have been instrumental in undermining the institutional bases of political participation in this way.

NGOs serve to aggregate and moderate political demands through which disputes can be negotiated and dissipated. As Feldman (1997) argues with respect to Bangladesh, the growth of foreign-funded NGOs is merely 'institutionalising representation', channelling all protest and citizen participation into the NGO sector which has come to represent a 'legitimised...controlled, organised arena of public debate with institutional and financial support from the donor community, (which) has come to speak on behalf of the citizenry' (1997: 59). Moreover, 'the participation of the donor community in NGO initiatives corresponds to a move toward the privatisation of resource distribution and forms of production away from locally initiated and locally controlled development activities' (*ibid.*).

Governments are distrustful of NGOs, especially if the sector sows seeds of political discontent and provides organisational channels through which opposition can be mounted against the incumbent regime (used for popular mobilisation against the state by other political parties). Governments also scrutinise NGOs when separatist agitation is at play or when NGOs try to maintain autonomy from the state through foreign funding. This highlights some of the important debates in the sector on NGOs' political role or alignment to political parties. This might be a central issue in collaboration and might raise significant questions regarding NGO leadership. Can the NGOs influence their much less like-minded government to expand the social and political space for people's organisations?

NGOs have little substantial influence at the policy formulation stage in either relatively undemocratic Pakistan or relatively democratic Sri Lanka. It is clear that the room to manoeuvre that is left at the implementation stage is typically insufficient to enable implementing agencies to have a significant impact on policy outcome, even if they have been shut out of the earlier stages of policy-making.

Accountability of NGOs

Generalisations are dangerous. Not all NGOs are effective; some talk a fine patter but do little for the vulnerable. In India and Bangladesh, NGOs congregate close to big cities such as Dhaka, Mumbai or Kolkata but are less common in the poverty belts of Bihar and Rajasthan. There are important questions of accountability and legitimacy, regarding for example their claim to 'speak for the poor' (Edwards and Hulme 1995). But the sector as a whole plays important and increasingly powerful roles. This may include challenging decisions of states, but in ways which shift governments to better serve their responsibilities vis-à-vis their populations, and therefore enhance their legitimacy. Legitimate governments, therefore, have no need to fear a well-functioning and responsible NGO sector.

Indeed, the question of NGO accountability to, and legitimacy among, the grassroots, given their increased funding, has become a dominant theme within the recent literature (Edwards and Hulme 1995). As discussed earlier, increasing financial support to NGOs has encouraged the proliferation of particular types of organisations with distinctive social and geographical characteristics. Getting the balance right is not easy, and will depend on the legal tradition and other aspects of the country in question. Governments should not seek to manage NGOs – this would undermine NGOs and inappropriately stretch government capacity. Instead they should create conditions that encourage effective self-regulation of the sector. NGOs which seek benefits from the state or the public should be expected to be transparent and accountable, proportionate to the scale of these benefits.

Accountability is now a key concept for NGOs. Currently, debates are less likely to be anti-state or pro-state, but rather to address the question of how people engage with the state and public affairs for accountability, in mainstream policy issues (Staudt 1998). Whatever the limitations of representative democracy, the state is at least nominally accountable to its citizens through the process of elections, and even law. By contrast, NGOs espousing participatory approaches to change in matters affecting their clientele, rarely permit participation by the poor in influencing the size, structure and objectives of their own organisation.

Also NGOs are just as susceptible to failure as governments. NGOs need new skills and capacities to mediate these linkages. This new relationship calls for major changes in NGO roles, relationships, capacities and account-abilities. The policy implication is that NGOs need to become much better at documentation, analysis and at learning from their own experience, and much better at understanding the local contexts of their projects; that they should pay much more attention to financial monitoring and to analysing costs in relation to benefits.

In all South Asian countries NGOs are characterised as either selflessly dedicated to the poor or self-interested charlatans; as supporting where government provision falls short or as agents of foreign powers; able to act effectively with a minimum of bureaucracy or a bunch of amateurs with funds far beyond their managerial skills. Corruption of NGOs will be the political game in the years ahead. One of the ways NGOs can avoid corruption is to develop systems for performance, monitoring and accountability.

Pressures on NGOs towards more formal procedures are likely to increase as the official donors look towards direct funding of NGOs. As they develop into institutions, NGOs almost inevitably take on some of the conservatism, the measures for self-protection and perpetuation, that are characteristic of state agencies. In many cases in South Asia, NGO staff modify their aims and programmes so as to accommodate the existing power structures, rather than attempt to challenge them.

Problems of NGO co-ordination

The fragmented nature of the NGO sector has ramifications for NGO co-ordination. Co-ordination among Indian NGOs is generally weak (Robinson 1993): despite the efforts made by NGOs to establish the Voluntary Action Network India (VANI), the unity of the Indian NGO sector remains relatively remote due to deep ethnic, religious, political and regional cleavages in civil society.

Strategic networking and shifting from service delivery to advocacy have been under way for some time in South Asia, although it remains difficult to ascertain to what degree these shifts have occurred. NGOs need to be selective about their foreign partners; their policy impact may also depend on selecting and building coalitions with key government deci-sion-makers or using what some NGO leaders describe as 'acupuncture' – targeting the weak points of a sick system.

State–NGO relations have taken a new form with the emergence of issue-oriented campaigning by networks of NGOs. These networks are formed around popular issues such as environment, women's rights, pavement dwellers and housing. A good example of such a campaign is the *Narmada Bachao Andolan* (save Narmada agitation) initiated by various developmen-tal NGOs, GROs, activists and intellectuals to oppose the Narmada Valley project (referred to by Chapman in Chapter 4 of this book).

If we look at NGOs which are engaged in the more routinised roles of service delivery, solidaristic group formation and advocacy within the societies of South Asia, there is a need to link discussions of co-ordination with greater levels of participation in the formation of social capital and institutional development at the grassroots.

On a wider scale, NGOs have much to learn from each other. As the densities of NGOs operational at field level increase, the need for institutionalised means of communication (and, where appropriate, co-ordination) among NGOs and between NGOs and government has become all the more urgent.

Conclusions

The state welcomes NGO activity which is welfare-oriented, especially if it substitutes for or supplements the state's own efforts. On the other hand, the Indian and Bangladeshi states have not encouraged, or even tolerated, empowerment-oriented NGOs. It is clear that while complementarities have often been acknowledged between NGOs and government agencies, far less thought has gone into the practical aspects of creating linkages. Many of the documented collaborations have been on an ad hoc basis, formed from local, spontaneous action or from higher-level contacts achieved around the efforts of key personalities. In general, there has been less attention paid to the question of building trust and informal alliances or the sharing of decision-making. While both NGOs and governments cannot ignore each other, the formal mechanisms for creating mutually agreed partnerships rarely exist.

NGOs' actions and their relations with the state are determined to a large extent by the political, economic and cultural context in which they operate. In these contexts lies a source of diversity. The particular political conditions and bureaucratic procedures of government authorities experienced by the NGO control many of the opportunities of NGO activity, and set very specific parameters within which NGOs can operate. All this has involved NGOs in a series of trade-offs. They can promote certain activities as opposed to others. They can focus on certain groups at the expense of others, perhaps with quite strong implications for the overall degree of apparent success achieved. They can engage intensively with a small number of communities, or more superficially with larger numbers.

At a deeper level, there are worries about the long-term impact of NGO service provision on the sustainability of national health and education systems (rather than programmes) and access to quality services for all. Robinson (1993) points out that large, influential and well-funded NGOs may be able to 'concentrate resources in regions and sectors that might not be most important for national development', with a 'patchwork quilt' of services of varying quality emerging without any overview of overall needs. There are echoes here of debates about the public/private mix in social services in countries such as the UK and USA. The evidence from these debates suggests that such fears are well founded, with unequal access and spiralling costs already a reality.

NGOs have traditionally defined themselves in relation to the state. As South Asia's experience shows, they operate in an enormously varied range of geographical and political contexts. At one extreme, the Maldives and Bhutan are among the world's smallest states, while in terms of population India, Bangladesh and Pakistan are among the world's largest. Similarly, the countries of South Asia have experienced political systems ranging from political dictatorship to democracy. NGOs thus have to adapt to an enormous range of social, economic and political conditions. As the state is transformed by the wider agendas of political change, democratisation and a reduction in resource flows to governments, NGOs are required to rethink their role (Bebbington and Farrington 1993). Political circumstances are changing rapidly, but it would be a mistake to conceive of this change as a linear movement towards greater freedom. If NGOs do not improve the efficacy of political participation, they can hardly be said to have strengthened civil society. There is no widespread evidence that NGOs do improve the efficacy of political participation. Some NGOs have succeeded in widening political debate and political participation, championing the causes of the poor and disenfranchised, and challenging the state on policy or legislative matters. However, such evidence is generally forthcoming from only a handful of NGOs.

NGOs face an unresolved dilemma in most societies – how to engage in the political process in order to achieve fundamental changes in the distribution of power and resources without becoming embroiled in partisan politics and thereby losing their independence. It is necessary to be creative and flexible in setting up appropriate mechanisms that will enable NGOs to engage directly in partisan politics with minimum risk to their credibility, independence and survival. This seems a very tall order indeed. In the absence of any obvious solutions to this dilemma, most NGOs have tended to concentrate on supporting the democratisation of informal politics through training grassroots activists, local institutional development, micro-policy reform, and education for citizenship.

Active partnerships are those built through ongoing processes of negotiation, debate, occasional conflict, and learning through trial and error. Dependent partnerships, on the other hand, have a 'blueprint character', with relatively rigid assumptions about comparative advantage, and are often linked to the availability of funding. Partnerships may bring extra costs which are easily underestimated, such as new lines of communication requiring demands on staff time, new responsibilities for certain staff; and the need to share information with other NGOs. Building partnerships is likely, therefore, to be difficult. In order to promote a healthy policy environment, governments can provide NGOs with information about state programmes and policies for dissemination to their constituencies and gathering feedback. NGOs can help governments strengthen citizen consultation.

Governments can offer opportunities for operational collaboration, commissioning NGO activities that complement their programmes and strengthen the NGOs. Governments can co-ordinate – or encourage co-ordination – between the various agencies (non-governmental, governmental, donors and private sector) who work in a common field. This is an important element of the 'Comprehensive Development Framework' approach currently being promoted and piloted by the World Bank (Wolfensohn 1999). It may be that South–South learning between NGOs is an important priority, since exchanged knowledge and experience drawn from comparable contexts may be more relevant (Lewis 1997).

Governments can involve NGOs in policy debate and public consultations on new policies or major government projects. For example, NGOs can help orchestrate effective public hearings about infrastructure projects. The World Bank now *requires* public hearings, including civil society, for all projects it funds that may have a significant environmental impact.

Training should be given to government officials to improve their skills to work with local groups, understand methodologies and techniques of community participation, and communication and interpersonal skills.[8] It is important to recognise that not only institutional restructuring but also changes in organisational culture are both crucial to sustain an ongoing commitment to NGO/government collaboration and people's involvement in programmes, otherwise most of the schemes will only exist on paper.

In all these areas there is potential for conflict and difficulties. NGOs may be at odds with governments; collaboration by some may appear to undermine advocacy efforts of others. Government agencies may seek to co-opt (and possibly corrupt) the NGOs it works with, or NGOs may find their own agendas get lost.

Notes

1. Many South Asians still believe in a notion of voluntarism which is essentially romantic, inspired by self-initiative and social commitment. Gandhi's voluntarism emphasised empowerment and transformation of the society and acquired a political content in place of the issue-based voluntarism of the nineteenth century. During the freedom struggle, dedicated workers from the Indian National Congress undertook rural development programmes as well as participating in India's independence movement.
2. For further discussion of economic reforms in India, see Corbridge and Harriss, this volume.
3. Growth in numbers has been rapid in North and South. Hulme and Edwards (1997) report a growth in the number of development NGOs registered in OECD countries from 1,600 in 1980 to nearly 3,000 by 1993 and a growth in expenditure from US$2.8bn to $5.7bn. Clark (1991) reports a doubling in resources over the decade from the mid-1970s to the mid-1980s, mainly as a consequence of increased official contributions rather than private generosity.
4. Proshika has organised 773,400 people in 44,400 groups across the country and claims to reach nearly 4 million individuals (Lewis 1997). BRAC is an NGO with more than 12,000 staff and its rural development programme has so far reached 1.42 million households (Lewis 1997). BRAC generates 31 per cent of its income from its own business sources, and is seeking to consolidate this with the recovery of 'user costs' from its group members. Its size means that it is effectively independent of any single donor.
5. Korten (1990) refers to these as public service contractors.
6. The Seventh Five-Year Plan defined NGOs as politically neutral developmental organisations, which would help the government in its rural development programmes (Government of India 1985). This would exclude empowerment-oriented NGOs.

7. In the early 1990s, there were 14 organisations working in Bombay for leprosy control and 156 centres have been established where free clinical examination and free treatment was available (Source: Alert India Report July 1993). Traditionally the field of leprosy control and eradication was seen as one where the voluntary sector could have much greater impact, because of the highly sensitive nature of the ailment and the social stigma still attached to its detection and subsequent treatment.
8. In the year 1989–90 in Bombay, all the Health Post Staff of about 600, had undergone a one-week intensive training input in 'community work' organised by the Department of Urban and Rural Community Department, Tata Institute of Social Sciences, Mumbai.

References

Aguirre, F. and **Namdar-Irani, M.** (1992) *Complementaries and Tensions in NGO–State Relations in Agricultural Development: The trajectory of AGRARIA (Chile).* Agricultural Research and Extension Network, Network Paper No 32, Overseas Development Institute, London.

Batley, R.A. (1999), 'Introduction' and 'The new public management in developing countries', *Public Administration and Development,* **11**: 755–65.

Bebbington, A.J. (1991) 'Sharecropping agricultural development: the potential of GSO-government cooperation', *Grassroots Development,* **15**: 20–30.

Bebbington, A. (1997) 'New states, new NGOs? Crises and transitions among rural development NGOs in the Andean region', *World Development,* **25**: 1755–65.

Clark, J. (1991) *Democratizing Development: The role of voluntary organizations.* Earthscan, London and Kumarian Press, West Hartford, CT.

Clark, J. (1995) 'The state, popular participation and the voluntary sector', *World Development,* **23**(4): 4.

Clarke, G. (1998a) *The Politics of NGOs in South-East Asia: Participation and protest in the Philippines.* Routledge, London.

Clarke, G. (1998b) 'Non-governmental organisations (NGOs) and politics in the developing world', *Political Studies,* **XLVI**: 36–52.

Desai, V. (1999) 'Urban NGO–government relationship: experiences from Bombay', in Graham P. Chapman, Ashok K. Dutt and Robert Bradnock (eds), *Urban Growth and Development in Asia: Living in the cities,* Vol. II. Ashgate, Aldershot: 368–87.

Dunleavy, P. and **Hood, C.** (1994) 'From old public administration to new public management', *Public Money and Management,* July/September: 9–16.

Edwards, M. (1999) 'NGO performance – what breeds success? New evidence from South Asia', *World Development,* **27**: 361–74.

Edwards, M. and **Hulme, D.** (eds) (1992) *Making a Difference: NGOs and development in a changing world.* Earthscan, London.

Edwards, M. and **Hulme, D.** (eds) (1995) *NGO Performance and Accountability: Beyond the magic bullet.* Earthscan, London and Kumarian Press, West Hartford, CT.

Faaland, J. (ed.) (1981) *Aid and Influence: The case of Bangladesh.* Macmillan, London.

Farrington, J. and **Bebbington, A.** (1993) *Reluctant Partners? Non-governmental organisations, the state and sustainable agricultural development.* Routledge, London.

Farrington, J. and **Lewis, D.J.** (eds) (1993) *Non-governmental Organisations and the State in Asia: Rethinking roles in sustainable agricultural development.* Routledge, London.

Feldman, S. (1997) 'NGOs and civil society: (un)stated contradictions', *Annals of the American Academy of Political and Social Science,* **554**: 46–65.

Fowler, A. (1991) 'The role of NGOs in changing state–society relations: perspectives from Eastern and Southern Africa', *Development Policy Review,* **9**: 53–84.

Government of India (1992) *Eighth Five Year Plan, 1992–97.* Vol. I, *Objectives, Perspectives, Macro-Dimensions, Policy Framework and Resources.* Government of India, Planning Commission, New Delhi.

Heyzer, Noeleen, Riker, James and **Quizon, Antonio** (eds) (1995) *Government-NGO Relationships in Asia: Prospects and challenges for people-centred development.* Macmillan Press, London and St Martin's Press, New York.

Hodson, R. (1997) 'Elephant loose in the jungle: the World Bank and NGOs in Sri Lanka', in D. Hulme and M. Edwards, (eds), *NGOs, States and Donors: Too close for comfort?* Macmillan, London: 168–90.

Hulme, D. and **Edwards, M.** (eds) (1997), *NGOs, States and Donors: Too close for comfort?* Macmillan, London.

Karim, M. (1996) 'NGOs in Bangladesh: issues of legitimacy and accountability', in M. Edwards and D. Hulme (eds), *Beyond the Magic Bullet: NGO performance and accountability in the post-Cold War world.* Kumarian Press, West Hartford, CT.

Kooiman, J. (1999), 'Socio-political governance: overview, reflections and design', *Public Management,* **1**(1): 67–92.

Korten, D.C. (1990) *Getting to the Twenty-First Century: Voluntary action and the global agenda.* Kumarian Press, West Hartford, CT.

Larbi, G.A. (1999) *The New Public Management Approach and Crisis States.* Discussion Paper No. 112, United Nations Research Institute for Social Development, Geneva.

Lewis, D. (1997) 'NGOs, donors and the state in Bangladesh', *Annals of the American Academy of Political and Social Science,* **554**: 33–45.

Main, H. (1999) 'Environmental improvement in Calcutta: non-governmental organisations and community-based organisations', in G.P. Chapman, A. Dutt and R.W. Bradnock (eds), *Urban Growth and Development in Asia,* Vol. II: *Living in the Cities.* Ashgate/SOAS Studies in Development Geography, Aldershot, Chapter 24.

Minogue, M. (1998), 'Changing the state: concepts and practices in the reform of the public sector', in M. Minogue, C. Polidano and D. Hulme (eds), *Beyond the New Public Management: Changing ideas and practices in governance.* Edward Elgar, Cheltenham: 17–37.

Närman, A. (1999) 'Getting towards the beginning of the end for traditional development aid: major trends in development thinking and its practical application over the last fifty years', in D. Simon and A. Närman (eds), *Development as Theory and Practice: Current perspectives on development and development co-operation,* Longman, Harlow: 149–80.

Pathak, P. (1999) 'Two decades of urban poverty reduction through income generation: what have we learnt?' in G.P. Chapman, A. Dutt and R.W. Bradnock (eds),

Urban Growth and Development in Asia, Vol. II: *Living in the Cities.* Ashgate/SOAS Studies in Development Geography, Aldershot, Chapter 16.

Robinson, M. (1993) 'Governance, democracy and conditionality: NGOs and the New Policy Agenda', in A. Clayton (ed.), *Governance, Democracy and Conditionality: What role for NGOs?* INTRAC, Oxford.

Robinson, M. (1997) 'Privatizing the voluntary sector: NGOs as public service contractors', in D. Hulme and M. Edwards (eds), *NGOs, States, and Donors: Too close for comfort?* St Martin's Press, New York: 59–78.

Sen, Siddhartha (1999) 'Some aspects of state–NGO relationship in India in the post-Independence era', *Development and Change,* **30**: 327–55.

Sethi, H. (1988) 'Trends within', *Seminar,* **348**: 21–4.

Sharma, Kalpana (1995) 'Chronicle of a riot foretold', in S. Patel and A. Thorner (eds), *Bombay: Metaphor for modern India.* Oxford University Press, Bombay: 268–86.

Staudt, Kathleen (1998) *Policy, Politics and Gender: Women gaining ground.* Kumarian Press, West Hartford, CT.

Tvedt, T. (1998) *Angels of Mercy or Development Diplomats? NGOs and foreign aid.* James Currey, Oxford.

UNDP (United Nations Development Programme) (1997) *Reconceptualizing Governance.* Discussion Paper 2, UNDP, New York.

White, Sarah (1999) 'NGOs, civil society and the state in Bangladesh: the politics of representing the poor', *Development and Change,* **30**: 307–26.

Wolfensohn, J. (1999) *A Proposal for a Comprehensive Development Framework.* World Bank, Washington, D.C.

Wood, G. (1997) 'States without citizens: the problems of the franchise state', in D. Hulme and M. Edwards (eds), *NGOs, States and Donors: Too close for comfort?* Macmillan, London: 79–92.

World Bank (1992) *Governance and Development.* World Bank, Washington, D.C.

World Bank (1996) *Pursuing Common Goals: Strengthening relations between government and NGOs in Bangladesh.* Dhaka University Press, Dhaka.

World Bank (1997) *World Development Report 1997: The state in a changing world.* Oxford University Press, New York.

CHAPTER 8

The Eppawela phosphate controversy in Sri Lanka
Development in the context of a 'war for peace'

Yolanda Foster

Massive demonstrations by Sri Lankans; foreign investors reportedly fed-up with a government stalling for time; and allegations of a huge sell-out. These are some of the intrigue and controversy surrounding a proposal to mine rock phosphate at Eppawela.
(Frederica Jansz, *Sunday Times* 5/9/99).

The Eppawela phosphate mining issue is the biggest and most controversial foreign investment deal struck in post-colonial Sri Lanka. Eppawela raises several important issues regarding power and authority in relation to resource use. In 1994, the Sri Lankan government signed a deal with US-based Freeport McMoran to mine phosphates at Eppawela.[1] The handling of the Eppawela controversy problematises public accountability in the sphere of national development. The variety of responses to the project highlight the variety of interests at play in the alliance against the Eppawela mine, which is located about 15 km south of Anuradhapura in the dry zone of northern Sri Lanka. In addition, the role of the American-based company Freeport McMoran brings out the ambivalence of the globalisation process. What is also interesting about the Eppawela controversy is that it marks a shift from state-sponsored development projects imbricated with nationalism, to understanding what development means in the context of a 'war for peace'.

In the past, state-sponsored development projects in Sri Lanka were constrained by a majoritarian Sinhala Buddhist state and the intrusion of party politics. This is well documented by Tennekoon who explored nationalist resonances in President Jayewardena's *Mahaweli* project (Tennekoon 1988). More recently, James Brow has explored the role of party politics in a government housing project in Kukuwela (1996). Both authors foreground the politics of nationalism in contemporary Sri Lanka. From Independence in 1948 until 1956 the governing United National Party had had representatives not only of the majority Sinhala community but of the minority Tamils. The Sri Lanka Freedom Party fought the 1956 election largely on the language issue, 'promising to make Sinhala the national language and to support Buddhism' (Farmer 1993: 99), thereby in their own eyes excluding Tamils from full and equal status as Sri Lankan citizens. Although the subsequent declaration of Sinhala as the only national language was ultimately reversed, the issue of ethnic identity has remained central to the political conflicts in the island, as, after an anti-Tamil pogrom in 1983, Tamil nationalism has taken increasingly extreme forms, notably through the guerrilla secessionist war waged by the Liberation Tigers of Tamil Eelam, the LTTE. In the past, the allure of nationalist imaginings has been deployed by various state leaders to secure support for development projects and boost popular appeal. What is important to explore are shifts in the context of development.

After more than 11 years of varying intensity of civil war, in 1994, the People's Alliance (PA) government came to power on a peace and reform mandate. The PA started peace negotiations with the LTTE, but after a few months the LTTE broke them off. The Sri Lankan government then opted for a 'military' solution. Jaffna was occupied by the army in December 1995, but the war went on. At the moment of writing, it is fought in the jungles south of Jaffna and a part of the island called the Vanni. Violence is part and parcel of development. Analysts need to recognise that the Sri Lankan state, by definition, claims the right to control the means of violence.

Sri Lanka: complex political emergency

A new approach to understanding state-sponsored development locates itself in an emerging literature which argues that conflict in Sri Lanka now has its own war economy dynamic (Rajasingham 1999). This emerging literature raises questions concerning how we locate and situate analyses of development. Since 1983, Sri Lanka has been associated with 'ethnic' conflict and high levels of political violence.[2] The anti-Tamil riots of 1983, known as Black July, mark the beginning of a permanent civil war between the Sinhala-dominated government and the LTTE (Kloos 1997). One of the key failures of the post-colonial state is the failure of a politics of accommodation, or the ability to manage diversity – a centralised state being a key problem.

Today, some scholars read Sri Lanka's conflict as a 'complex political emergency' (CPE) and have started to explore the role of the war economy in shaping and influencing development projects (Foster 2000; Rajasingham 1999). The term 'complex political emergency' has been used to describe a situation in which one aspect of emergency is that the state is contested or collapsed (Goodhand and Lewer 1999). One of the legacies of a protracted conflict is a transformation in the rules and practices of politics. War between the Sri Lankan government and the LTTE recommenced in April 1995 after the collapse of peace talks. Since this period there has been an increase in the political weight of those who control the means of violence (the security forces; the LTTE; politicians with connections to thugs and paramilitaries). In the past, the allure of nationalist imaginings has been deployed by various state leaders to secure support for development projects and boost popular appeal. An understanding of the Sri Lankan conflict as a complex political emergency requires that we reimagine the way we analyse development projects.

In the case of Sri Lanka, dominant ideologies of development that emphasised the role of the state have multiplied the state's power. The state sector was first promoted by the Left under the guise of socialism (see Yapa, this volume). The mushrooming of NGOs since the 1970s is a legacy of the awareness of tasks that needed to be carried out. Dependence on the state to deliver goods for development has now extended beyond a rhetoric of the Left. Political parties across the spectrum have looked to the central state as an engine of growth and as a mechanism to deliver social justice (Bastian 1999). While many writers have celebrated the democratic trad-itions of Sri Lanka, we must bear in mind that the practice of democracy is often inflected with populist and majoritarian notions of democracy.[3] Sri Lankan democracy has witnessed two insurgencies from the extreme left in the Sinhalese-dominated southern parts of the island, orchestrated by the Jatika Vimukthi Peramunna (the JVP, or 'People's Liberation Army'), as well as facing a separatist war in the north from the LTTE. The JVP was crushed as a militant force in 1989, but the Sri Lankan state has a peculiar north–south dynamic. Modernisation is going on in the south and yet the country has a long-term civil war. A deceptive 'normality' pervades the capital and yet people in the Vanni region of the north may have problems receiving food aid. Sri Lanka remains in the grip of a ruinous war which threatens to undo all the major advances in education and health, which it won over previous decades (HDC 1999: 3). These challenges affect the way the state relates to development.[4]

Despite the particular challenges of a complex political emergency, civil society activists as well as writers interested in 'governance' have advocated constitutional reform as the solution to many of Sri Lanka's problems. What must be recognised is that these institutions work in specific histori-cal circumstances. The legal and normative basis of institutions cannot themselves democratise society, protect human rights and ensure the rights of minorities. My own research in Sri Lanka has explored whether a majoritarian state can imagine diversity – what shapes and limits political identities? Despite the fact that the newly elected 1994 PA government led by Chandrika Bandaranaike Kumaratunga was the first government to advocate a peace and reform mandate seriously, the context for my research journeys is a 'war for peace'. Before discussing the Eppawela project in detail, it is important to situate the project in its political context. Seven years on from the August 1994 PA victory, the political crisis has deepened. Political life continues to be marked by deep uncertainty. Part of the failure of the state to imagine diversity stems from the ongoing military conflict. The 1994 government 'for peace' found itself in the ambivalent position of 'waging a war for peace'. The slender victory gained by Mrs Kumaratunga's govern-ment in the 2001 elections has not yet done anything to resolve the conflict.

Despite rhetorical gestures towards constitutional reform and devolu-tion, the government 'for peace' has become an increasingly authoritarian regime. Jayadeva Uyangoda has commented on the 'reform-resistant char-acter' of the Sri Lankan state (Uyangoda 2000). A flagpost of this deepen-ing 'regime authoritarianism' was the imposition of the Public Security Ordinances Act on 4 May 2000. The Act allows for increased censorship, the banning of public demonstrations and adds to the tensions of a nation suffering from protracted conflict. A notable post-Independence pheno-menon in Sri Lanka, as elsewhere in the South, is that 'in what appears to be the standard response to all social upheavals, the ruling elites strengthen the coercive capacity of the state' (Uyangoda 2000: 96). When we analyse the Eppawela project, it is worth understanding that the state needs money to fuel the ongoing military conflict. This is one reason why questions of nationalism have been less important than the appeal of a quick cash injection – despite lack of long-term financial gains from the project. The corruption record of successive Sri Lankan governments, and of public administration in general, has facilitated the collapse of a reform mandate to pragmatic management of the 'war for peace'.

The Eppawela project: background

The Eppawela phosphate mine is located in the northern interior of Sri Lanka, in the Anuradhapura district (Figure 8.1). The Eppawela project – the only known phosphate resource in Sri Lanka – was conceived in 1992 by the former United National Party (UNP) government but stalled when the PA swept parliamentary and presidential polls in mid-1994.[5] President Kumaratunga made an election promise in 1994 not to go ahead with the project but subsequently went back on her word. The monk spearheading protest at Eppawela notes: 'she promised us. She told me at an election meeting in Eppawela itself that she would abandon the project. But see what has happened' (Mahamankadawala Piyaratana Thera, Buddhist priest and Chief Incumbent of the Galkande Purana Temple: interview with the author, 2000).

The mine was to be a joint venture project by Freeport McMoran Resource Partners, IMC Agrico and Japan's Tomen Corporation. The $425 million Eppawela phosphate mining and Trincomalee fertiliser production joint venture contract was touted by government officials as a showpiece for the country's manufacturing sector. A deal should have been signed in the run-up to golden jubilee celebrations on 4 February 1998. Public protest as well as environmental concerns, however, stalled the development. The mine was to cover an area of 56 km² in opencast mining, and would displace about 12,000 families.[6] These villages have repeatedly expressed their opposition to the scheme because of the ensuing loss of not only their lands but also their very means of livelihood.

Lack of public accountability in national development

One of the key problems with the Eppawela phosphate proposal is lack of public accountability. Indeed Eppawela could be seen as a case study in the lack of people's participation in Sri Lanka's decision-making process. Involving the public at the planning stage would have averted some of the conflicts now being played out. Such an involvement would have given the

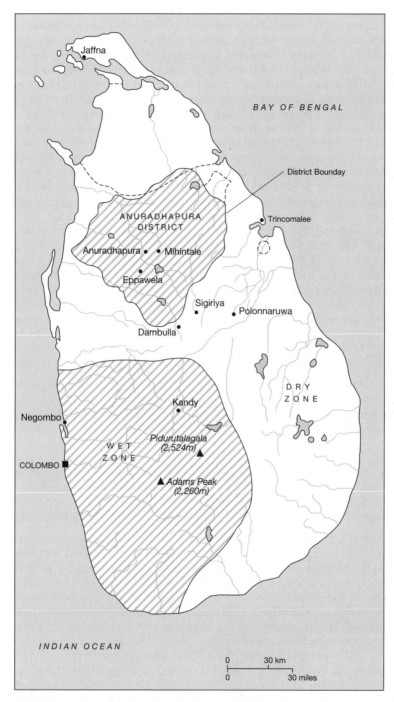

Figure 8.1 Eppawela phosphate mine location

government a clearer understanding of all the issues and concerns considered important by various interest groups. The idea of consulting 'stakeholders' was absent; however, in July 1999, Lord Avebury, speaking out in support of Eppawela protesters, expressed his concern about the lack of consultation. He explained that there is a need to consult people affected by such a large project and a need to abide by World Bank guidelines requiring the operators of projects to consult local people (*Sunday Times* 25/7/99).

The 30-year Eppawela mining deal was cleared by the Sri Lankan Cabinet in August 1997 with no public consultation.[7] Government officials justified the Eppawela project on the grounds that the country lacks the know-how, the capital and machinery to tap high-risk industry such as large-scale mining. The question of whether large-scale mining was actually appropriate for Sri Lanka's national development seemed absent from planning. Subsequent government preoccupation with financial and other benefits of the phosphate project made villagers suspicious and they started opposing the project. The National Science Foundation and the National Academy of Science also published a report warning of environmental hazards of the projects.

My interest in this chapter is not in the scientific and technical questions arising from the project.[8] These are well documented by Mendis (2000) and show clear environmental concerns. It is appropriate, however, to quote a short extract from the Managing Director of IMC-Agrico, Udaya Borelessa. His response to the National Science Foundation Report was not an attempt to outline the benefits of the mine but rather to lash back at the National Science Foundation. He claimed, among other matters that:

> the wide publicity given was a devilish face, [the idea] that the US venture is a total destruction of cultural heritage, [would lead to] displacement of 12,000 families, several towns and public buildings, rivers and lakes going dry is an utterly false picture.

He continued:

> my standing on the National Science Foundation (NSF) report on the project is [that it is] a highly biased report, meant to divert the good intentions of the government and push through another interest. (Quoted in a report by Janaka Perera, *The Weekend Express*, 9–10 October 1999).

The type of language deployed indicates the atmosphere surrounding dialogue of the Eppawela issue. Given the serious financial and environmental costs, what interests did the state have in pursuing the project?

Over the years, the Sri Lankan state has played different roles in development. In the period 1956–77 the SLFP government was keen to embark on new policies, with direct intervention by the state promoted to control and manage the economy. During this period the public sector expanded on an unprecedented scale (Ranugge 2000). The opening up of the economy in 1977 marked a new phase. The new government – the UNP – pledged itself to introduce radical economic changes. In addition, a new Constitution with an Executive Presidential system was promulgated in 1978. The civil service was, however, further politicised, vesting all the powers of appointments, promotions and transfers of public servants in the hands of the Executive Presidency and Cabinet of Ministers. Political considerations were the rule rather than the exception in this period. After 1977 and increased liberalisation, the role of the state was supposed to be narrowed down to keeping law and order and providing and maintaining infrastructure facilities to facilitate private sector development. However, despite several attempts to restructure the civil service it remains immune to fundamental structural reform. As a result of the deep politicisation of the bureaucracy, some writers have noted that public servants have 'become more passive in reacting to what they are told to do rather than proactive in performing their managerial functions'. A public servant who shows too much interest in his/her work will probably be transferred. In 1998, several divisional secretaries were transferred through fax messages to the Ministry of Public Administration despite complaints from the Sri Lanka Administrative Service, of which they were members, which was able to challenge the legality of those transfers before the Supreme Court (Ranugge 2000: 57). As filing a fundamental rights plea is very expensive, this course of action is the exception rather than the rule. The politicisation of the bureaucracy is relevant to the Eppawela issue as it means that even if certain public officers had reservations about the project, in the current reform-resistant bureaucracy, it is difficult to file a complaint.

The turnaround by the PA government concerning its support for Eppawela would not have been challenged by public servants. The turn-around also reflects the regime's growing trend during the period 1994–9 to forgo long-term planning in favour of short-term gain. Widely seen as failing to deliver on her 1994 election platform pledges, President Chandrika Bandaranaike Kumaratunga may have seen the Eppawela mining scheme as a mechanism to inject badly needed dollars into the nation's coffers. During the period 1994–9, the government's military expenditure has doubled in order to fund ongoing war in the north and east.[9] In the context of a 'war for peace', the role of public protest in contesting the state is to be especially commended. During the period 1994–2000 regime authoritarianism is marked by a series of measures to deepen censorship. What the 'war for peace' also requires is that analysts of development take into account the compromises that come with managing a 'complex political emergency'.

Role of different interest groups

Concerned parties directly affected by Eppawela phosphate mining

Protests around the Eppawela issue reflect the multidimensional aspects of collective action. The issue which unites different groups is the problematic way in which development is being pursued by the state. On the surface, it appears that the Eppawela issue is an ecological concern, but within the spectrum of protest there exist different ideologies of how people, land and development are represented. The organisation and diversity within the movement reflect the complex and often contradictory nature of resistance.

The Buddhist monk Mahamankadawala Piyaratana Thera and the 'Committee for the Protection of the Phosphate Deposit'

Those people who are directly affected by the Eppawela issue are villagers in surrounding areas. Protests by the villagers are spearheaded by Mahamankadawala Piyaratana Thera, a Buddhist priest and Chief Incumbent of the Galkande Purana Temple. The monk is head of the 'Committee for the Protection of the Phosphate Deposit'. His aim is to support villagers in the area who believe that the project will lead to the loss of their homes. Veteran trade unionist, Bala Tampoe, notes: 'In my entire career of over 50 years, this monk is the most unique person whom I have come across. He has expressed his readiness not to kill but to die and he has inspired the villagers of Eppawela in that spirit.'[10] Mahamankadawala Piyaratana Thera himself has said: 'we will not leave; the government will have to use soldiers to remove us from our homes' (*Sunday Leader* 17/1/99). Inspired by the monk's commitment, three mass demonstrations took place in June 1997, February and August 1998, when thousands marched to the Sacred Bodhi Tree in Anuradhapura to protest against the plans of Freeport McMoran (interview with Reverend Piyaratana, October 2000). In the shadow of Pospet Kanda or phosphate hill, stands an ancient Buddhist temple where the chief monk has himself led demonstrations, protests and even hunger strikes.

In the same interview, Piyaratana stressed his belief that 'Land belongs to the people, not the government. The government has no right to push forward what it calls development, when we, the people of Eppawela, will suffer (interview with author, October 2000)'. Reverend Piyaratana was also cynical of the fact that the government has anything other than short-term interests in mind. Having considered the lack of long-term financial gain from the Eppawela project, questions were raised concerning tenders and the lack of proper consultation on the signing of the Minerals Investment Agreement. Concerns of lack of transparency were echoed by the Minister of Science and Technology, Batty Weerakoon, who noted the problematic shift in negotiations from Freeport McMoran to IMC-Agrico.

The environmentalists

Supporting the protests organised by the Venerable Piyaratana are environmentalists concerned about Eppawela. The Environmental Foundation Ltd warns that the project will bring untold environmental and infrastructural degradation to Zone H of the multi-billion rupee Mahaweli Development Scheme.

> If mining takes place as proposed, up to a depth of about 100 metres, within 20 years, it is most likely that this area will become uninhabitable. Due to the possibility of irrigation and other water resources getting affected, people of this area will face a problem regarding their daily requirements of water, both for drinking and other purposes.

Other allegations are that strip mining would destroy the ancient irrigation system of Kalaweva Jayaganga. This water management system in the

174

Rajarata Kingdom in the dry zone generated the agricultural surplus to support the Anuradhapura Buddhist culture from the 3rd century BC until the 11th century AD, and continues to function today. UNESCO, through its International Committee for Monuments and Sites, ICOMOS, at a meeting in Ottawa in 1994, had recommended that the Kalaweva Jayaganga in Sri Lanka, and the Grand Canal in China, should be recognised as cultural heritage sites and monuments under their programme. Unfortunately this recommendation had not been taken up by the Sri Lankan government in respect of the Kalaweva Jayaganga. If that had been done, the Kalaweva Jayaganga would have been a protected cultural landscape, and such a proposal as the mining to destruction of the Eppawela phosphate rock deposit could not have been considered (Dipek Gywali, *Kathmandu Post*, 31/10/1999).

The Ceylon Mercantile Union

Another interest group which has persistently protested against the Eppawela mine is the Ceylon Mercantile Union (CMU).[11] The General Secretary of the CMU, Bala Tampoe, is particularly concerned about the short-term interests of transnationals:

> We are convinced that the American transnational company will stop at nothing to get a mining licence from the government to plunder the valuable national asset of the Eppawela deposit wholesale for its short term commercial gain, estimated at over five billion American dollars

> (Bala Tampoe, quoted in the *Sunday Times* 25/7/99).

In March 1999, Bala Tampoe went to Britain to present the Eppawela case before the Permanent People's Tribunal on Global Corporations and Human Wrongs. The Permanent People's Tribunal investigates human rights violations by multinational corporations. The primary aim of the tribunal is to expand on a 'people's jurisprudence' and to provide a forum where victims of multinationals could voice their grievances. Mr Tampoe said that three cases were presented at the tribunal sessions. The transnational corporations concerned were Union Carbide, responsible for the still unresolved Bhopal disaster, Monsanto Corporation and Freeport McMoran/Rio Tinto. The tribunal acknowledged that the role of Freeport McMoran in Papua New Guinea was highly problematic. Where the CMU differs from other interest groups is in their concern to use Eppawela to problematise the liberalisation process in Sri Lanka. The CMU link up the local to global struggles and question whether capitalism can have a human face.

What is interesting about the different responses to the Eppawela issue is that they reflect different interest groups. The environmentalists are concerned with what they call the irreversible harm which could be caused by the project; archaeologists and historians are attempting to protect the invaluable cultural and historical aspects of the area; scientists are focusing on the use of non-renewable resources while the trade union movement is concerned about rights. Although these groups are right to criticise lack of public accountability, some of the slogans used are problematic.

The rallying call to protect Rajarata

The Eppawela mine is located at the heart of what Sinhala nationalists refer to as 'the ancient Rajarata water and soil conservation ecosystem'. From the top of the excavations that currently exist at Eppawela, the ruins and stupas of Anuradhapura are clearly visible. One reason why the Eppa-wela controversy has generated so much support is a valorisation of an imagined Sinhala–Buddhist rural idyll:

> Villagers there worship the upper boughs of the Sri Mahabodhi,12 that ancient and previous natural relic of Sri Lankan and Buddhist glory that can be seen from the summit. The water of the Jaya Ganga (Yoda Ela), that two millennia old irrigation wonder, where water flows along a negligible gradient, courses through Eppawela. For centuries it has watered the fertile hands of the region, and the preciseness of its construction continues to baffle and astonish modern engineers reliant on their computerised aids. Beginning at the Kala Wewa, the Jaya Ganga is one of the main arteries through the heart of the famed irriga-tion network of the North, feeding old and new tanks, and continuing to connect the farmer with that most precious of all commodities, water
>
> (Bala Tampoe, *op. cit.*).

While there is nothing wrong with valorising the environmental beauty of the area, some of the tracts generated by Eppawela protesters share a certain style of writing with some more problematic Sinhala nationalist groups. Various members of the Buddhist *sangha*[13] in Sri Lanka supported Eppawela protesters as a response to fears that globalisation means cultural colonialism in a new form. This can be observed in media writing on the subject in which the *Mahanayakes* (head monks) condone the loss of sovereignty or writers demonise the American foreigner. Perhaps as misguided cultural resistance, the concept of 'national' space and a sacred geography transforms itself into a discourse of 'national authenticity' which overwrites a long tradition of intermarriage and cultural mixing. The last Kandyan kings were, after all, descendants of the south Indian *Nayakkar* dynasty. Buddhist traditions have certainly enriched Sri Lankan culture. However, the collapse of traditions into 'national pride' bodes ill for imagining diversity in Sri Lanka.

Sri Lanka becomes represented as *dhammadipa* (*dhamma* refers to the Buddha's teachings, *dipa* to island). In a particular narrative on 'national authenticity', Sri Lanka becomes the chosen land of Theravada Buddhism and it is incumbent upon the Sinhalese to protect it. The destruction of Eppawela is thus seen as a blow to the heart of the Sinhala–Buddhist rural idyll. The question is what kind of protection is needed: is it possible to preserve cultural traditions without using the language of 'national authenticity'?[14]

While the people of Eppawela are justified in feeling an attachment to their homes, this should not be used by other interest groups to put forward a particularised view of 'national authenticity'. Today's encounter with global discourses encompasses new potentials as well as radical doubt. While it is important to probe the role of transnational corporations – like Freeport McMoran in Sri Lanka – it is also important to embrace the positive aspects that globalisation could offer such as education and exchange (Simon and Närman 1999; Bastías Gonzales 1999). There has been a trend in recent years in Sri Lanka, for groups who wish to reassert particularised identities to tout the fear of globalisation as a scapegoat to divert attention from their own political motives. The type of discourse produced on ancient kingdoms should therefore be treated with some degree of scepticism. Protests at the local level should have connected up with demands for reforms in the public sector which would compel development to be more transparent.

This is not to say that globalisation presents no problems. Fears of the McDonaldisation of the world, or 'global blanding', first emerged in 1989. The *National Interest* published *The End of History*? by Francis Fukuyama, then a senior official at the US State Department. In that comparatively short but extremely controversial article, Fukuyama speculated that liberal democracy may constitute the 'end point of mankind's ideological evolution' and hence the 'final form of human government'. The article sparked a global debate on the consequences of liberal democracy as an end point. What would happen to 'national authenticity' if Fukuyama's thesis were true? The current debates in Sri Lanka for control over who gets to define culture and the nation are a good example of the struggles that globalisation can bring. The Buddhist *sangha* and the political parties argue over whose cultural values count. Locking oneself into fixed notions of authenticity, however, may be the wrong approach to the challenges of globalisation. Although globalisation adds its own spin to the development of religious identity, Buddhist practices have changed across the centuries. There is no reason why cultural values should be eroded in a global world. After all, the exchange of monks between Thailand and Burma and Sri Lanka has gone on since the thirteenth century, and the transfer of ideologies from region to region is not simply a matter of the modern era of globalisation.

The problem of invoking concepts like Rajarata (land where the Royal Kandyan kings reigned) in protest against the lack of consultation, is that this type of protest does not build social capital across the whole of Sri Lankan society. It reflects the promotion of an interest group rather than the beginning of a coalition to improve public dialogue and consultation. What about the voices of other people who are suffering in Sri Lanka? Do those who rally for Rajarata speak out for the rights of the internally displaced? Who is speaking out for the land rights disputes of the *veddahs* (indigenous tribal people) in Mahiyangana, on the plains some 50 km east of Kandy? Compared to some of the groups in the Alliance, the CMU has a broader platform than others since it also represents labour rights. The CMU have been keen to highlight the fact that Eppawela also affects people in Trincomalee (where refuse was to be drained) as well as villagers in Eppawela. The CMU articulates the downside of the transnational process and the state's complicity with this. 'The company will get $5 billion from looting the phosphate deposits', notes Bala Tampoe. From this perspective, the CMU challenges the 'dominant class' interests of the state.

Recent developments

On 20 March 2000, a large demonstration was held outside the Colombo Fort Railway Station.[15] The Committee for the Protection of the Eppawela Phosphate Deposit was present as well as farmers, the CMU and various other interest groups. Among the thousands of protesters, one could observe many women. One woman, a farmer, noted: 'Leave? We will never leave this land even if the government and the US company go ahead with the project. We have grown up here. This land belongs to our children and future generations. We are prepared to die for it,' said Kanthi Ambathale.[16] Although leadership of the Eppawela protests has been dominated by men, women have played an active role in grassroots activism. Women will be directly affected by the McMoran project as farmers and as mothers. A teacher at the demonstration noted that the state is just giving away land and it is the women who have consistently attended marches and rallies. The Committee to protect the Eppawela phosphate see development as something being done to them by a state that has given up caring for its citizens (see also Simon 1999: 23–9). Women at the march were very cynical, saying that Eppawela is just a way for politicians to make money. The CMU challenges the state more directly over issues of power and authority in relation to resource use. They highlight failures of post-colonial development strategy and criticise pro-liberalisation policies. Members of the *sangha* who use the discourse of ancient kingdoms and sacrilege tend to use the Eppawela issue to front their own visions of Rajarata. What should be noted, however, is the variety of Buddhist responses to this issue. Undoubtedly the local monk, Mahamankadawala Piyaratana Thera, Chief Incumbent of the Galkande Purana Viharaya, has been pivotal in enthusing commitment for the Eppawela issue. He speaks with great local knowledge and a commitment to the area. Whether the Venerable Piyaratana continues to champion other environmental causes will be something to observe in the future.

Success story?

The interesting thing about the Eppawela controversy is that despite the multidimensionality of interests in the alliance, it appears to be a success story of collective action. On 2 June 2000, the Sri Lankan Supreme Court ruled in favour of the petitioners of the Eppawela fundamental rights case against the government. In a judgement that vindicated the long-running objections by residents of Eppawela, as well as Sri Lanka's scientific community, the Sri Lankan Supreme Court ordered the government to desist from going ahead with its controversial phosphate mining project. The Supreme Court judgement reads:

> This BOI [Board of Investment] approved project, which was to be carried out by 'Sarabhumi Resources (Pvt) Limited', a company with a ninety per cent foreign shareholding, envisaged the exploitation of the known rock phosphate reserves at Eppawela over a mere thirty-year period, in a high intensity mining operation that its detractors claimed would devastate the local environment.

The success of Eppawela can be compared to Enron's Dabhol project in India in which the Supreme Court dismissed the petitions filed against Dabhol.[17]

The fundamental rights case was filed by six owners of agricultural land and the Viharadhipathi of the Galkanda Purana Viharaya, all within the exploration area of the project. They claimed that they were in danger of losing their lands and livelihood as a result of this government-sanctioned project which, they said, was not for a public purpose but for the enrichment of a private company. They also claimed that the project was being entered into in a manner that circumvented the environmental laws of this country, and that clauses in the agreement binding the government to assist the company to obtain all necessary approvals meant that any environmental impact assessment conducted thereafter (in which they as citizens were entitled to participate) was likely to be biased and not conducted in good faith.

The petitioners accordingly claimed an imminent infringement of their rights under Article 12(1) of the Constitution – right to equality before the law and equal protection of the law – and Articles 14(1)(g) and (h) – right to choose their place of residence and carry on their livelihood. In its final judgement, arrived at after lengthy arguments and written submissions by the contesting parties, the Supreme Court Bench headed by Justice A.R.B. Amerasinghe, with Justices Wadugodapitiya and D.P.S. Gunasekera in agreement, held that the petitioners had established an imminent infringement of their fundamental rights. The Sri Lankan state was directed to desist from entering into any contract until a comprehensive exploration and study had been carried out with regard to the locations, quantity and quality of the rock phosphate deposits in consultation with the National Academy of Sciences and the National Science Foundation (*The Island*, 11 June 2000).

Motives for the Eppawela mining project?

The way in which the state managed the Eppawela issue reflects a lack of vision for long-term planning. It is important to contextualise state management of the project in the context of a 'war for peace'. The Sri Lankan government has slid towards regime authoritarianism during the period 1995–9 as a means to cope with the ongoing military war in the north.[18] Observing the Eppawela controversy from the outside, it is relatively easy to identify the main actors and their intentions. The way in which the Eppawela project was handled by the state in Sri Lanka was less transparent. A lack of dialogue and a deliberate attempt to conceal damaging information tended to make the Eppawela issue a local or environmental issue rather than something which reflects state policy on development. Given ongoing emergency regulations and censorship, the state electronic media can control what type of information they wish to give out. Yaju Dharamaraja notes that the Eppawela mine would have been an incredibly hard deal for Sri Lanka, which would get a mere 10 per cent of the total share equity.[19] There are no safeguards to the outflow of pollutants from phosphate and the government is using Trincomalee as dumping ground for this waste, under cover of the ongoing war in the north-east. Consequently, Dharamaraja notes that the infrastructural development – including a railway, bridges and docking facilities – that would come in the wake of the project, would be minimal in comparison to the destruction the project would cause. What interest did the state then have for wanting to push the Eppawela mine project forward?

Conclusion

The Eppawela project can be seen as part of a spectrum of social movement activity in the South Asian region (including the Chipko struggle of Uttaranchal, and the anti-dam protests in Narmada and elsewhere – see Chapman, this volume). It reflects the evolving relationship between communities and the state. Issues like Eppawela problematise contemporary 'feel good' development discourse on empowerment. The idea of local people as stakeholders does not seem feasible in the context of a reform-resistant public sector. The real successs of the Committee for the Protection of the Eppawela Phosphate Deposit is the way in which popular attitudes and expressions of disillusionment came to be expressed in the realm of the public arena. The Eppawela issue allowed connections to be made between different organisations and ideologies resulting in popular participation. The use of public space to articulate claims is a fraught area in Sri Lanka as a history of the policing of demonstrations would show. The fact that Eppawela protesters did take their message to the streets is to be especially admired. Perhaps the involvement of monks facilitated public protest. This is not to homogenise the range of actors involved in the Eppawela issue. However, in the context of a 'war for peace' in which war is often prosecuted through essentialised categories, it was positive to see communities emerging to take on issue-based action rather than identitarian action. Some groups did echo the language of Sinhala nationalists but in the end the Supreme Court judgement reflected the main issue, which was a failure of the state to consult its citizens. Awareness of an intrusive state is vital at a time when the transnational attenuation of 'local' space is at its peak. Demonstrators at the protest on 30 March 2000 were all too aware of problematic claims on the nation made by a state which was not prepared to consult the very communities for whom they claim to speak.

Notes

1. Freeport McMoran has been accused of being the number one polluter in the USA. The US mining company has also been questioned with regard to its international dealings. One controversy is over the Grasberg mine in the Indonesian province of Irian Jaya. This is where the Amungwawe people live. When some Amungwawe people resisted Freeport's expansion, Indonesian troops drove them to the mountains. According to the *Far Eastern Economic* review of July 1994, natives of Irian Jaya have been protesting against Freeport McMoran since 1977.
2. Chronologies of the conflict are inevitably controversial. We can discuss many strands of the Sri Lankan conflict – for brevity's sake I will note here that the July 1983 pogrom could be read as a key moment in the outbreak of hostilities between the Sri Lankan state and the LTTE – see Gunasinghe (1984).

3. Sri Lanka has many of the formal institutions of democracy, but media censorship, lack of the right to free and fair association as well as widespread political thuggery require that the type of democracy in Sri Lanka be better adjectived.

4. In 1998, the Defence Ministry closed down an international non-governmental organisation (NGO), Peace Brigades International (PBI). PBI provided accompaniment to those at risk of political violence. The Defence Ministry demanded that PBI should submit its situation reports from eastern Sri Lanka for 'editing' before publication and provide the names and addresses of PBI's clients and contacts. PBI said that agreeing to these demands would make its non-partisanship meaningless (*Sri Lanka Monitor* No. 122, March 1998, p. 1). This event can be regarded as a sign that the space for humanitarian work under the Sri Lankan government became restricted. It is also part of the strategy of the government, of which censoring newspapapers forms part, to hide its failure to deal effectively with the war (Kloos 1997: 32).

5. The Eppawela phosphate rock (apatite) deposit was discovered by the Geological and Mineralogical Surveys Department in 1971. The rock had been quarried under a Divisional Development Councils project in the Ministry of Planning and Economic Affairs. The gradual extraction of the insoluble apatite was planned to meet some of the needs of local plantation agriculture. Since 1974, the huge phosphate deposit in Eppawela has been used to provide phosphate fertiliser for local agriculture. Then in the 1990s the idea of a joint foreign venture was conceived.

6. These claims were reported in the Xinhua News Agency, 28 July 1999.

7. Under the deal, IMC and Tomen will respectively control 65 and 25 per cent of equity in the joint venture, with the balance, 10 per cent, going freely to state-run Lanka Phosphate Ltd. They will be allowed to mine and export up to 3.6 Mt of phosphate – in rock form – for the first 12 years, after 18 months and 15 months of exploration and feasibility study. Mining industry officials note that, given the increasing international demand for phosphate, the mineral needed to fuel an ever-expanding international fertiliser market as world populations soar, these foreign investors should turn enough profits on rock export sales to repay bank loans and recover the bulk of their outlay – under the present arrangement, 35 per cent of the $425m. The deal also allows them to explore over an area of 56 km^2, and scour a 'buffer zone' extending to 10 km from the Eppawela deposit boundary. According to a report made by a five-man Cabinet-appointed negotiating team on the final round of talks conducted with the investors, the 'Mining Area' is loosely defined as 'the land within the exploration area which is covered by the mining licenses issued to the company by the Geological Survey and Mines Bureau' (report by Imran Vittachi, *Sunday Times*, 28/12/97).

8. Two scientists, Prof. Dahanayake and Prof. Gunawardena, were in a small committee appointed by the President to look into the Eppawela matter. They submitted a report which noted that

> it (is) in the interest of Sri Lanka to have foreign collaboration, but with 'the national interest' as the priority. Our local scientists are saying that we have a good quality deposit, but we should go for a process which is not harmful to our national interest. They have proposed setting up of a single super phosphate plant. Although I have said earlier, theoretically you can even do that in your own backyard, on a larger scale we should get the experts, foreign aid, but do it for our own benefit.

Another scientist opposed to the Eppawela mining plans said that the executive committee of the PA – including ministers – had agreed prior to the December 1999 presidential poll to abandon the project (Professor Tissa Vitharana, senior advisor to the Ministry of Science and Technology). 'But there's pressure from hidden quarters to sign the agreement. That's why this protest is taking place.' Controversy over lack of transparency comes in the wake of other problematic business deals in Sri Lanka such as the Air Lanka and Channel 9 scandal. Prof. Vitharana added that: 'under the proposed deal, the Sri Lankan government will be paid five dollars per tonne of phosphate extracted, while the mining companies will export it at world market prices currently at between 40 and 70 dollars per tonne. In simple economic terms, Sri Lanka is the loser.'

9. The defence budget was 1 billion rupees in 1981–2. In 1994 it was 24 billion rupees. In 2000 it was 51 billion rupees (Interview with Harry Goonetilleke, Former Air Vice-Marshal, 30/9/2000). The burden of defence spending in Sri Lanka as a percentage of education and health spending is 107 per cent (HDC 1999: 18). The only other South Asian country to be over 100 per cent is Pakistan.

10. Bala Tampoe is the Ceylon Mercantile Union General Secretary. He spoke at a seminar at ICES, Colombo, in 1999.

11. The London-based human rights organisation, Amnesty International, joined the environmental and civic action groups opposing the sale of Eppawela phosphate mines to a foreign firm known for its alleged disregard for environment and labour laws. Lord Avebury spoke out about Freeport McMoran (*Sunday Times* 25/7/99).

12. The Mahabodhi refers to the sacred Bo Tree at Anuradhapura. The Bo Tree is believed to be a cutting from the tree at Bodh Gaya under which Gautama, Lord Buddha, gained enlightenment.

13. The *sangha* is the organisational and institutional structure for the three Buddhist orders.

14. Lakshman Yapa (this volume) raises parallel questions when considering the geographies of alternative development as embodied in Sri Lanka's Sarvodaya Shramadana Movement.

15. I was present at the demonstration. The fact that so many protestors gathered at a time when Colombo has increasing security checks is evidence of the protestors' commitment.

16. InterPress Report, 30 March 2000.

17. 'Enron's Dabhol Project. Maharashtra State', *Frontline*, 2000: 84.

18. This is a term used by Jayadeva Uyangoda, Lecturer in Political Science at Colombo University.

19. Yaju Dharamaraja is a postgraduate student who is exploring the role of the CMU in the Eppawella protest. Seminar at ICES, Colombo, July 1999.

References

Bastian, S. (1999) *The Failure of State Formation, Identity Conflict and Civil Society Responses – the case of Sri Lanka*. Bradford Peace Studies working paper.

Bastías Gonzales, M.V. (1999) 'Popular organisation, local power and development', in D. Simon and A. Närman (eds), *Development as Theory and Practice: Current perspectives on development and development co-operation*, Longman, Harlow: 247–65.

Brow, J. (1996) *Demons and Development*. University of Arizona Press.

Farmer, B.H. (1993) *An Introduction to South Asia*, 2nd edn. Methuen University Paperbacks, London.

Foster, Y. (2000) *Sri Lanka. Donor policy in a complex political emergency*. Working Paper for International Alert.

Goodhand, J. and **Lewer, N.** (1999) 'NGOs and peace-building in complex political emergencies', *Third World Quarterly*, **20**(1): 69–87.

Gunasinghe, N. (1996) *Selected Essays*, edited by Sasanka Perera. SSA, Colombo.

HDC (Human Development Centre) (1999) *The Crisis of Governance in South Asia*. Oxford University Press, Karachi.

Kloos, P. (1997) 'The struggle between the lion and the tiger. The relevance of inter- and intra-ethnic conflict for the construction of ethnic identities in Sri Lanka', in C. Govers, and H. Vermeulen (eds), *The Politics of Ethnic Consciousness*. Macmillan, London: 223–49.

Mendis, D.L.O. (2000) *Eppawela–Destruction of Cultural Heritage in the name of Development*. Vishwa Lekha Printers.

Rajasingham, Darini (1999) 'Democracy and the problem of representation: the making of bi-polar ethnic identity in post/colonial Sri Lanka', in Joanna Pfaff-Czarnecka *et al.* (eds), *Ethnic Futures. The State and Identity Politics in Asia*. Sage, New Delhi: 99–134.

Ranugge, S. (2000) 'State, bureaucracy and development', in S. Hettige and M. Mayer (eds), *Sri Lanka at the Crossroads*. Macmillan India, New Delhi: 50–62.

Simon, D. (1999) 'Development revisited: thinking about, practising and teaching development after the Cold War', in D. Simon and A. Närman (eds), *Development as Theory and Practice: Current perspectives on development and development co-operation*, Longman, Harlow: 17–54.

Tennekoon, S. (1988) 'Rituals of development: the accelerated Mahaweli Development Program of Sri Lanka', *American Ethnologist*, **15**: 294–310.

Uyangoda, J. (2000) 'A state of desire? Some reflections on the unreformability of Sri Lanka's post-colonial polity', in S. Hettige and M. Mayer (eds), *Sri Lanka at the Crossroads*. Macmillan India, New Delhi: 92–118.

Sri Lankan newspapers consulted

Daily News
Island
Sunday Times
The Weekend Express

Part Three Social inequality and social change

CHAPTER 9

Rethinking poverty in Sri Lanka

Lakshman Yapa

This chapter presents a case study of the general argument that the process of economic development cannot eradicate poverty because it is implicated as a causative agent of the problems that poor people face.[1] The most common approach to studying poverty first identifies a 'poverty sector' viewed as the least developed part of the larger underdeveloped economy, and then investigates it through such instruments as 'poverty surveys' within that sector. All contemporary poverty discourses in Sri Lanka, be they conservative, liberal or radical, believe that there is a distinct, bounded poverty sector in the economy whose problems can be eradicated through economic development. But the origins of scarcity do not lie within the so-called 'poverty sector' because they are diffused throughout the larger society. It is my view that we cannot usefully address issues of poverty in Sri Lanka without first critically engaging with the development discourse that is a central part of the problem. Of course the case I make is a very general one and is not confined to the specific situation in Sri Lanka, as Simon and Närman have made clear in their recent discussion of development theory and practice (1999).

The chapter is divided into five parts. The first describes a new model of poverty, building on a concept called 'the end-use of a commodity'. The second part gives a brief history of post-colonial economic development and poverty alleviation in Sri Lanka that included a variety of approaches – conservative, liberal, Marxist and others. Despite some profound differences, they all agreed that economic development was the answer to eradicating poverty in Sri Lanka. The third part carries a critical evaluation of the programmes of poverty alleviation described in the previous part. Since the state has played a very dominant role in poverty alleviation in Sri Lanka, I use the fourth part of the chapter to cite the work of the Sri Lanka Sarvodaya Movement as an example of a non-governmental approach to the problem. But Sarvodaya is not entirely free of the 'development philosophy' I criticised in state-sponsored approaches. The final section uses 'the end-use model' laid out in the first part to present a brief narrative of food in Sri Lanka as a concrete example of the proposed new way of seeing poverty.

A new model of poverty

When viewed in the context of the larger society, the material deprivation that poor people suffer is a form of 'socially constructed scarcity'. This section of the chapter will define the term 'scarcity' and propose a hypothesis about its 'constructed' nature. Scarcity can be understood in two ways. First, it can refer to outright physical scarcity caused by conditions such as natural disasters or war. Second, economists use the term to describe a world where consumers have unlimited wants, but only limited resources to satisfy them. In this case, something could be abundant, but still defined as scarce relative to the demand for it. My own analysis begins at the point where scarcity comes into being at the interface of both demand and supply. The point of departure for economists in understanding scarcity is the commodity, but to understand the notion of socially constructed scarcity we need to move beyond the commodity to a concept that I wish to call the 'end-use' of a commodity, by which is meant its 'final end-use'.

Given the end-use of a commodity, we can determine the availability of alternative ways in which that particular end-use can be met. Now we can understand scarcity inhering not in the demand for a single commodity, but within a system of commodities of varying value, all of which satisfy a desired end-use. Thus, the demand for a commodity may expand because the other ways in which that end-use can be met are unavailable. More technically, assume there are different sources of supply for a particular end-use of a given commodity – what economists call substitutes. The demand for that commodity can shift (creating scarcity) because alternative sources of supply of that end-use have contracted. Consider the example of chemical fertiliser as a commodity the end-use of which is supplying nitrogen to the soil. Next, consider alternative and complementary sources of soil nitrogen. These include crop rotation, polyculture, companion planting, green manure, compost, human and animal waste, agricultural residues, slurry from the anaerobic production of biogas and so on. Through programmes of economic development, chemical fertilisers usually become the primary commodity for supplying nitrogen in the soil. Here the expanded demand for chemical fertiliser is partly a result of the contraction of alternative sources of supply or the non-availability of substitutes. The history of economic development is replete with instances where one commodity (or source) has come to dominate the satisfaction of a particular end-use to the exclusion of other sources, a process described by the philosopher Illich (1973: 51–2) as the creation of 'radical monopolies'.

Increased demand for a particular commodity can also result from expanding end-uses beyond its original use-value. For example, even though the original use-value of an automobile is derived from transport, the demand for it can come from other end-uses associated with success, status, glamour and even sex. These added end-uses might prevent or discourage substitution of one commodity for another, thereby constructing scarcity of the preferred commodity despite the presence of available substitutes.

In addition to substitution there is another connection among commodities called complementarity that leads to construction of scarcity.

Complementarity refers to commodities that are technically linked, so that you cannot use one without using the other. An example is the cultivation of high-yielding seeds which requires a number of complementary inputs – such as irrigation, commercial fertiliser, chemical pesticides and agricultural extension services – in order to fulfil the end-use of the original commodity.

My argument about the constructed nature of scarcity, and therefore, of poverty, is built on the three premises of end-use given above: contraction of alternative sources of supply of the end-use; expansion of end-uses of a commodity beyond the original use-value; and demand for complementary goods created by the purchase of a given commodity. The next step in understanding scarcity through end-use analysis is to consider the commodity within a 'nexus of relations' which refers to the bundle of relations associated with a commodity and the mutual interactions among these relations. For purposes of this chapter I shall recognise only six broad categories of relations – technical, social, cultural, ecological, political and academic (Yapa 1996, 1998). The term 'technical relations' refers to physical attributes of the commodity, a concept similar to Marx's forces of production in which he included raw materials, resources, labour and technology used in production (Marx 1989 [1869]). The term 'social relations' is used in a manner identical to its use in Marxian economics, where it refers to matters of ownership of the means of production (Marx 1989 [1869]). Production and consumption of commodities also require a myriad of interactions with the biophysical environment. These are 'ecological relations' related to inputs of matter and energy, and outputs of products, waste materials, waste heat, chemicals and so on. 'Cultural relations' refer to the interaction of production with the ways of life of social groups as embodied in shared meaning, beliefs, values, symbols, signs, language, lifestyles and consumption. 'Political relations' include interactions between the state and society in the organisation of economic activity. Finally, 'academic relations' show the interactions among discourse, practices and the development of the commodity.

This concept of academic relations is central to my argument about poverty. Academic relations are of two kinds: internal and external. Internal relations refer to epistemology, that is, how we know what we know. They describe the understanding that science has of itself in the production of knowledge. By defining, naming and describing social issues such as poverty and racism, scientists divide the world into two realms, that of the problem and of the non-problem. People who are not poor, and even those who study poverty, are located in the realm of the non-problem. But this binary view of the world turns out to be very unhelpful, emphasising commodities originating from the area of the 'non-problem' and negating alternative ways of satisfying an end-use. If, as I have argued, economic development is part of the reason why poverty persists then the people who produce, propagate and implement that development discourse are implicated as causative agents of the problem.[2] External academic relations refer to the discourses that are produced at other sites in the nexus – technical, social, cultural, political and ecological – where our understanding of technology, property, culture, the state and nature is mediated through social theories produced about these matters. Here discourse is not conducted under rules of free inquiry; it is constructed out of, and constrained by, the very material circumstances that it studies. Thus each node of the nexus is the site of both discursive and non-discursive practices.

I shall briefly summarise the implications of this new model of poverty: first, we need to establish a new understanding of scarcity as socially constructed by building on end-use principles. Second, the concept of a poverty sector as a set of households falling below an income threshold, though helpful in targeting social welfare, does not help us to find out why those households are poor. We need a methodology, which recognises that the explanations exist in a larger nexus of relations diffused throughout the larger society. Third, even though economic development is viewed as the way out of poverty, it is important to understand how that process itself is implicated in a parallel process of socially constructed scarcity embedded in a nexus of relations. The poverty discourse in Sri Lanka is built on two key elements – the concept of a poverty sector and the promise of economic development, both of which are serious impediments to improving the material conditions of poor people.

There has of course been extensive debate about the poverty issue stretching back over a century, a debate which is still active. Both the UNDP (2000) and the Oxfam Poverty Report (Watkins, 1995) have recently engaged in extensive analyses of continuing poverty across the world. Both these documents work off the concept of social justice and equality. In that sense, however, they are still a part of the conventional development paradigm. Of course the Oxfam Poverty Report gives a large number of very concrete examples of the destruction caused by development, adjustment policies and debt. But neither of these reports has a theoretical framework that implicates development itself as a causative agent of poverty. The argument which follows is constructed around discourse theory, the social construction of scarcity, and the very irrationality of the hegemonic development paradigm. I have argued that the inner logic of development itself is irrational. While I have been aided by the empirical examples provided by these two publications, they make no reference to many of the postmodern tools I employ below such as discourse theory, dualism and deconstruction.

Poverty alleviation in Sri Lanka

This section contains a brief history of the economic development of Sri Lanka to provide a context for my critique of the poverty discourse.[3] Sri Lanka is cited as an exemplary case of direct poverty alleviation because of a long history of social welfare and high values in quality of life indices. Despite a very low per capita income, Sri Lanka has achieved remarkably high levels in quality of life indices such as life expectancy, infant mortality and literacy. Many scholars believe these statistics reflect the benefits of sustained government intervention in social welfare (Gunatilleke *et al.* 1992; Sen 1981). Indeed Sri Lanka is often cited as a 'test case' of the efficacy of direct public intervention in poverty alleviation (Anand and Kanbur 1995: 228). The country has a long, sustained history of government anti-poverty programmes, beginning in colonial times. And yet, recent estimates show that over a fifth of all households do not consume the required minimum in caloric food energy (Ratnayake 1998). It is not my intent to criticise anti-poverty measures in Sri Lanka because it is certain that the poor would have been worse off without such programmes. The national leadership is to be commended for setting up safety nets for the poor despite the prevailing dominance worldwide of economic liberalism, increased austerity and reduced social welfare. As necessary as it is, however, providing welfare for poor people is not the same thing as 'solving the poverty problem'. Anti-poverty measures are (in)formed by the way policy-makers think about poverty. Because these policies are founded on an academic discourse of development that perpetuates the very problem we wish to solve, anti-poverty measures cannot solve the poverty problem in Sri Lanka. Using the changing role of the state as the criterion, I shall recognise three distinct periods in the history of post-colonial poverty alleviation in Sri Lanka.[4]

Post-Independence: 1948–56

In 1948, when Sri Lanka became politically independent from the British, it was not unlike any other former colony, a nation of dependent capitalism the control of which had passed into the hands of a class of people comprised of owners of export–import houses and plantations. The country's principal revenue came from the export of three plantation crops – tea, rubber and coconut. There was no base of manufacturing industries. Subsistence farming had lagged behind the more capitalised plantation economy: rice and subsidiary foodstuffs accounted for most of the imports of the country (de Silva 1981: 504). Post-independence, the new government placed great emphasis on peasant agriculture and made large investments in irrigation and land settlement. Paddy output expanded greatly as a result of the extension of the cultivated area, and a scheme of generous guaranteed prices for the farmers at rates well above the world market prices (de Silva 1981: 506). The surplus in export earnings the country enjoyed at the end of the Second World War soon turned into a deficit. Nevertheless, expenses on social welfare, which absorbed 56.1 per cent of the government budget in 1947, remained high (de Silva 1981: 495). Food subsidies and rationing, first introduced in 1942 by the British as a wartime relief measure, continued into the post-war period, becoming a cornerstone of post-colonial welfare services. After independence, expenditures on public health, hospitals, free drugs, free education and subsidies for imported essential foods continued despite a deteriorating foreign account balance.

Throughout this period, the government was in the hands of a conservative group that had been in the forefront of the independence movement. The country was enjoying prosperity resulting from high prices received for its rubber during the Second World War. A strong welfare economy was in place. There were no explicitly stated plans for economic or industrial development, although a great deal of attention was paid to peasant agriculture. The opposition consisted of several small warring factions whose leader in Parliament was a Trotskyite. The Marxist discourses of this period paid scant attention to issues of poverty or agriculture. Their primary objective was the organisation of trade unions among white-collar clerical workers and among urban and plantation workers.

The rise of state capitalism: 1956–77

The parliamentary elections of 1956 brought a populist coalition government into power under Solomon Bandaranaiyake, a left-of-centre liberal who subscribed to a vague amalgam of nationalism, socialism and Sinhala Buddhist chauvinism. Several Marxists were prominent ministers in the coalition government. This period saw the beginning of modern state capitalism in Sri Lanka. Influenced by the policies of the Economic Commission for Latin America (ECLA), the government inaugurated a programme of import substitution based on state-owned manufacturing enterprises in steel, cement, paper, petroleum refining and so on. A major boost to the growth of state capitalism came from the takeover of all private bus companies. The Soviet model of development was much admired by both liberals and Marxists in the coalition government. As a result, the process of national planning was strengthened, and there was a new emphasis on large-scale industrialisation.

Politically, the comprador classes continued in power, sharing it with a new, more broad-based, petty bourgeoisie consisting of small traders, small landowners, teachers, Ayurvedic physicians (practitioners of indigenous medicine) and Buddhist priests. The basis of the new class alliance was the rise of religio-linguistic nationalism among the Sinhala Buddhists. The Official Language Act of 1956 made Sinhalese the sole official language of the state, much to the dismay of the Tamil population. This laid the foundation for the ethnic conflict of the 1960s and 1970s, and the deadly civil war of the 1980s and 1990s.

After the assassination of Bandaranaiyake in 1959, his widow, Sirimavo, was elected to power. The new government of 1960 was also a coalition of liberals and Marxists who continued the tendency towards state capitalism with strong restrictions on imports and state control of foreign exchange. In 1965, a new right-of-centre government was elected, but it continued the broad policies of the previous government with some relaxation of controls on exchange and imports. The previous left-of-centre 'socialist' coalition returned to power in 1971 and governed until 1977, during which time state capitalism expanded even further, ending foreign ownership of plantations, brokering firms and agency houses. Ceilings were placed on ownership of land and houses. Food imports were restricted in order to encourage local production. Despite several changes in political leadership between 1956 and 1977, however, the broad outline of poverty policy remained intact with social expenditure on education, health and food subsidies taking up almost 10 per cent of the GNP each year (Anand and Kanbur 1995).

For the purpose of this chapter I wish to highlight three themes from this period. First, poverty alleviation took the form of continued social welfare, which included subsidised prices on essential food items, free health care and tuition-free education. These programmes were funded entirely out of the government budget, and proved to be very vulnerable to cuts in the subsequent period of the open economy. Second, the state greatly expanded its role in the economy as a result of the takeover of bus companies, plantations, agency houses, petroleum refining and distribution, and schools, in addition to the start-up of several state-owned industrial enterprises. Even before this period the state was the leading employer in Sri Lanka, but now it enjoyed a near monopoly in this respect. With high levels of unemployment throughout this period, it was not easy to get a coveted government job without the patronage of a Member of Parliament. Similar projects of economic development in Third World countries have invariably increased the power of the state vis-à-vis civil society because of its access to project funds, foreign aid, bribery, corruption and the liberal use of the power of eminent domain. In Sri Lanka the exercise of such power attained exaggerated proportions due to the extent of the welfare economy and the exercise of state control through state capitalism. The expanded role of the state inhibited private, individual and non-government initiatives. Third, during this period the Tamil population became increasingly alienated because the state control of the economy was accompanied by the continued rise of Sinhala Buddhist influence inside the governing party. This conflict prevented a large section of the population and a large part of the territory from contributing to the nation's economy. Moreover, the conflict gave rise to the civil war that began in 1983 which continues to haemorrhage the financial resources of the country. In 1996, defence absorbed nearly 6 per cent of the GDP (Nithiyanandam 2000).

The open economy: 1977–2000

The year 1977 marked a watershed in the post-independent economic history of the island when the new right-of-centre government introduced sweeping reforms: the institution of an open economy; elimination of exchange controls; liberalisation of imports; promotion of exports; encouragement of foreign investment; the building of a free-trade zone; massive borrowing in international credit markets; and the construction of a highly capitalised multi-purpose river basin development centred on the River Mahaweli. Yolanda Foster has explored the role of a phosphate mining project undertaken as part of this liberalisation process (see Chapter 8 in this volume). The new model for development was the newly industrialised countries – Hong Kong, Singapore and Taiwan. The new policies produced a massive increase in the size of the foreign debt. Throughout the 1980s the total foreign debt was over 5 per cent of GNP, and in 1987 that debt represented 23.2 per cent of export earnings (World Bank 1999). This period represented a setback to the new class of small-scale capitalists and to local farmers, among whom were the Tamil onion and chilli producers of Jaffna who had previously catered to a protected domestic market.

At the same time, expenses on social welfare were reduced. In 1979, the food subsidy scheme was eliminated and replaced with an American-style food stamp programme. Besides the traditional welfare measures in food, education and health, there were other new programmes of poverty alleviation, for example the National Housing Programme, the Janasaviya Programme (People Empowerment) and the Integrated Rural Development

Programme (IRDP). The objective of the National Housing Programme was to increase home ownership among the poor. The Janasaviya Programme gave small grants to the poor for consumption, and for starting up small-scale enterprises. It received substantial funds from international development agencies, including the World Bank, because its logic was compatible with the new development philosophy of free markets, investment in human capital, enterprise development and safety nets for the poor. The IRDP started in 1979 was an effort to organise economic planning on a decentralised district-wise basis. By 2000, 18 out of the 25 districts had district development programmes with support from a number of countries and multilateral agencies like the World Bank. For example, the rural districts of Hambantota and Monaragala received funds from Norway to implement a series of small, decentralised projects in health, sanitation and micro-credit for start-up businesses.

In broad outline, the People's Alliance government, which came into power in 1994, continued the policies of the open economy initiated by the previous government. In 1995, a new poverty alleviation initiative called the Samurdhi Development Programme was begun to replace the Janasaviya Programme of the UNP. The programme reflects the government's continued commitment to provide a welfare safety net for the poor. Modelled on the logic of the Grameen Bank, it provides credit for micro-enterprises, but the main emphasis of the programme is on consumption grants. For example, during the years 1996 and 1997, 80 per cent of the programme budget was spent on consumption grants to the poor (Samurdhi Authority of Sri Lanka 1998).

Conservatives on the one hand, and liberals and Marxists on the other, disagreed on the size of the welfare budget, and on the extent of the state's role in alleviating poverty. But there was a remarkable consensus among these groups on the need to recognise a distinct, bounded poverty sector whose problems could be solved through economic development. The poverty alleviation NGO, the Sarvodaya Sharamadana Movement, needs special mention because its logic and mode of operation are different from the state-sponsored initiatives I have described so far. I shall consider the work of Sarvodaya in the fourth part of the chapter.

A critique of the conventional approach to poverty

Notwithstanding the long history of programmes for direct poverty alleviation in Sri Lanka, the population below the poverty line remains significant. Using the income required for minimum nutrition as the criterion, the Central Bank of Sri Lanka estimated that in 1987 23.6 per cent of the population was poor (Gunaratne 1987). A World Bank poverty assessment in Sri Lanka published in 1995 reported the proportion of individuals living below the poverty line in 1990 to be about 22 per cent (World Bank 1995). Using a more generous poverty line, the fraction of people deemed poor was estimated at 40.6 per cent in 1985–6 and at 35.3 per cent in 1990–1 (World Bank 1995: 7).[5] Interestingly, it is the higher estimate that is reported in the World Bank's widely read *World Development Report 2000/2001: Attacking poverty* (World Bank 2000: 281).

At a certain level of generalisation there is a remarkable continuity in the history of poverty alleviation programmes of Sri Lanka because they all proceed from two basic premises. First is the idea that poverty can be understood by studying characteristics of households in poverty. Second is the axiom that general economic growth will eradicate poverty. In fact, the World Bank (1995) report on poverty assessment in Sri Lanka that can be viewed as the 'official word' on the subject is built on this two-pronged logic. I will reflect critically on the poverty discourse by organising my comments under three headings: dualism, internalism and economism.

As mentioned above, the poverty discourse begins by dividing all households into two sectors: the poor and the non-poor. The poverty 'problem' is seen to exist within the poverty sector. That gives rise to another dichotomy: the problem and the 'non-problem'. Those who are not poor, and those who study, help and administer programmes for the poor are seen as distinct and separate from the poor. Further, as part of the non-problem they see themselves as part of the solution. Despite its appeal and seemingly commonsense logic, that dualism represents a very unhelpful way of seeing poor people, and policies based on that logic have proven ineffective in alleviating poverty.

The official analysis of poverty begins with the construction of a poverty profile – a description of who the poor are, what the extent of their poverty is, where they live, and a list of characteristics that explains why they are poor; this strategy is based on the belief that a study of the poor will reveal why they are poor. According to the World Bank's *Poverty Reduction Handbook* (1993: 13–17), 'The poverty profile portrays the extent and nature of poverty and the distinguishing characteristics of the poor...[it] is a snap-shot of the poor.' The poverty profile can be visualised as a data table with rows and columns. The rows represent individuals, households or places. The columns contain two types of information about the poor: poverty measures and 'distinguishing characteristics of the poor'. Poverty measures include such things as income status on a poverty line. Distinguishing characteristics are variables such as location, urban/rural, gender, race, ethnicity, family size, marital status, employment, occupation, education, assets and access to markets (World Bank 1995: 95–120). In most analyses the distinguishing characteristics of the poor are the independent variables that 'explain' the magnitudes of the poverty measures. Once the poverty sector is defined as a distinct, measurable, bounded entity, then the analysis examines the poor households to look for causes of poverty. This practice is what I mean by internalism, with the implicit assumption that the sector of the non-poor is not implicated in creating poverty.

Economism is the idea that lack of food, shelter and health care is an economic problem that calls for economic solutions. According to the Governor of Central Bank, 'There is no doubt that success in poverty alleviation programmes largely hinges on the ability to maintain sustained economic growth...[to] enable the poor to get more jobs, higher incomes, and higher labour force participation' (Central Bank 1987: 4). There are two aspects to this economism. First is the notion that poor people have characteristics that prevent them from fully participating in the economy. Second is the notion that the overall economy itself needs development because it is now underdeveloped. Those who emphasise the first aspect seek 'direct' solutions to poverty, and others look to development for 'indirect' solutions. The Janasaviya and Samurdhi Programme are examples of the former, state capitalism an example of the latter.

An element of flawed dualistic thinking is present even in the decentralised area-based IRDP. Using a variant of the core–periphery model of uneven development, the concept has become deeply embedded in official planning thinking in Sri Lanka, documents coming out of the Planning Ministry referring to it routinely. It has commonly been argued that the city of Colombo and its surroundings represent the regional core, while districts like Hambantota, Monaragala and Badulla represent the marginalised periphery of Sri Lanka. The conditions of the so-called periphery are problematised because several economic indicators show values lagging behind those of the metropolitan area of Colombo. According to this model, places like Hambantota are evaluated not in terms of their internal needs, but by what they are not – 'Hambantota is not like Colombo.' The IRDP was financed primarily through bilateral foreign aid where donor nations such as Norway and the Netherlands funded particular districts. The Ministry of Plan Implementation in Sri Lanka is replacing the IRDP initiative with a new one called the Regional Economic Advancement Programme (REAP). Examinations of their programme documents (Ministry of Plan Implementation 1997) show that this is a recycling of the regional core–periphery logic with a greater emphasis on the role of private capital, as in many other countries of the South, in keeping with the current liberalisation of the economy.

There are three serious problems with the core–periphery logic of the IRDP and REAP development programmes. First, it is difficult to see an investment programme being put into place where regional inequalities in the values of economic indicators will begin to disappear. In fact, regional differences in economic indicators persist throughout the world even in wealthy countries. Second, it is not clear why the city of Colombo is held up as a standard for the rest of the country to emulate. Incomes in Colombo are higher partly because it has the highest cost of living of any city on the island. The region consumes far more resources per person than any other region. It has more congestion, pollution and crime per capita than any other region. It produces very little by way of food. It is for the most part a parasitic region. In addition, the city is home to a large number of disease-ridden slums with no sanitation, piped water or electricity. Its schools continue to produce a mass of 'educated youth' who are socialised to believe that getting a white-collar job and consuming a middle-class basket of goods is the epitome of social success. Given its self-referential and unique conditions, Colombo should not be a model or standard of comparison for any other region. Third, the doctrine of regional inequality is a prescription for according a status of permanent inferiority to districts such as Hambantota. Knowing that 'catching up' with Colombo is not feasible in the foreseeable future, we continue to use it as a standard for judging other places, which are inevitably found to be wanting and inferior to Colombo. Such regional 'envy' also prevents us from imagining creative solutions to the real problems that poor people face in the so-called periphery.

The Sarvodaya Shramadana Movement

The Sarvodaya Shramadana Movement, founded by Ariyaratne in 1958, is the largest non-governmental organisation engaged in development and poverty alleviation in Sri Lanka outside the formal framework of the state. Its philosophy and methods make it very different from other poverty programmes, and therefore I examine it separately from the others. The movement, inspired by the philosophy of Mahatma Gandhi and Vinoba Bhave has been adapted to make Buddhism in Sri Lanka function as a 'social gospel'. The term 'Sarvodaya' comes from two Sanskrit words *sarva* (universal) and *udaya* (awakening). Ariyaratne uses the word in two ways, to mean the awakening of all people and/or the total awakening of individuals. The term *shramadana* is also derived from two Sanskrit words, *shrama* (labour) and *dana* (gift), i.e. the gift of labour. In the Sinhalese language, the two words together, *sarvodaya shramadana*, have come to mean 'the sharing of one's time, thought, and energy for the awakening of all' (*Dana*, February 1987: 15).[6] Through this modest economic philosophy, Sarvodaya hopes to satisfy 10 basic human needs: (1) a clean and beautiful environment; (2) an adequate supply of safe water; (3) minimum requirements of clothing; (4) a balanced diet; (5) simple housing; (6) basic health care; (7) communication facilities; (8) energy; (9) total education related to life and living; and (10) fulfilment of cultural and spiritual needs.

From its beginnings in 100 villages more than four decades ago, the movement today has become active in 8,600 of Sri Lanka's 23,000 villages, and it commands the services of a large number of paid and volunteer workers (Sarvodaya.org). They believe that development should be more than material growth. Sarvodaya holds that social change requires personal change, a notion that is emphasised in its methods and activities (Macy 1985). So the Sarvodaya process involves psychological, moral and spiritual dimensions as well as social, economic and political ones. The explicit goal of Sarvodaya is to eradicate poverty, but it believes that this cannot be achieved through the creation of wealth. The movement promotes the vision of a 'no-affluence, no poverty' society, following the Buddhist logic of the 'middle way' which advocates avoiding the extremes of unlimited consumption or ascetic austerity. Several years ago the famous economist E.F. Schumacher (1973: 57), who was also an associate of Ariyaratne, expressed the idea in his essay titled, 'Buddhist economics' in the following words:

> From an economist's point of view, the marvel of the Buddhist way of life is the utter rationality of its pattern – amazingly small means leading to extraordinarily satisfactory results.... For the modern economist this is very difficult to understand. He is used to measuring the 'standard of

living' by the amount of annual consumption, assuming all the time that a man who consumed more is 'better off' than a man who consumes less.

Sarvodaya's material critique of development arises from its explicit rejection of affluence; its most celebrated slogan is 'no affluence, no poverty'. A similar plea has been advanced by Rana P.B. Singh in his call to recognise the importance of an 'alternative world-view of interconnectedness and holism' based on 'the ancient spiritual world-view of integral living' (1999: 55). The rejection of a capital-intensive, high-technology, open economy is clearly at odds with prevailing government policies. Sarvodaya hopes to build a no-affluence economy founded on self-reliant intermediate technology. Its advocacy of limiting consumption following the Buddhist middle way weakens one of the principal driving forces of the market economy, namely the constant acceleration of consumption. No major political party in Sri Lanka has ever advocated 'no affluence', an economic philosophy that goes against the very grain of the prevailing political economy philosophies.

This chapter looks at Sarvodaya separately because I believe the concept of no affluence can form the basis of a serious social theory critique of development where striving for affluence is one of the principal causes of socially constructed scarcity. However, Sarvodaya has weakened this argument by stating it entirely within a framework of personal morality and individual choice. The desire to acquire material goods, which Sarvodaya calls 'craving', is presented as a psychological constant, an intrinsic human condition that must be overcome through moral advancement, a point of view taken directly from Buddhist theology. But what Buddhists and Sarvodaya call 'craving' is seen by economists as consumer demand, a necessary element for the functioning of a modern market. So-called 'craving' as it relates to the desire to consume material goods takes shape within the consumerist socialisation of modern capitalism.

In Sri Lanka the size and composition of the market basket of goods are determined within a nexus of relations comprising of class, status, work, success, family, socialisation, and a culture of image making and advertising. It is unrealistic to advocate the voluntary limiting of consumption through the exercise of personal moral choice without recognising the power of myriad forces that have profoundly transformed consumption into a universal ethic by which we define human dignity and use it to rank order individuals, communities and even nations. The now large literature produced by Sarvodaya contains no serious effort to unpack secular discourses and understand mechanisms of socially constructed scarcity. Apart from a superficial dismissal of economics and development theory, Sarvodaya theorists have not produced a coherent argument to engage critically with the logic of growth, free markets and globalisation. As a result, many social scientists in Sri Lanka that I have interviewed over the years do not regard Sarvodaya as a serious social movement, and some even dismiss it as theoretically naive and lacking a historical and structural analysis of conditions in Sri Lanka.

Since the spiritual message of Sarvodaya comes from Buddhist theology, it is difficult to see it as a secular movement in which Tamils and Christians may feel welcome and included. Sinhalese identity and the perception of 'self' are intimately tied to Buddhism and its social history and mythology. Conversely, the Sinhalese perceive the Tamils as the non-Buddhist 'other'. As in a mirror image, Tamils, in turn, perceive the Sinhalese as 'the other' and Buddhist. Many Tamils argue that the privileged status accorded to Buddhism by the state is one of the mechanisms through which they are discriminated against. And it is also true that many prominent Buddhist priests have been at the forefront of anti-Tamil sentiment and protest (Tambiah 1992). Notwithstanding the central role accorded to tolerance and non-violence in Buddhist philosophy, the recent history of Sri Lanka has been marred by much violence with overtones of linguistic and religious chauvinism. Given the significance of Buddhism to the main parties of the conflict in Sri Lanka it is important to see the contradictions of a social movement that uses Buddhist ideology for non-violent social change and poverty alleviation.

While Sarvodaya wants to solve problems of the poor by valorising the 'simple life', there is no simultaneous reflection and critical theory on the life of the non-poor and the rich who are directly implicated in creating scarcity for the poor. In fact, Sarvodaya leaders and officials themselves do not adhere to simple lifestyles. They educate their own children to be professionals and push them to win scholarships abroad (Gombrich and Obeyesekere 1988: 249). Added to this dualism is another contradiction: while Sarvodaya advocates the ideal of self-reliance, the organisation itself is heavily dependent on Dutch, German and American aid. Those funds have enabled Sarvodaya to build up a vertical bureaucracy taking on the attributes of a service-delivery organisation rather than a social movement of the poor and by the poor.

Conclusions: rethinking poverty

Understanding the idea that scarcity of basic goods is socially constructed is the key to a new view of poverty. Since the poverty threshold in Sri Lanka is defined as the amount of income required for a nutritionally adequate diet we can ask the direct question: Why do large numbers of people not have enough to eat? I am not satisfied with the answer that it is because the poor do not have enough money to buy food. Why not look at food directly as a way of alleviating poverty? This is also consistent with our 'end-use' logic because the end-use of income for the poor is the purchase of basic goods such as food. The limitations of approaching poverty indirectly via economic growth are very evident in the World Bank's (1995) official understanding of this topic as seen in the report *Sri Lanka Poverty Assessment*. I wish to call attention to this document because it was a joint effort between the World Bank and the leading poverty theoreticians in the country. The report does give attention to the topic of nutrition (World Bank 1995: 19–22), but this is done as part of a discussion on trends in indicators of human welfare. In Sri Lanka the poverty line is defined as an income below which a person is unable to obtain basic nutritional requirements. The poor devote nearly four-fifths of their income to food, with rice alone accounting for a quarter of total consumption expenditure (World Bank 1995: 28). Sri Lanka is a tropical country with copious rainfall, abundant sunshine, much arable land and plentiful labour. Why so much hunger? There is nothing in the physical geography of the island to indicate that this must be so. The *Sri Lanka Poverty Assessment* report has little to say on the myriad ways in which an average individual can obtain 2,000 calories of food energy with 50 g of protein. Presumably it is not the task of a poverty report to address issues best left to agriculture, nutrition and food scientists! After all poverty reports should be written by poverty experts. But this gets to the crux of my argument. We must not separate the topics of poverty and nutrition. Finding inexpensive, creative ways of getting adequate nutrition is itself a strategy for fighting what we call 'poverty'. Not doing so contributes to social construction of scarcity, and thus perpetuates poverty. I shall illustrate this argument with reference to food. But it should be clear that the same logic can and should be extended to other topics such as health, housing, energy and transport.

So why are there large numbers of people in Sri Lanka who do not get enough to eat? Here is my answer: the scarcity of food is socially constructed within a nexus of relations that is diffused throughout the larger society. That is why we cannot find what causes poverty (and presumably eliminate those causes) by looking inside the so-called poverty sector. Instead, we should explore this question in detail, relation by relation and site by site. A detailed exploration of why food is scarce will not only explain exactly what is going on, but will simultaneously provide practical answers to what we can do to alleviate poverty. Indeed, we will not find a solution unless we first work through the substantive details of the problem. I shall offer such an analysis of the scarcity of food by way of a brief example.

In the first part of the chapter I presented a model using the concept of end-use of a commodity where scarcity occurs through: (a) the contraction of alternative sources of supply; (b) expansion of end-uses beyond the original use value; and (c) the generation of demand for complementary goods. Only one or two of these mechanisms may function in any practical instance of constructed scarcity. In the scope of this chapter I shall consider the food drive undertaken by successive governments in Sri Lanka from 1970 onwards. Their main focus was on the cultivation of high-yielding rice that required a comprehensive package of inputs: commercial seed, irrigation, fertiliser, pesticides, herbicides and credit. By 2000 rice production had increased by an impressive 300 per cent since the early 1960s (FAO 2000). But as we have seen, there is a persistent scarcity of food in the country. To understand this scarcity, this production increase must be compared against several aspects of increased costs, the records of which are now well publicised. First, the new high-yielding seeds do not reproduce well so the farmers have to purchase new seeds at planting time. Even though the monetary cost of seed is not very high it has increased the dependence of farmers on seed suppliers. Second, the new seeds are genetically uniform and their low natural resistance to diseases has created a need for pesticides. Third, in the absence of heavy doses of fertiliser the seeds do not yield well. Besides such direct costs to the farmer there are also a large number of indirect costs. The use of chemical fertiliser and pesticides have contaminated water in wells and streams which are often the only sources of water for drinking and bathing. The use of pesticides has also destroyed the natural enemies of pests leading to a still greater use of pesticides in what is called 'the treadmill of chemical farming'. In other ecological relations, pesticides have eliminated fish and crab that used to flourish in paddy fields – all useful sources of protein.

The dominance of the chemical commodities has further contributed to scarcity and dependence in poor populations. Between 1961 and 1998 the total outlay on imports of manufactured fertiliser to Sri Lanka was US$1,299 million (FAO 2000). During that time there were no systematic efforts to develop and use alternative sources of plant nutrients, including agricultural waste, green manure, the cultivation of leguminous crops, animal manure, human waste, compost, intercropping, companion planting and the use of organic waste from methane digesters. Between 1961 and 1998 the total cost of chemical pesticide imports was US$243 million (FAO 2000). Despite the existence of several effective cultural, genetic and biological methods of pest control it was the chemicals that received the most amount of official support and sponsorship. And in Sri Lanka we have not even begun to calculate the cost to the economy from the loss of natural enemies of crop disease, and the effect of chemicals on pollinating species. Turning to consumption, the food balance sheets show that over 10 per cent of calories in the average Sri Lankan diet comes from imported wheat flour consumed in the form of bakery bread (FAO 2000). All this at a time when foods such as manioc, yams, kurakkan, jak and breadfruit have declined in importance. The end result of this nexus of relations was a massive inflation in food prices: the index of food prices in 1995 was 5.2 times that of 1980 (Central Bank 1995: Table 25).

This brief survey of some aspects of the food system in Sri Lanka illustrates the concepts of socially constructed scarcity and its relation to the poverty discourse. First, it shows how food scarcity in Sri Lanka is literally constructed by social policies, practices and discourses. Among other reasons, the high cost of rice production is related to loss of genetic diversity, neglect of organic fertilisers, and lack of alternative methods of pest control. This example illustrates two themes of the model of scarcity based on end-use of commodities, namely, the contraction of alternative sources of supply, and the generation of demand for complementary commodities. Second, it shows the impact of values and norms on food choices: preference for imported wheat flour and the neglect of foods such as yam and jak. Such added use-values (or devaluations) further demonstrate the role of cultural relations in the nexus in creating scarcity. Third, this analysis of food scarcity shows a history of agricultural policy that emphasised high-yielding rice and neglected other food initiatives such as the improvement of home gardens, promotion of freshwater aquaculture and food preservation. This is an example of how the number of sources for satisfying an end-use has been systematically narrowed by agricultural research and policy (academic and political relations of the nexus) designed to alleviate poverty but implicated instead in the construction of scarcity.

To summarise the argument: since the lack of access to a diet sufficient in calories is the criterion for measuring poverty in Sri Lanka, I posed the question of why a tropical country well endowed with natural resources is unable to provide enough food for large numbers of its people. The answer to that question cannot be found through an internal examination of the so-called poverty sector. My brief discussion shows that the scarcity of food is socially constructed in a nexus of relations diffused throughout the wider society. Since there are multiple ways in which food scarcity is socially constructed, they also provide opportunities or sites at which scarcity can be overcome. In fact, a detailed investigation of the food sector along the lines I have indicated will reveal a large number of possible actions that Sri Lanka can undertake to overcome the scarcity of food.

I believe that there is a much larger story to tell about social construction of scarcity by looking at the circumstances surrounding all other basic needs as housing, transport and health care. By situating each basic good within a nexus of relations we can 'uncover' how scarcity is socially constructed at each site of a nexus – technical, social, cultural, political, ecological and academic – in a network of relations diffused throughout the wider society. Hence each site is also a locus of 'opportunity' to creatively engage the very forces that create scarcity. In this alternative approach the questions we ask, the answers we get and the actions we can take are very different from those associated with the conventional approach. Poverty discourses in Sri Lanka, be they liberal, neo-liberal, Marxist or populist, are all firmly committed to the belief that economic development will eradicate poverty. That prevents them from seeing how development is implicated in the social construction of scarcity. This in turn hinders the adoption of realistic, practical, routine, everyday measures to alleviate poverty.

Notes

1. Longer versions of the argument appear in Rahnema (1997), Sachs (1992) and Yapa (1996, 1998).
2. Despite some exceptions, I believe the claim that academic discourse is implicated as a causative agent of poverty is generally true. See for example, the influential World Bank document, *World Development Report 2000/2001: Attacking poverty*, in the light of my critique.
3. This section of the chapter has drawn on the work of an earlier paper (Yapa 1998).
4. For a similar scheme of periods created for a different purpose, see Nithiyanandam (2000).
5. The survey excluded the north and the east where data could not collected due to the ongoing civil war.
6. *Dana* is a monthly journal of the Sarvodaya Shramadana Movement in Sri Lanka.

References

Anand, S. and **Kanbur, S.M.R.** (1995) 'Public policy and basic needs provision: intervention and achievements in Sri Lanka', in J. Dreze, A. Sen and A. Hussain, (eds), *The Political Economy of Hunger*. Clarendon Press, Oxford: 298–331.

Central Bank (1987) 'Welcome address to the symposium', in *The Alleviation of Poverty in Sri Lanka*. The Central Bank, Colombo: 1–6.

Central Bank (1995) *The Annual Report of the Central Bank of Sri Lanka.* The Central Bank, Colombo.

De Silva, K.M. (1981) *A History of Sri Lanka.* University of California Press, Berkeley, CA.

FAO (Food and Agricultural Organization of the United Nations) (2000) *Statistical Databases* (Website: www.fao.org).

Gombrich, R. and **Obeyesekere, G.** (1988) *Buddhism Transformed: Religious change in Sri Lanka.* Princeton University Press, Princeton, NJ.

Gunaratne, L. (1987) 'The poorest of the poor in Sri Lanka', in *The Alleviation of Poverty in Sri Lanka.* The Central Bank, Colombo, Sri Lanka: 243–75.

Gunatilleke, G., Perera, M., Wanigaratne, R.A.M.C., Fernando, R.E., Lakshman, W.D., Chandrasiri, J.K.M.D. and **Wanigaratne, R.D.** (1992) 'Rural poverty in Sri Lanka: priority issues and policy measures', *Asian Development Review*, **10**: 164–98.

Illich, I. (1973) *Tools of Conviviality.* Harper & Row, New York.

Macy, J. (1985) *Dharma and Development: Religion as resource in the Sarvodaya Self-Help Movement.* Kumarian Press, West Hartford, CT.

Marx, K. (1989 [1869]) *A Contribution to a Critique of Political Economy.* International Publishers, New York.

Nithiyanandam, V. (2000) 'Ethnic politics and Third World development: some lessons from Sri Lanka's experience', *Third World Quarterly*, **21**: 283–311.

Rahnema, M. (ed.) (1997) *The Post-development Reader.* Zed Books, London.

Ratnayake, R.M.K. (1998) *Sri Lanka: Poverty sector study.* Samurdhi Authority of Sri Lanka, Colombo.

Sachs, W. (ed.) (1992) *The Development Dictionary: A guide to knowledge as Power.* Zed Books, London.

Samurdhi Authority of Sri Lanka (1998) *Samurdhi: National Programme for Poverty Alleviation.* Samurdhi Authority of Sri Lanka, Ministry of Youth Affairs and Sports, Colombo.

Sarvodaya Shramadana Movement in Sri Lanka (2000) (Website: www.sarvodaya.org).

Schumacher, E.F. (1973) *Small is Beautiful: Economics as if people mattered.* Harper & Row, New York.

Sen, A. (1981) 'Public action and the quality of life in developing countries', *Oxford Bulletin of Economics and Statistics*, **43**: 287–319.

Simon, D. and **Närman, A.** (eds) (1999) *Development as Theory and Practice: Current perspectives on development and development Co-operation.* Longman, Harlow.

Singh, R.P.B. (1999) 'Rethinking development in India: perspective, crisis and prospects', in Simon and Närman (eds), *op. cit.*: 55–75.

Tambiah, S.J. (1992) *Buddhism Betrayed?: Religion, politics, and violence in Sri Lanka.* University of Chicago Press, Chicago.

UNDP (United Nations Development Programme) (2000) *UNDP Poverty Report: Overcoming human poverty.* UNDP.

Watkins, K. (1995) *The Oxfam Poverty Report.* Oxfam Publishing, Oxford.

World Bank (1993) *Poverty Reduction Handbook.* World Bank, Washington, D.C.

World Bank (1995) *Sri Lanka Poverty Assessment, Report No. 13431-CE.* World Bank, Washington, D.C.

World Bank (1999) *1999 World Development Indicators on CD-ROM.* World Bank, Washington, D.C.

World Bank (2000) *World Development Report 2000/2001: Attacking poverty.* World Bank, Washington, D.C.

Yapa, L. (1996) 'What causes poverty? A postmodern view', *Annals of the Association of American Geographers,* **86**: 707–28.

Yapa, L. (1998) 'The poverty discourse and the poor in Sri Lanka', *Transactions of the Institute of British Geographers,* **23**: 95–115.

CHAPTER 10

'A fist is stronger than five fingers'
Caste and dominance in rural north India[1]

Craig Jeffrey

Craig Jeffrey: So what happened next, how did Bed Vir respond?
Pause
Yeshpal: He became cheeky, so we punched out his teeth.

(Field notes, Daurala, August 1997)

Society is like a web which touched at any point shivers through the whole.

(Thomas Hardy)

Introduction: trouble at mill

In July 1997 in Daurala, a small town built around a sugar mill in Meerut District, western Uttar Pradesh (UP), an argument broke out between Yeshpal, a rich farmer belonging to the Jat caste and Bed Vir, a poor Scheduled Caste (SC) labourer.[2] Yeshpal is one of the largest farmers in the settlement. He owns 5 ha of irrigated land and maintains close social links with local politicians and police officials.[3] Bed Vir is the head (*pradhan*) of the local government representative body (the *panchayat*) in Daurala. Prime Minster Jawaharlal Nehru established *panchayats* in the immediate post-independence period as a means of increasing the participation and representation of rural people in local government. In the early 1990s, measures were introduced to reserve the headship and positions on the *panchayat* committees for various disadvantaged sections of society, including SCs.

On a hot day, early in July 1997, several rich Jat farmers, led by Yeshpal, approached Bed Vir to complain about the condition of the roads in a Jat portion of the settlement, close to the sugar mill. Yeshpal began by asking Bed Vir why the drainage in 'their' section of Daurala had not been improved by the *panchayat*. This task, Yeshpal pointed out, should be the 'public duty' of the local *pradhan*. Bed Vir responded indignantly. When Jats had been *pradhans* of the village, he said, they had done nothing to improve SC parts of the village, 'so why should we help them now?' At this, Yeshpal became enraged (he 'boiled with anger' as an SC informant

205

put it) and hit Bed Vir in the face very hard. Bed Vir was then dragged to the ground and kicked and beaten. Bed Vir was badly injured after the assault. 'We taught him a lesson' as a Jat recalled. 'What lesson?' I asked in an interview:

> We showed him and the other Scheduled Castes that one Jat is the same as ten Scheduled Castes in strength and that we have the power [*shakti*] in the village. We showed them that just because they are pampered by the government does not mean that they rule the village (Yeshpal's brother, Daurala, August 1997).

Scheduled Castes are a collection of formerly untouchable castes identified by the British in the 1930s as requiring special government assistance and listed on so-called 'schedules'. In western UP, as in many other areas of India, they are generally confined to menial, unskilled or semi-skilled wage labour (Mendelsohn and Vicziany 1998). It is important to distinguish SCs from so-called 'Other Backward Castes' (OBCs), identified as economically and socially deprived by the Mandal Commission that reported to the Indian government in 1980. The Jats are a locally powerful caste that controls the majority of land and well-paid employment in and around Daurala. They claim Kshatriya (warrior) caste status (second beneath the Brahmins) in the *varna* hierarchical ranking of castes from which SCs are excluded.[4] They are not classed as OBCs in UP.

Caste conflict and struggle are a common feature of contemporary rural life in India. This chapter provides an introduction to how caste works in rural north India, how it is changing and how it relates to other axes of power. My principal argument is that while caste as a religiously sanctioned system of resource transfer is declining in importance in India, caste remains a powerful structuring principle in society. In particular, caste is significant as a political identity, symbolic resource and basis for social organisation. In combination with class, caste operates to reinforce established patterns of social exclusion and inclusion and therefore contributes in important ways to the poverty, powerlessness and immiseration of the rural poor in many parts of South Asia.

I frame my empirical insights around the work of the Indian sociologist M.N. Srinivas (1987) on caste dominance, and a critique of Srinivas's 'dominant caste theory' provided by Oliver Mendelsohn (1993). Mendelsohn (1993) argues that caste dominance has been eroded in rural India as powerful castes have lost their authority at the local level. He also maintains that caste has declined as a socially relevant identity. I shall argue – *contra* Mendelsohn – that caste identity continues to be an important basis for social mobilisation and organisation at the level of party, movement-based and 'everyday' politics. I shall also demonstrate the continuing authority and dominance of the Jat caste in Meerut District, western UP.

The chapter is divided into a further five sections. The first discusses the changing importance of caste in rural India with particular reference to Srinivas's (1987) dominant caste theory, and Mendelsohn's (1993) critique of this theory. The subsequent two sections investigate the relevance of this work for understanding social differentiation, political activity and poverty in rural Meerut District, western UP. These sections draw extensively on my own doctoral research and focus on the Jats. I argue that Jats, and particularly rich Jats, have been fairly successful in reproducing their economic and political privileges in the post-independence era at different political levels. The fourth section of the chapter returns to the story of Bed Vir and Yeshpal in order to consider possible threats to Jat dominance and the limits to these threats. In the final section of the chapter, I reflect upon the implications of my argument for understanding the social geography of caste dominance in South Asia.

The discussion refers throughout to the economic, social and political connections between Jats and SCs in rural north India. The high proportion of Jats and SCs in the extreme western part of UP, and large difference in wealth between these two sections of rural society, make this a sound basis for an analysis of dominance. But I must acknowledge obvious 'silences' in this account. It does not consider, for example, the relationship between Jats and other upper castes or the relationship between Jats and OBCs, who have recently been granted reserved places in government employment (see Balagopal 1991; Galanter 1991; Heuzé 1991; Béteille 1992; Srinivas 1996). Nor do I refer in any detail in this chapter to how caste is internalised by rural people (see Ramanujan 1989; Fuller 1996; Parish 1997), or the complex links between caste and gender (see, for example, Vatuk 1972; Sharma 1993; Chowdhry 1994, 1997; Lambert 1996; Palriwala 1996; Dutta 1997). Rather, I use social and political practice as a starting point for assessing *some* of the relationships between caste, identity, spatial practice and economic difference with reference primarily to men.

Casteism exists in most parts of South Asia, even in non-Hindu societies, but the strength and nature of caste feeling vary across the region: from state to state, district to district and even village to village. Commentators often draw a broad distinction between the south and west of India, where caste discrimination appears to have been particularly oppressive historically, and the northern and eastern parts of the subcontinent, where rules of caste purity and pollution were relatively relaxed (Mendelsohn and Vicziany 1998: Chapter 1). Since 1947, new patterns of caste discrimination emerging in South Asia appear to be connected to differences in the willingness and capacity of State governments and social movements to promote equitable economic development and implement anti-caste legislation. Thus, for example, the Marxist governments in Kerala and West Bengal have been more effective in reducing caste discrimination than successive government regimes in UP and Bihar, where (until very recently) successive State regimes have done relatively little to promote socially inclusive development or reduce casteism (see Corbridge and Harriss 2000: Chapter 9). Other factors that affect local and regional variations in caste feeling are the relative organisational and political power of elite and non-elite castes, the nature of political mobilisation and degree of overlap between caste and class. For example, Lerche (1995) explains the persistence of caste and class oppression in western UP relative to eastern parts of the State with reference to the greater political power and organisation of dominant castes in the western region. The forms of caste solidarity and discrimination discussed in this chapter clearly exist in many regions of India, but the specific nature of caste relations discussed here should not be generalised.

Caste and dominance

The Jats constitute only around 2 per cent of the population of UP as a whole (Hasan 1989), but in rural Meerut District they correspond in certain ways with what the renowned Indian sociologist M.N. Srinivas (1955) has called a 'dominant caste'. Srinivas defined the dominant caste as follows:

> A caste may be said to be 'dominant' when it preponderates numerically over the other castes, and when it wields preponderant economic and political power. A large and powerful caste group can more easily be dominant if its position in the local caste hierarchy is not too low (Srinivas 1955: 8; quoted in Mendelsohn 1993: 806).

The term 'caste' is being used here by Srinivas to connote *jati*:[5] a local kin network within which marriages are arranged. Each *jati* was formerly associated with a single occupation. According to Srinivas, the dominant *jati* would usually belong to one of the upper three *varnas*: Brahmin, Kshatriya or Vaishya.

In Srinivas's schema, dominance is partly a function of economic wealth. Drawing on fieldwork conducted in the 1950s in south India, Srinivas identifies dominant castes that monopolise agricultural land, educational opportunities and lucrative white-collar and professional occupations. As principal landowners, the dominant caste enjoyed a privileged position within a system of exchange of goods and services between castes, known as the *jajmani* system. In this system, castes specialising in particular occupations rendered their services or supplied goods to other villagers, and particularly to the dominant caste landholders, in return for an annual wage paid in kind. These asymmetrical economic relationships were legitimised with reference to culturally entrenched norms of purity and pollution and appear to have survived over several generations, at least up until the 1950s (see Wiser and Wiser 1963; Raheja 1988). Although patron–client relations often extended outside the village, the *jajmani* system focused economic and social struggles on the village arena.

Srinivas (1987) argues that the social and political power of the dominant caste is most fully expressed in the institutional form of caste or village councils. Since few villagers contemplated referring local conflicts to bodies outside the village, these councils assumed a judicial function in almost all village disputes. Srinivas is careful to distinguish these 'traditional' councils from the official government *panchayat* bodies (ibid.: 102–3). Unlike the official *panchayats*, the traditional institutions are active in the settlement of disputes, draw on patron–client cultural norms and manifestly operate to neutralise threats to the power and hegemony of dominant castes. 'The patrons of the dominant caste tend to support, if not to create, local structures of authority' (Srinivas 1987: 113).

In Srinivas's (1987) account, social and political dominance is also secured through more informal means. Access to local toughs appears to be crucial: 'the capacity to "field" a number of able-bodied men for a fight and a reputation for aggressiveness are relevant factors in determining the position of a caste vis-à-vis the other castes' (*ibid.*: 101). Dominant castes are also better placed to use caste solidarity, or other forms of influence, to obtain favours from politicians and other officials outside the village. This is a theme that Srinivas (1996: xi) has emphasised more strongly in subsequent work. In an introduction to a recent edited volume on caste, Srinivas approvingly quotes one of the contributors: 'The membership of a caste implies that a person becomes part of a person-based social network which controls insiders' information about economic opportunities, transmits skills, and provides varied types of human and material support' (Panini 1996, quoted in Srinivas 1996; xi).

Srinivas also notes that the efficacy of caste as a support base extends beyond the boundaries of the *jati* to encompass other castes, usually of a similar economic and ritual status. This 'elasticity' of caste, or 'horizontal stretch', is said to be most visible in the case of dominant castes (Srinivas 1987; xiii).[6] Given that each *jati* comprises only a very small proportion of the overall population of each Indian State, a broad coalition of *jatis* based on 'elastic' conceptions of caste identity is usually required to secure power in State or central elections.[7]

In stressing that rural life in India tends to revolve around a single strong *jati*, Srinivas was making an important point. Srinivas showed that the *varna* system of caste ranking did not necessarily, or even often, coincide with the local distribution of economic and political power (Mendelsohn 1993: 807). Brahmins, for example, are often not the most powerful caste locally despite being of the highest ritual status, a point also acknowledged, in a rather different way, by Louis Dumont in his classic account of the hierarchical principles underlying caste (Dumont 1970).[8]

In the present intellectual climate, Srinivas's focus on local relations of dominance and subordination is significant for at least three reasons. First, it warns against romanticised notions of 'the community' contained within certain post-colonial analyses of the rural Third World (see Vanaik 1997: 180–7). Accounts of rural relations in poorer countries, influenced by postmodernism and Foucauldian analyses, frequently posit a state versus 'local population' dichotomy that obscures class, caste, gender and ethnic divisions (Jeffrey and Lerche 2000). The strength of thinking in terms of caste dominance is that it reminds us of the relational nature of caste: that there are dominators and dominated.

Second, Srinivas's account of dominance demonstrates how everyday social struggles in rural India throw light on wider social and political structures. Srinivas encouraged his students to take 'the worm's eye view' of social and political change in South Asia. As Mendelsohn (1993: 807) points out, village-based ethnographic studies containing detailed descriptions of everyday practice are no longer viewed as providing a privileged window on an 'authentic' India. But many political economists of the last three decades have moved too far in the other direction and ignored the rich ethnographic material available on local practices of rural dominance in India and the experience of poverty (see also Gupta 1989).

Third, Srinivas's account hints at the intellectual value and increasing importance of perceiving caste as a social or symbolic resource that is brought into service within networks of social relations. This interesting aspect of caste power has been discussed in detail by David Rudner in his historical work on the organisation of banking among the Nattukotai Chettiars of south India (Rudner 1994). Rudner shows how this south Indian merchant-banking caste built up a commercial empire, spanning much of South and South-East Asia, through a complex network of inter-dependent family businesses, tied together to form a unified banking system (*ibid.*: 89). Caste-defined social relationships of trust, based on business territory, residential location, descent, marriage and common cult membership, and symbolised by gift-giving, allowed Chettiar organisations to loan and deposit money with one another effectively (*ibid.*: 234).

The general importance of social contacts in the economic success of social groups is emphasised in recent accounts of 'social capital' within social science research (Bourdieu 1992; Coleman 1990; Putnam 1993). Bourdieu (1984, 1986), in particular, highlights the links between a social stratum's economic dominance and their capacity to recycle wealth effectively into the accumulation of useful social contacts based upon shared cultural values and tastes. Rather more than Srinivas (1987), I shall explore how the ability to maintain networks of social contacts within and outside a rural settlement has become a crucial aspect of dominance.

Oliver Mendelsohn (1993) has recently questioned the contemporary relevance of Srinivas's dominant caste model and argued that caste is becoming much less important as a basis for social organisation and differentiation. Drawing on his own field observations in Behror village, that lies in an advanced rural area of northern Rajasthan, Mendelsohn maintains that the relationship between the dominant caste and less powerful castes has undergone a fundamental change. Mendelsohn notes that by the late 1980s, the traditional system of *jajmani* (patron–client) relations in Behror had become marginal to the overall economy of the village (1993: 821).[9] There appears to have been a parallel decline in an awareness of caste as a religiously sanctioned institution and in traditional councils dominated by a single caste and claiming authority over other villagers. Mendelsohn links the decline of the *jajmani* system and traditional councils to a sharp rise in rural people's economic activity outside the village. The decline of the *jajmani* system, allied to the growth of non-agricultural employment outside the village, is said to have provided labourers in Behror with a new independence from the formerly dominant Ahirs (*ibid.*). The erosion of *jajmani* relations and rise in off-farm employment have been noted in many parts of India, including Gujarat (Breman 1993), western UP (Jeffery and Jeffery 1997; Gupta 1998), West Bengal (Harriss 1993; Williams 1997), Madhya Pradesh (Mayer 1996), Tamil Nadu (Harriss 1991; Harriss-White 1996), Andhra Pradesh (Upadhya 1988) and Karnataka (Epstein 1973; Shivalingappa 1997). Nevertheless, these authors are generally much more cautious in their assessment of how far the rural poor have become independent of powerful local castes.

Mendelsohn makes a second claim in his account of social relations in Behror. The expansion of opportunities to enter business or employment outside the village is said to have increased economic differentiation among Ahirs. This has eroded a sense of collective interest within this caste. Caste solidarity has been replaced by 'pragmatic individualism or at least family-centredness' (Mendelsohn 1993: 824–5). The declining social relevance of caste for Ahirs has also led to a decline in forms of collective action based on caste, with the notable exception of marriage, which continues to be in-caste (endogamous). The influence of caste as a form of political and social organisation is also said to be on the wane. This individualistic orientation among the Ahirs is linked in Mendelsohn's account to their lack of interest in village affairs and orientation to a wider space economy embracing nearby towns and cities. Mendelsohn's two central claims – that the power of the formerly dominant caste has declined and that caste is increasingly irrelevant in the social life of the village – are clearly linked. An absence of caste solidarity and organisation within a formerly dominant caste is likely to weaken their grip on the rural poor. Wealthy individuals belonging to the formerly dominant caste have lost both the inclination and capacity to police the boundaries of their privilege along caste lines within the arena of the village. My account of the economic and social standing and strategies of rural Jats in western UP casts doubt on the direction and content of Mendelsohn's argument.[10]

Caste and class differentiation in rural Meerut District

My analysis of social reproduction in western UP draws upon 12 months' field research that I conducted in the sugar township of Daurala and two villages of Khanpur and Masuri in Meerut District, between December 1996 and December 1997 (Figure 10.1). The three settlements are located within 25 km of Meerut City and were selected as being broadly representative of dynamic rural settlements in this part of western UP. Jats comprise between a third and a half of the total population of these settlements.[11]

I collected information regarding the assets, education, employment and political contacts of 250 male household heads belonging to four Jat lineages[12] within the three settlements. These extended kin groups were broadly representative of the wider Jat population in the three villages in terms of their wealth and occupational profile. I also interviewed a random sample of 30 Muslim, SC or Most Backward Caste (MBC)[13] household heads and 40 Jat and lower-caste women within Daurala, Khanpur and Masuri (combined). In addition, I interviewed politicians, policemen, lawyers, land officials and other official notaries living in or close to Meerut City. The interviews were conducted in Hindi.

Jats possess over 90 per cent of the agricultural land in each of the three settlements in which I worked. Of the 250 households within my sample, 12 per cent possess more than 5 ha. In contrast, of the 73 SC households in the sample, only 33 per cent own agricultural land and none possess more than 2.5 ha (Jeffrey and Lerche 2000). Such inequalities of landownership are fairly typical of rural western UP (Singh 1992; Wadley 1994; Jeffery and Jeffery 1997; Srivastava 1997; Jeffrey and Lerche 2000).

The Green Revolution in Indian agriculture, beginning in the mid-1960s, had a major impact on the wealth and lifestyle of Jat farmers (Jeffrey 1997).

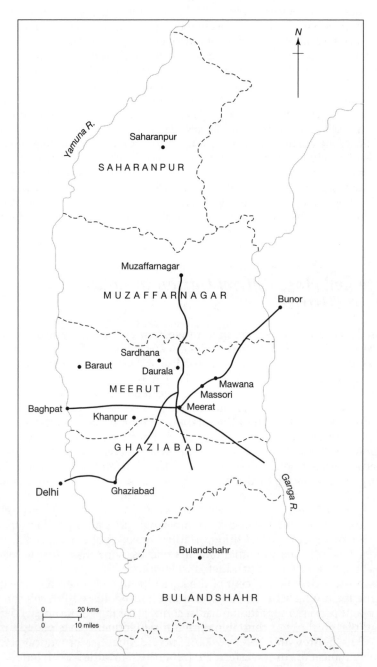

Figure 10.1 Meerut District, Uttar Pradesh

From the mid-1960s, Jat farmers in Daurala, Khanpur and Masuri invested in tractors, electric tube wells, chemical fertilisers and high-yielding varieties of wheat and sugar cane. These technologies increased agricultural yields markedly. The vast majority of Jat farmers in rural Meerut District cultivate a rotation of sugar cane and wheat for the market. Jats possessing more than 5 ha usually hire farm servants on contracts of between six and nine months. Farmers cultivating more than 2.5 ha of land tend to employ temporary workers for the most labour-intensive agricultural activities such as the cutting, tying and weeding of sugar cane and harvesting of wheat. Many of the temporary labourers and farm servants are SC men or women.

The rapid subdivision of land associated with the prevailing system of partible inheritance encourages Jats to seek off-farm incomes, especially salaried employment for their sons. The UP State government, like other north Indian State administrations, has failed to create a large number of jobs in the private sector (Srivastava 1995; Dube 1998; Hasan 1998). As a result, efforts to secure salaried jobs tend to focus on public-sector employment opportunities that expanded relatively quickly between 1947 and the early 1990s (Hasan 1998). Rural people attach great importance to being able to secure even low-ranking service employment, such as a job as a bus conductor, or peon in a local bank. These jobs are secure, provide a regular wage, offer relative economic independence (relative to agricultural labour, for example) and may provide opportunities for obtaining large illegal incomes 'on the top' (*uparli*).

In my research I distinguished between Jats possessing more than 5 ha and those possessing less than 5 ha as a basis for examining social differentiation within the caste. This corresponds with local definitions of what constitutes a 'big' farmer, as opposed to a 'medium' or 'small' farmer. It is also generally only farmers possessing more than 5 ha who employ labourers on long-term (six to nine month) contracts.

Jats living in rural Meerut District have traditionally engaged in agricultural production, although a small number obtained employment within the British system of colonial administration (Chowdhry 1994; Dutta 1997). Of the 108 male Jats born before 1946 for which I have occupational information, 20 per cent (22) entered non-agricultural employment. Table 10.1 shows that Jat men in my sample born between 1947 and 1971, who by 1997 had completed the transition to employment or non-employment, have effected a more pronounced move out of farming. This has been achieved through investing in the schooling of male, and more recently female, children, particularly within English-medium private schools. In addition, intra-caste arranged marriages are acting as an imperfect, but occasionally successful, means for rich landowners to 'buy' their way into professional and urban circles of influence.

Table 10.1 points to intra- and inter-caste variations in the pattern of off-farm diversification. As this table suggests, rich farmers have tended to enter the most prestigious professional and white-collar employment and have, with one exception, avoided labouring work. Poorer Jats have been less successful in entering lucrative and high-ranking non-agricultural employment, but have managed to obtain lower-ranking professional or white-collar work, for example as policemen, teachers, managers and clerks.

Table 10.1 Principal occupation of male Jats in the four lineages, born between 1947 and 1971 by landholding of guardian, excluding Jats whose guardian was a higher professional or businessman

	Farming % (no.)	Business % (no.)	Higher professional % (no.)	Lower professional or white-collar % (no.)	Labour % (no.)	Unemployed % (no.)	Total % (no.)
Jats owning over 5 ha	52 (38)	7 (5)	12 (9)	27 (20)	1 (1)	0 (0)	99 (73)
Jats owning 5 ha or less	31 (45)	6 (8)	2 (3)	38 (56)	19 (28)	4 (6)	100 (146)
Total Jats	38 (83)	6 (13)	6 (12)	35 (76)	13 (29)	3 (6)	101 (219)
SCs, MBCs and Muslims	1 (3)	13 (28)	2 (4)	12 (26)	68 (144)	3 (6)	99 (211)

Notes: Percentages may not add up to 100 due to rounding. 'Higher professional' employment refers to employment as a doctor, lawyer, engineer, tertiary-level teacher, Grade I or Grade II government administrator or senior army official. These forms of employment are usually better paid than other professional and white-collar occupations and are often accorded a higher status in Jat society. 'Lower professional' employment refers to work in government service, and managerial, white-collar or professional private employment.
Source: Sample surveys conducted between December 1996 and December 1997 in Khanpur, Daurala and Masuri.

The evidence of a move into lucrative professional employment among the Jat elite, and the simultaneous shift among poorer Jats into lower-ranking jobs, suggests that, among the Jats, class differences based on land are being reproduced in the quest for jobs outside the village (Jeffrey and Lerche 2000). This ties in closely with Mendelsohn's (1993) thesis that class and caste no longer overlap in rural north India. What we are seeing, it seems, is the emergence of classes within castes, and, in the case of my research, within families.

Table 10.1 suggests that SCs, MBCs and Muslims have been far less successful in entering skilled non-agricultural employment and the majority remain in labouring work. It appears that SCs have been largely unable to take advantage of reserved places in public-sector employment and educational institutions introduced in 1950 to promote the economic mobility of formerly untouchable castes. As a friend of Bed Vir's said: 'don't talk to me about reservations! Without money and social contacts, we just can't get a job!' (Daurala, September 1997). The competition for jobs among SCs in the reserved quota, and among Jats and other middle and upper castes in the so-called 'general quota', is said to have become privatised. A 'government job' comes to those with influence in the bureaucracy and the money to bribe recruitment officials or brokers.

The proportion of work that labourers perform outside rural settlements and agriculture appears to have increased quite rapidly since the mid-1960s in rural Meerut District (Sharma and Poleman 1994). This is largely a result

214

of the spread effects of capitalist agricultural growth. The Green Revolution led to an increase in agricultural processing, transport and storage enterprises, and an attendant rise in consumer incomes has stimulated the expansion of a range of more diverse rural industries. Bed Vir is typical of the mass of SC labourers in terms of his employment. He spends roughly six months of the year employed on the farm of a rich Jat farmer, two to three months in construction work in and around Meerut City and occasionally helps a relative run a phone booth on the main road. Of the 144 SC, MBC and Muslim male labourers in the sample, 36 per cent were principally involved in agriculture, 13 per cent in factory work and 11 per cent in construction. The remaining 40 per cent were involved in miscellaneous labouring activity, usually outside the village and not under the supervision or employment of local Jats. I was unable to obtain similar figures for SC women, and it is likely that they remain more closely tied to waged work for local Jats. Holding in mind that the area in which I worked was a fairly advanced part of rural western UP, this pattern of off-farm diversification is in line with other research in the region (see Lerche 1999). It also ties in with Mendelsohn's (1993) observation of increased off-farm employment among the rural poor in a similarly advanced rural area of northern Rajasthan. Diversification out of traditional caste occupations among the rural poor has led to the decline of former *jajmani* relationships.

Nevertheless, certain relationships of obligation continue to exist between the rural poor and dominant Jats. For example, the Jats are often able to coerce the poor into planting sugar cane in return for the provision of a feast, known as *makali*. They have also been able to institutionalise a system whereby labourers receive only the leafy sugar cane tops (*gole*) in return for cutting the sugar cane (see Lerche 1999). Women remain particularly tied to forms of exhausting and poorly paid work, for example in making dung cakes (a common fuel) for rich Jats who have withdrawn their wives from this activity.

The dependence of the rural poor on the Jats relates not only to their continued partial reliance on labouring work. The poor also depend on the Jats for credit, access to fields to defecate and cut grass for their cattle, and assistance in negotiations with government officials. Jats can therefore apply a range of sanctions to discourage recalcitrance among SCs. These sanctions are spatial in nature and include banning SCs from walking through Jat fields, restricting their use of wells in the village and intimidating SCs as they walk through Jat-dominated areas of the village (see Singh 1992).

The Green Revolution in Indian agriculture, in combination with an expansion in public-sector employment for much of the post-colonial period, appears to have strengthened the economic dominance of Jats in Meerut District and promoted class differentiation within this caste. The incursion of the state into people's everyday lives and the uneven impact of capitalist agricultural development are creating new forms of dependence and sustaining some older relations of dominance.

The social and political geography of Jat dominance

This section outlines the nature of Jat political strategies since the mid-1960s at three political levels: that of party political competition, the level of movement-based politics and an informal level of political intrigue based around obtaining assistance from government officials. At each level, caste has been important as a basis for social organisation and as a symbolic resource. The following analysis seeks to move between a quite detailed account of the tactics and nature of social action and attention to broader geographies of political transformation and social change. This represents an effort to situate a worm's eye view of caste dominance, or what Geertz (1983) would describe as 'thick description', within the wider political economy of caste in UP.

Within party politics, the power of the Jats has been closely associated with the success of the Jat politician and intellectual Charan Singh. During the 1950s and 1960s, as a politician within the Congress Party, Charan Singh resolutely sought to protect the interests of the richer Jat peasantry. In 1967, Singh broke from the ruling Congress Party in frustration at the policies of the party leader of the time in UP, C.B. Gupta. Singh then established his own political party: the Bharatiya Kranti Dal. During the period between 1967 and 1986, Singh built on the loyalty of members of his own Jat caste to galvanise the votes of middle and rich peasants across UP (Hasan 1989; Duncan 1997). Between 1969 and 1985, Charan Singh's party consistently polled between 28 and 34 per cent of the votes in the UP Legislative Assembly elections in Meerut District and captured 43 and 48 per cent of the parliamentary election vote in 1980 and 1984 respectively (Singh 1992: 130–5).

As a member of the Cabinet in the State and central parliaments, Singh represented the interests of his rich peasant supporters. Most noticeably, he raised subsidies on agricultural inputs and government prices for key cash crops and opposed the introduction of radical redistributive land reform and effective forms of agricultural taxation (Byres 1988; Hasan 1989, 1998). The coherence of a middle/rich peasant vote declined quite rapidly in the late 1980s and early 1990s (Duncan 1997). But Charan Singh's political career had a lasting impact on the ability of middle and rich cultivators to obtain a surplus from Green Revolution agriculture in western UP. As Yeshpal himself noted at the end of an interview about politics, and presumably with no intended reference to the Mafia: 'for the Jats, he was a Godfather figure' (Daurala, August 1997).

Charan Singh strenuously denied accusations of casteism (Byres 1988). Nevertheless, Singh clearly played on his specifically Jat peasant origins in fostering electoral support, and Jats loyally supported Singh (Singh 1992). Occasionally, this Jat politician's specific interest in assisting his fellow caste members became clear. For example, Singh's effort to extend the principle of positive discrimination in government employment to 'kisan (farming) communities' (Hasan 1989: 185–6), represents a thinly veiled attempt to promote Jat interests. Furthermore, the wider political consensus nurtured by Charan Singh appears to have depended upon implicit distrust or suspicion of the extremes of the caste hierarchy: Brahmins and untouchables (Byres 1988: 163–5). It has often been suggested that there was a connection between the rise of Charan Singh's political party within UP politics and violence targeted against SCs (Brass 1985: 330; Byres 1988: 165).

This description has implications for our understanding of caste and caste dominance. Mendelsohn acknowledges that at elections, and in defence of a political party, corporate caste interest and solidarity tend to assert themselves, even among castes that may be moving towards a more secular, 'Western' outlook (Mendelsohn 1993: 828). As a number of commentators have noted, the spread of democracy and democratic institutions across India has reinforced rather than undermined the importance of caste, particularly since the mid-1960s (Kothari 1970; Fuller 1996; Srinivas 1996). Caste continues to be central in the calculus of political parties in UP and most other States of India. In addition, several authors have shown that caste organisation and associations may be important building blocks for political parties, factions or coalitions (Kothari 1970). The building of political factions and coalitions has been one of the primary ways in which particular *jatis* have extended their reach and influence outside the village arena.

The link between the party political power of a caste and its influence over local government institutions has been the subject of less research. Mendelsohn maintains that the 'occasional position' of the Jats as the principal political power of the UP State should not be mistaken for local dominance: 'once caste dominance has been lost at the village level it cannot functionally be replaced at a higher political level' (1993: 834). It is possible to interrogate this claim with reference to the Bharatiya Kisan Union (BKU) farmers' movement that came to prominence in western UP in 1987.

Under the leadership of the charismatic Jat, Mahendra Singh Tikait, the BKU undertook its first large-scale demonstration in January 1987, in protest against a government-imposed hike in the electricity tariff on irrigation pumps. At this rally, held in Shamli, Muzaffarnagar District, western UP, the main aims of the BKU were voiced in an 'oral charter'. The chief demands were for a more efficient and cheaper supply of key inputs to capitalist agriculture and higher government purchasing prices for cash crops (Bentall and Corbridge 1996: 32). Drawing on Gandhian imagery, the BKU declared war on urban 'India', which was identified as the enemy of an 'authentic' mother India, *bharat* (the Hindi word for India). The BKU proceeded to organise large rallies in Meerut City and Delhi in 1988 and 1989 to publicise its grievances. These demonstrations lasted for several days and were militant in nature (Hasan 1995: 166). In Muzaffarnagar District, Jats were able to establish their own 'village republics' characterised by farmers' rule (*kisan raj*) (Hasan 1995; Bentall 1996). Roads were blocked, villages closed to government officials and farmers refused to pay taxes and electricity bills or to clear their interest from banks and credit co-operatives (Hasan 1995).

Jim Bentall's fieldwork in western UP, conducted in 1992–3, suggests that Jats were at the forefront of BKU activity (Bentall 1996; Bentall and Corbridge 1996: 40–1). This is reflected in the significance of Jat ideologies of direct cultivation, plain living and brotherhood in BKU rhetoric (Bentall and Corbridge 1996: 39–40; A. Gupta 1997). The Sisauli *panchayat*, headed by Tikait and a fulcrum of the grassroots decision-making structure, was an extension of a traditional Jat clan council (Hasan 1995: 183; see also Lerche 1995). Like the traditional councils, the BKU *panchayat* had no well-defined agenda and involved long sittings in large male gatherings, elaborate feeding arrangements and a degree of sacrifice from the participants, who had to leave the villages for days or weeks (Madsen 1996: 89). This ties in with evidence of the continuing salience of Jat kinship and caste councils in the resolution of disputes over marriage (Madsen 1996; Chowdhry 1997). In western UP, breaches of marriage norms are frequently referred to local or regional caste councils that assume responsibility for administering punishments.

This level of Jat political co-ordination casts doubt upon some of the conclusions drawn by Mendelsohn (1993), particularly the close relationship he posits between the extension of powerful castes' spatial imaginations outside the locality and the rise of individualism. Social capital, institutionalised within caste councils, provides an important means for Jats to exert pressure on the government and to police the boundaries of their privilege.

Occupational diversification among the Jats also appears to have strengthened, rather than undermined, their interest in maintaining informal networks of power that bridge rural and urban areas. This power is expressed and reinforced through cultivating 'good relations' with key social contacts in nearby large towns and cities. Here I refer to the core features of these social networks as a means of understanding the role of caste in structuring access to local government bureaucracies and in order to stress the connection between social capital, spatial practice and the subordination of the rural poor.

The support and assistance of the local police service are viewed as an important potential means for all rural people to defend themselves from attack or seek redress after a crime. In this context, police assistance has become almost wholly privatised. Police protection in rural western UP is best viewed as a prize to be won, rather than as a right to be demanded (Brass 1997). To obtain police help, for example after being the victim of theft or violence, it is usually necessary to work through a chain of intermediaries, each of whom may demand a bribe.

Working through a concrete example will illustrate the social dynamics and spatiality of the process of seeking assistance from the police. A rich Jat farmer named Giriraj was aggrieved by the encroachment of a neighbour on to his land. His immediate reaction, one that is very common, was to approach a fellow caste member, Rishipal, who was known to have a 'good relation' or 'link' with a politician. Giriraj had to convince Rishipal of the urgency and seriousness of the matter at hand. 'This is a matter of playing the game', Giriraj confided in me afterwards. At their meeting, held in Rishipal's *baithak* (male sitting area on the edge of a courtyard), a plan of action was agreed over long drafts on the *hookah* (pipe), a ritual that is part of Jat culture in rural western UP (D. Gupta 1997). The pair then sped off on Giriraj's scooter to visit Rishipal's political contact, a Member of the Legislative Assembly (UP Parliament) living in Meerut. At this meeting, the politician listened carefully to the details of the case and the appeals of Giriraj and Rishipal for help. In this instance, the politician was a Muslim, and this foreclosed the opportunity to appeal to common Jat or Kshatriya (warrior) caste solidarity. But the politician realised that it was important to maintain a 'good relation' with an influential Jat such as Rishipal in order to raise funds and canvass votes before elections in Daurala. The politician therefore agreed to help the two Jats with the case and informally discuss the matter with a senior police official living in Meerut. He reported back to Rishipal and Giriraj that the police would act, but that it would be necessary to 'sweeten the palm' of the senior police official to ensure rapid action. The sweetener proved to be 12,000 rupees (roughly £200 in 1997). Rishipal maintained that such a sum was typical in this sort of case. It is equivalent to over double the yearly wage of an agricultural labourer in rural Meerut District.

Small wonder then that people belonging to the SCs, such as Bed Vir, lose out in disputes over land, property and violence. Considerations of social and economic capital intersect to deny poor SCs access to state help. I happened to mention the bribe that was required in the Giriraj case to an SC labourer and lamented that most SCs would find it impossible to pay this sum. The labourer replied that if *he* had approached the politician for help, he would have been forced to pay a larger bribe to the politician to ensure that the work would be completed: 'I can't think about the 12,000 rupees, for me it might have been 24,000!' (Bedhu, Daurala, September, 1997). Without the political clout of Rishipal, the politician would have demanded a substantial fee for his role as intermediary. In addition, SCs lack the private transport that is necessary to move rapidly between key contacts. Significantly, SCs employ spatial metaphors to describe the superior social contacts of the Jats. The dominant caste men are said to possess a 'setting' or are referred to as 'mobile' or 'wandering' men.

Although informal local political networks link people of different castes and religions, a sense of Jat caste solidarity is often significant in securing help from state officials. This becomes clear if we emphasise the performative dimensions of local corruption (Gupta 1995). In order for Rishipal and others to maintain good relations with officials, who generally belong to the upper castes, it is necessary for him to regularly reaffirm the link through organising a meeting. These meetings often occur in the Jat's own home. One of the most common forms of entertainment laid on for visiting notables are 'chicken and whisky' parties that frequently involve shooting, songs, arm-wrestling competitions and heated card games. These parties often celebrate the Jat self-image of physical strength and martial prowess and draw on a tradition of Jat wrestling and army service (Dutta 1997). In other cases, they affirm a wider warrior caste identity. This is testament to the 'elasticity' or 'horizontal stretch' (Srinivas 1996; xiii) of upper-caste masculine identity.

Scheduled Castes tend to lack the money required to lay on ostentatious parties for visiting notables. Jats and OBCs frequently ridicule male SCs for a style of dress, manner and insouciant air that are said to betray low-caste desperation for influence and recognition. SCs maintain that in conversations with state officials, most of whom belong to the forward castes, they often feel uncomfortable and lack the easy style that wins respect: 'we can stand around and look strong, but it makes no difference' (SC labourer, Masuri, August 1997). It appears therefore that SC access to networks of influence is limited not only by their poverty and lack of local political clout, but also by their inability to master appropriate and largely caste-specific styles of masculine behaviour that characterise interactions between state officials and those seeking patronage. Caste matters. Or as Rishipal put it:

> When someone comes to me for help in some matter, first I think is he my friend, is he of my caste? If he is a Jat, of course we think we should help each other. Jats have a saying: a fist is stronger than five fingers (Rishipal, Daurala, August 1997).

Threats to Jat power

Caste relations between Jats and SCs in rural UP have recently been influenced by the emergence of the Bahujan Samaj Party (BSP) – literally 'the majority of society' party – within UP politics. The BSP seeks to increase the political visibility, standing and security of SCs in UP. Political efforts aimed at assisting SCs in local struggles go back at least as far as the 1870s (Singh 1992; Lerche 1999). Nevertheless, the BSP is the first political party in post-independence UP solely committed to helping SCs. During the 1990s, the BSP held power at the State level three times, in each case in a coalition with another party. It has increased SC representation in the local administration and tightened legislation relating to caste violence (Pai and Singh 1997; Duncan 1999; Lerche 1999). It has also embarked upon an ambitious drive to raise the profile and pride of SCs through the construction of parks, statues, libraries and other 'civic' monuments, often dedicated to the SC hero, Dr Ambedkar. This effort appears to have increased the self-confidence and political purchase of SCs at the local level in many parts of rural UP (Jeffrey and Lerche 2000). Nevertheless, it has failed to release SCs from their dependence on Jats in western UP. A vibrant identity politics on its own provides an insufficient basis for effectively reducing inequalities in the distribution of material and social resources. The moderate improvement in SC political confidence but continuity in Jat dominance are demonstrated in the following account of the aftermath of the assault on Bed Vir.

The Bed Vir/Yeshpal denouement sent a ripple of rumours, gossip and anxiety through the fabric of Daurala society. Bed Vir obtained the *pradhanship* of the *panchayat* on a reserved seat, but he commanded respect and support from many poorer sections of local society. When SCs in Daurala became aware of the attack on Bed Vir, they first tried to bring the aggressors to justice. Scheduled Castes in Daurala organised a two-pronged response. One party of SCs created a *rasta roko* (roadblock) on the main road close to Daurala in an effort to publicise the act of violence. Another party took the bus to Meerut City and reported the incident to the Sub-Divisional Magistrate (SDM). This direct approach to seeking assistance from an influential State official seemed at first to have borne fruit. The SDM, who was a member of the SCs, listened carefully to the story of the assault. The official then told the SCs to return to Daurala and apologise to the Jats. The SDM's argument appears to have run along the following lines:

> You must learn to live better with the Jats. You must not antagonise them. Without good relations with the Jats you will find it difficult to get work, to get fodder for your cattle and move around the village. So go back and

apologise to the Jats and do not trouble me with this again (field notes, August 1997, based on reports of several SCs present at the meeting).

The SCs returned to the village and Bed Vir went to apologise to Yeshpal and his cronies. Bed Vir later acknowledged the force of the SDM's arguments, adding that without social 'links', or a *jan paychan* (known and recognised figure) in the judiciary, it would be very expensive to pursue a legal case against Yeshpal.

Such examples of caste-based violence and subsequent official inaction are a depressingly common feature of the social geography of post-independence rural India (see Samata Sanghatana 1991; Ilaiah *et al.* 1991; Bhatia 1997; Breman 1997). In many areas the judicial role and authority of traditional councils have weakened, or, as in Behror, Rajasthan, disappeared altogether (Mendelsohn 1993). But this decline has not been accompanied by a reduction in the effective power of the dominant caste. The fact of the rural poor's rural residence, the vulnerability of their wives (many of whom continue to work for the Jats), and their dependence on Jats for cattle fodder and intermittent employment, militates against the emergence of an independent rural poor. SCs also lack the material and symbolic resources to build and maintain effective 'social capital', figured locally as 'links' and referring, in particular, to contacts within government bureaucracies.

What makes Bed Vir's case rather unusual is the fact that he was the *pradhan* (head) of the local village *panchayat* at the time of the fracas. Bed Vir has actually been quite successful in using his position to improve the access of SCs to resources in Daurala, such as government scholarships for SC children and grants to buy buffaloes. This is a significant change. For most of the period since their introduction in the early 1950s, the headship of the *panchayats* has been controlled by the richer Jats in Daurala, as in other settlements in western UP (Lieten 1996; Gupta 1998). The Jats used their position as *pradhan* to control government development resources that entered the village. These could be used for personal enrichment and chan-nelled to supporters in the settlement. Bed Vir's humiliation can be inter-preted as a reassertion of the status quo in Daurala. In a letter I received after returning from fieldwork to the UK, I was told that Bed Vir is now understandably frightened of crossing the Jats again. The incident appears to have brought the settlement into line with many surrounding villages in which low-caste or female *pradhans* act as puppet leaders for powerful Jats (see Lieten 1996). In the absence of more substantial economic, social and political change, Jats are able to build rural–urban social capital that acts to isolate SCs, such as Bed Vir, who had taken advantage of reserved posts.

Conclusions

Caste is changing in rural India. As it changes, it poses new and excit-ing questions for geographers interested in the reproduction of social inequality and resistance to this reproduction. Caste as a religiously and culturally sanctioned system of resource transfer appears to be on the wane.

The *jajmani* system and associated norms of authority described by Srinivas (1987) for rural south India in the 1950s have been eroded. But caste as an identity, form of social organisation and basis for staking claims to valued resources remains significant. The discussion can be given greater clarity here by referring back to the two arguments distilled from Oliver Mendelsohn's (1993) work. First, Mendelsohn argued that caste dominance has declined in rural north India. *Contra* Mendelsohn, I have shown that, while political representation and increased off-farm employment opportunities have moderately improved the economic and political position of SCs in Meerut District, the rural poor continue to be economically dependent on the Jats and dominated by this caste.

Second, Mendelsohn argued that caste as an identity and form of social organisation has become less significant. In Meerut District, however, at a variety of political levels, forms of caste organisation and solidarity continue to be important in reproducing Jat power. The democratisation of party political competition from the late 1960s onwards in UP, capitalist agricultural advance and the increasing incursion of the state into people's everyday lives have collectively acted to politicise and invigorate caste identities and collectivities. My account of caste dominance differs somewhat from that of Srinivas (1987) in the attention it pays to the social resources and political links of the dominant caste. Dominance is perpetuated through spatialised social networks that link the rural and urban. The forms of individualism identified by Mendelsohn (1993) for the Ahirs of Behror coexist with other obligations and identities, including strong feelings of caste loyalty. Mendelsohn (1993: 827) acknowledges this point, but in emphasising 'atomisation' he fails to give due weight to the contemporary significance of caste solidarity and organisation.

These conclusions have wider relevance for the themes of this book. The rise of the BSP is representative of a much broader move across India towards a more open and plural competitive political democracy (Khilnani 1997). This offers the rural poor opportunities to obtain political representation, usually on the basis of caste rather than class (Robinson 1988; Mendelsohn and Vicziany 1997; Pai and Singh 1997). The position of the rural poor in UP, as in other parts of northern India, increasingly depends upon how far these parties match their emancipatory rhetoric with real efforts to alter the distribution of assets.

This chapter has also uncovered disturbing aspects of the state's failure to protect its citizens. These shortcomings cannot be swept under the carpet by referring to occasional examples of progressive governmental reform or assistance. They are failings that are likely to be particularly unsettling for those readers of this book brought up in the Western liberal tradition of believing in the redemptive capacity of the state. They are no less shocking to many of the rural people with whom I worked, who continue to believe in the viability of the state as a source of support (see also Pushpendra 1999). What we are witnessing when, for example, we recall the reported words of the SDM to the indignant SCs, is the failure of a particular 'state idea' in India. This refers to the state's responsibility to protect and enhance the lives of all members of the Indian population regardless of caste, creed or gender. The laudable aims of the Indian Constitution to guarantee Indian citizens freedom from ignorance, hunger and arbitrary violence are now being violated by precisely those who were expected to uphold people's rights: representatives of the state such as policemen, magistrates, government teachers and local government representatives. This chapter has demonstrated, with reference to ethnographic field research, how some of these failings impact upon the life chances and security of people in rural India.

Notes

1. An earlier version of this chapter appeared in the *Transactions of the Institute of British Geographers*, **26**(2), 217–36. I am grateful to the publishers for their permission to publish this version.
2. I employ pseudonyms throughout the chapter. The caste name 'Jat' is pronounced with a long 'a', as in 'cart'.
3. While for consistency all figures are quoted in metric units, the common unit of areal measurement in UP is still acres. The figures quoted in this chapter are rounded conversions from acres.
4. *Varna* literally means colour. The *varna* system of caste ranking originates in the Hindu sacred texts: the Vedas. In one of the earliest texts, the Rig Veda, society is linked to the divisions of the cosmic man 'Purusha'. The head corresponds to Brahmins, who perform a priestly function in society; the arms correspond to Kshatriyas, the warriors; the torso or stomach corresponds to the Vaishyas, the merchants; and the feet correspond to the Shudras, the menial workers. The untouchables, who traditionally performed work regarded by Hindus as defiling, were excluded from this hierarchy altogether. There is no direct translation of the word 'caste' in Indian languages. The word actually comes from the Portuguese word *'casta'* meaning 'pure breed'.
5. *Jati* literally means 'species' or 'type'. Where I refer to caste in the remainder of the chapter, I mean *jati*.
6. In pointing to the value of caste as a social or symbolic resource, Srinivas is in line with a wide range of Indian authors who have demonstrated that caste can provide a dynamic basis for capitalist enterprise. Most work has been conducted on castes primarily involved in merchant or business activity (Fox 1969; Timberg 1978; Béteille 1993; Gorter 1996; Harriss-White 1997). Nevertheless, caste solidarity and action have also been an important resource for upwardly mobile peasant castes seeking off-farm employment (Epstein 1973; Upadhya 1988; Rutten 1995).
7. In the remainder of the chapter, 'State' refers to the individual governments within the Indian federation and 'state' to the central Indian government in New Delhi.
8. See Quigley (1993) for a useful summary and critique of Dumont's work.
9. It is possible that in his determination to identify change, Mendelsohn may exaggerate the rigidity of caste structures in the past. I am grateful to Glyn Williams for pointing this out to me.
10. As in the area of northern Rajasthan studied by Mendelsohn (1993), western UP does not contain a large number of dominant Brahmin landowners. The dominant *jati* in most parts of the upper doab of western UP are of middling caste status (especially the Jats) (see Singh 1992; Lerche 1995). This links western UP to many areas of rural Bihar (Bhatia 1997) and central Gujarat (Rutten 1995; Breman 1997), where castes of middling or backward status are often dominant as landowners. A contrast can be drawn between these areas and areas where ritual and economic standing are more closely mapped on to one another. This includes the

areas of eastern UP studied by Lerche (1995) and Sharma (1978), where upper-caste Thakurs preside over local social and economic opportunities. It also includes parts of rural Madhya Pradesh (Mayer 1996), Tamil Nadu (Béteille 1965) and Karnataka (Srinivas 1976), where Brahmins have historically controlled land and the best employment opportunities.

11. The total population of the three fieldwork areas were: Daurala town 10,025 inhabitants, Masuri 4,373, Khanpur 4,913, according to the 1991 census (Office of the Registrar General 1991).
12. The genealogical tables for these groups stretch back at least 10 generations and were constructed from the memory of respondents and the records of the village genealogist for each settlement.
13. Most BCs refers to former service castes including Kumhars, Dhimars, Goswamis, Lohars, Telis and Julai (Pai and Singh 1997).

References

Balagopal, K. (1991) 'Post-Chundur and other Chundurs', *Economic and Political Weekly*, **26**(42): 2399–405.

Bentall, J. (1996) '"Bharat versus India": peasant politics and urban–rural relations in North-West India'. Unpublished PhD thesis, Department of Geography, University of Cambridge, Cambridge.

Bentall, J. and **Corbridge, S.E.** (1996) 'Urban–rural relations, demand politics and the "new agrarianism" in NW India: the Bharatiya Kisan Union', *Transactions of the Institute of British Geographers*, **21**(1): 27–48.

Béteille, A. (1965) *Caste, Class and Power: Changing patterns of stratification in a Tanjore village*. University of California Press, Berkeley, CA.

Béteille, A. (1992) *The Backward Classes in Contemporary India*. Oxford University Press, New Delhi.

Béteille, A. (1993) 'The family and the reproduction of inequality' in P. Uberoi (ed.)., *Family, Kinship and Marriage in India*. Oxford University Press, New Delhi: 435–51.

Bhatia, B. (1997) 'Massacre on the banks of the Sone', *Economic and Political Weekly*, **32**(51): 3242–5.

Bourdieu, P. (1984) *Distinction: A social critique of the judgement of taste*. Routledge and Kegan Paul, London.

Bourdieu, P. (1986) 'The forms of capital', in J.G. Richardson (ed.), *Handbook of Theory and Research in the Sociology of Education*. Greenwood Press, New York: 241–58.

Bourdieu, P. (1992) 'Ökonomisches Kapital–Kulturelles Kapital–soziales Kapital', in M. Steinrücke (ed.), *Die verbigene Mechanismen der Macht*. USA, Hamburg.

Brass, P.R. (1985) *Caste, Faction and Party in Indian Politics: Vol. I, Faction and Party*. Chanakya Publications, New Delhi.

Brass, P.R. (1997) *Theft of an Idol: Text and context in the representation of collective violence*. Princeton University Press, Princeton, NJ.

Breman, J. (1993) *Beyond Patronage and Exploitation*. Oxford University Press, New Delhi.

Breman, J. (1997) *Silencing the Voice of Agricultural Labourers in South Gujarat.* Kingsley Martin Memorial Lecture, University of Cambridge, Cambridge, November.

Byres, T.J. (1988) 'Charan Singh (1902–1987): an assessment', *Journal of Peasant Studies,* **15**(2): 139–89.

Chowdhry, P. (1994) *The Veiled Women: Shifting gender equations in rural Haryana 1880–1990.* Oxford University Press, New Delhi.

Chowdhry, P. (1997) 'Enforcing cultural codes: gender and violence in northern India', *Economic and Political Weekly,* **32**(19): 1019–28.

Coleman, J. (1990) *Foundations of Social Theory.* Belknap Press of Harvard University Press, Cambridge, MA, London.

Corbridge, S. and Harriss, J. (2000) *Reinventing India: Liberalization, Hindu nationalism and popular democracy.* Polity Press, Cambridge.

Dube, S. (1998) *In the Land of Poverty: Memoirs of an Indian family, 1947–1997.* Zed Books, London.

Dumont, L. (1970) *Homo hierarchicus: The caste system and its implications.* Weidenfeld and Nicolson, London.

Duncan, I. (1997) 'Agricultural innovation and political change in North India: the Lok Dal in Uttar Pradesh', *Journal of Peasant Studies,* **24**(4): 246–68.

Duncan, I. (1999) 'Dalits and politics in rural north India: the Bahujan Samaj Party in Uttar Pradesh', *Journal of Peasant Studies,* **27**(1): 35–60.

Dutta, N. (1997) 'Arya Samaj and the making of Jat identity', *Studies in History,* **13**(1): 97–119.

Epstein, T.S. (1973) *South India: Yesterday, today and tomorrow.* Macmillan, London.

Fox, R.G. (1969) *From Zamindar to Ballot Box: Community change in a North Indian market town.* Cornell University Press, Ithaca, New York.

Fuller, C.J. (1996) 'Introduction: caste today', in C.J. Fuller (ed.), *Caste Today.* Oxford University Press, New Delhi: 1–31.

Galanter, M. (1991) *Competing Equalities: Law and the backward classes in India.* Oxford University Press, New Delhi.

Geertz, C. (1983) *Local Knowledge: Further essays in interpretative anthropology.* Basic Books, New York.

Gorter, P. (1996) *Small Industrialists, Big Ambitions: Economic and political networks on a large industrial estate in Western India.* Oxford University Press, New Delhi.

Gupta, A. (1989) 'The political economy of post-Independence India – a review article', *Journal of Asian Studies,* **48**(4): 787–96.

Gupta, A. (1995) 'Blurred boundaries: the discourse of corruption, the culture of politics and the imagined state', *American Ethnonologist,* **22**(2): 375–402.

Gupta, A. (1997) 'Agrarian populism in the development of a modern nation (India)', in F. Cooper, and R. Packard (eds), *International Development and the Social Sciences: Essays on the history and politics of knowledge.* University of California Press, Berkeley, Los Angeles and London: 320–44.

Gupta, A. (1998) *Postcolonial Developments. Agriculture in the making of modern India.* Duke University Press, Durham, NC, London.

Gupta, D. (1997) *Brotherhood and Rivalry.* Oxford University Press, New Delhi.

Harriss, J. (1991) 'The Green Revolution in North Arcot: economic trends, household mobility, and the politics of an "awkward class"', in P. Hazell and C. Ramasamy (eds), *The Green Revolution Reconsidered: The impact of high-yielding rice varieties in South India.* Johns Hopkins University Press, Baltimore, MD: 29–57.

Harriss, J. (1993) 'What is happening in rural West Bengal? Agrarian reform, growth and redistribution', *Economic and Political Weekly,* **28**(24): 1237–47.

Harriss-White, B. (1996) 'The Green Revolution in south India', *Politica Internazionale,* **5**: 81–94.

Harriss-White, B. (1997 'The state and informal economic order in South Asia'. Paper to the Moscow School of Social and Economic Sciences Colloquium on Exploratory/Informal Economies: Substance and Methods of Study, January.

Hasan, Z. (1989) 'Power and mobilisation: patterns of resilience and change in UP politics', in F.R. Frankel and M.S.A. Rao (eds), *Dominance and Power in Modern India: Decline of a social order.* Oxford University Press, New Delhi: 133–95.

Hasan, Z. (1995) 'Shifting ground: Hindutva politics and the farmers' movements in Uttar Pradesh', in T. Brass, (ed.), *New Farmers Movements in India.* Frank Cass, Ilford, Essex: 165–94.

Hasan, Z. (1998) *Quest for Power: Oppositional movements and post-Congress politics in Uttar Pradesh.* Oxford University Press, New Delhi.

Heuzé, G. (1991) 'Troubled anthropologists: the controversy over employment quotas in India', *Anthropology Today,* **7**(6): 5–7.

Ilaia, K. *et al.* (1991) 'Upper caste violence: study of Chunduru carnage', *Economic and Political Weekly,* **26**(37): 2079–804.

Jeffrey, C. (1997) 'Richer farmers and agrarian change in Meerut District, Uttar Pradesh, India', *Environment and Planning A,* **29**: 2113–27.

Jeffrey, C. and **Lerche, J.** (2000) 'Stating the difference: state, discourse and class reproduction in Uttar Pradesh, India', *Development and Change,* **31**(4): 857–78.

Jeffery, R. and **Jeffery, P.** (1997) *Population, Gender and Politics: Demographic change in rural North India.* Cambridge University Press, Cambridge.

Khilnani, S. (1997) *The Idea of India.* Hamish Hamilton, London.

Kothari, R. (ed.) (1970) *Caste in Indian Politics.* Sangam, London.

Lambert, H. (1996) 'Caste, Gender and Locality in rural Rajasthan', in C.J. Fuller (ed.), *Caste Today.* Oxford University Press, New Delhi: 65–92.

Lerche, J. (1995) 'Is bonded labour a bound category? Reconceptualising agrarian conflict in India', *Journal of Peasant Studies,* **22**(3): 484–515.

Lerche, J. (1999) 'Politics of the poor: agricultural labourers and political transformations in Uttar Pradesh', in T.J. Byres, K. Kapadia and J. Lerche, *Rural Labour. Relations in India.* Frank Cass, London: 182–243.

Lieten, G.K. (1996) 'Panchayats in western Uttar Pradesh: "namesake" members', *Economic and Political Weekly,* **31**(39): 2700–5.

Madsen, S.T. (1996) *State, Society and Human Rights in South Asia.* Manohar, New Delhi.

Mayer, A. (1996) 'Caste in an Indian village: change and continuity 1954–1992', in C.J. Fuller (ed.), *Caste Today*. Oxford University Press, New Delhi: 32–64.

Mendelsohn, O. (1993) 'The transformation of authority in rural India', *Modern Asian Studies*, **27**(4): 805–42.

Mendelsohn, O. and **Vicziany, M.** (1998) *The Untouchables: Subordination, poverty and the state in modern India*. Cambridge University Press, Cambridge.

Office of the Registrar General (1991) *Census of India*. Uttar Pradesh volume, statistics prepared by the National Institute of Public Finance and Policy, New Delhi.

Pai, S. and **Singh, J.** (1997) 'Politicisation of Dalits and Most Backward Castes: study of social conflict and political preferences in four villages of Meerut District', *Economic and Political Weekly*, **32**(23): 1356–61.

Palriwala, R. (1996) 'Negotiating patriliny: intra-household consumption and authority in northwest India', in R. Palriwala and C. Risseeuw, (eds), *Shifting Circles of Support: Contextualising kinship and gender in South Asia and sub-Saharan Africa* Sage, New Delhi: 190–220.

Parish, S.M. (1997) *Hierarchy and its Discontents: Culture and the politics of consciousness in caste society*. Oxford University Press, New Delhi.

Pushpendra (1999) 'Dalit assertion through electoral politics', *Economic and Political Weekly*, **34**(36): 2609–18.

Putnam, R.D. (1993) *Making Democracy Work: Civic traditions in modern Italy*. Princeton University Press, Princeton, NJ; Chichester.

Quigley, D. (1993) *Interpretation of Caste*. Clarendon Press, Oxford.

Raheja, G.G. (1988) *The Poison in the Gift: Ritual prestation, and the dominant caste in a North Indian village*. University of Chicago Press, Chicago and London.

Ramanujan, A.K. (1989) 'Is there an Indian way of thinking? An informal essay', *Contributions to Indian Sociology*, **23**(1): 41–58.

Robinson, M. (1988) *Local Politics: The law of the fishes – development through political change in Medak District, Andhra Pradesh (South India)*. Oxford University Press, New Delhi.

Rudner, D.W. (1994) *Caste and Capitalism in Colonial India: The Nattukottai Chettiars*. University of California Press, Berkeley and Los Angeles, CA.

Rutten, M. (1995) *Farms and Factories*. Sage, New Delhi.

Samata Sanghatana (1991) 'Upper caste violence: study of the Chunduru carnage', *Economic and Political Weekly*, **26**(36): 2079–84.

Sharma, M. (1978) *The Politics of Inequality: Competition and Control in an Indian village*. University of Hawaii Press, Hawaii.

Sharma, R. and **Poleman, T.T.** (1994) *The New Economics of India's Green Revolution: Income and employment diffusion in Uttar Pradesh*. Cornell University Press, Ithaca, New York.

Sharma, U. (1993) 'Dowry in north India: its consequences for women', in P. Uberoi, (ed.), *Family, Kinship and Marriage in India*. Oxford University Press, Delhi: 341–56.

Shivalingappa, B.N. (1997) 'Mobility and change in the Western Ghats', in J.-L. Racine (ed.), *Peasant Moorings: Village ties and mobility rationales in South India*. Sage, London: 163–232.

Singh, J. (1992) *Capitalism and Dependence: Agrarian politics in Western Uttar Pradesh 1951–1991.* Manohar, New Delhi.

Srinivas, M.N. (1955) 'The social system of a Mysore village', in M. Marriott (ed.), *Village India.* Chicago University Press, Chicago.

Srinivas, M.N. (1976) *The Remembered Village.* Oxford University Press, New Delhi.

Srinivas, M.N. (1987) *The Dominant Caste and other Essays.* Oxford University Press, New Delhi.

Srinivas, M.N. (1996) 'Introduction', in M.N. Srinivas (ed.), *Caste: Its twentieth century avatar.* Penguin Books, New Delhi: ix–xxxviii.

Srivastava, R. (1995) 'India's uneven development and its implications for political processes: an analysis of some recent trends' T.V. Sathyamurthy (ed.), *Industry and Agriculture in India since Independence.* Oxford University Press, New Delhi: 190–210.

Srivastava, R. (1997) 'Change and resilience in producer strategies in Uttar Pradesh Agriculture', in J. Breman, P. Kloos and A. Saith (eds), *The Village in Asia Revisited.* Oxford University Press, New Delhi: 199–236.

Timberg, T.A. (1978) *The Marwaris: From traders to industrialists.* Vikas, New Delhi.

Upadhya, C.B. (1988) 'From kulak to capitalist: the emergence of a new business community in coastal Andhra Pradesh, India'. Unpublished PhD thesis, Yale University.

Vanaik, A. (1997) *The Furies of Indian Communalism: Religion, modernity and secularization.* Verso, London, New York.

Vatuk, S. (1972) *Kinship and Urbanization: White collar workers in North India.* University of California Press, Berkeley, CA.

Wadley, S.S. (1994) *Struggling with Destiny in Karimpur, 1925–1984.* Vistaar Publications, New Delhi.

Williams, G. (1997) 'State, discourse and development in India: the case of West Bengal's Panchayati Raj', *Environment and Planning A,* **29**(12): 2099–112.

Wiser, W.H. and **Wiser, C.V.** (1963) *Behind Mud Walls 1930–1960.* University of California Press, Berkeley, Los Angeles, CA.

CHAPTER 11

Informal settlements and urban sustainability in Pakistan

Arif Hasan

Informal settlements and sustainability in the South

Housing has been one of the most significant issues facing governments in the post-colonial countries of the South, including all the countries of South Asia. Hardoy and Satterthwaite (1997: 265) stated that

> Overall, the number of urban dwellers living in very poor conditions has grown rapidly, even though most nations are much wealthier than they were in 1950. An estimate for 1990 suggests that at least 600 million urban dwellers live in life threatening or health-threatening homes and neighbourhoods because of poor quality shelter, dangerous sites, and inadequate provision for safe water supplies, sanitation, drainage or health care.

India, Bangladesh and Pakistan, the three largest South Asian countries in terms of their population, have, with the active encouragement of the World Bank and other agencies, all adopted formal institutional approaches to meeting urban housing need. Yet the scale of the task remains daunting, and the failure of successive programmes to meet their objectives has resulted in a sense of continuing experiment in the field of urban housing.

The challenge has arisen partly from urban growth itself, and partly from the steady increase in the proportion of the population living in towns and cities across South Asia. In India, for example, the total urban population rose from 62 million in 1951 to over 300 million in 2001 (Census of India 2001), although the percentage of India's population classified as urban had only risen from 17 to 30 per cent during this period. While these rates of growth have been lower than in some parts of Africa or South America, leading some observers in the 1960s and 1970s to question why India's rate of urbanisation was so 'slow' (Turner 1962), the housing needs embedded in these rates of growth have represented one of the most outstanding challenges of urban planning and development of the post-colonial period.

Policy towards urban development has itself changed in each of the countries of South Asia during the last 50 years. In part these changes have reflected global shifts in policy goals, in part they are a response to local conditions. 'Informal housing' has played a major role in meeting urban housing need across the South, and is one of the most persistent features of contemporary urban development in South Asia. From the early 1950s to the early 1970s governments in South Asia saw informal housing, which went under a variety of local names – ranging from *cheris* in Madras (Bradnock 1989) to *bustees* in Kolkata or *jhuggies* in Delhi – as the problem in itself. Often designated as slums, the solution commonly proposed was their demolition and replacement with 'properly built', formal housing, serviced with the full range of water, electricity and sewerage services of the 'modern' urban sector. Despite massive investment in formal rehousing schemes through the 1970s, informal housing continued to multiply. Official government estimates in India suggested in the 1990s that as much as 40 per cent of India's urban population lived in informal housing, and both central and State governments were forced to re-examine urban development policies. From the 1970s governments across South Asia, with the support of international funding institutions, turned towards schemes designed to improve conditions in areas of informal housing rather than their wholesale replacement. While grandiose schemes of urban building such as the New Bombay Project (Jacquemin 1999) continued to promote large-scale investment in formal housing and urban development, many cities began to develop sites and service initiatives. In many cases such projects seemed little more than a recognition of existing informal processes of urban development. As is shown in the Pakistan case study below, the absence of formal urban physical environments may often conceal the tightly organised web of informal sector structures through which the informal sector has been developed. Despite the importance of region-specific characteristics which, as the Pakistan case study shows, continue to inform both problems and solutions, Maria Veronica Bastías Gonzales has shown in the South American context how important local focused approaches can be to the mobilisation of effective urban empowerment, itself a condition of improvement (Bastías Gonzales 1999: 257). This chapter explores the working out of urban development issues in Pakistan since independence with particular reference to the informal housing sector.

Urbanisation in Pakistan

In the 1951 census, the first after Partition, the urban population of Pakistan was 6.019 million or 17.8 per cent of Pakistan's total population. According to the 1998 census, the urban population is 42.458 million or 32.51 per cent of the country's total population. In the last 18 years, the urban population has grown at an average rate of 4.34 per cent against a total growth of 2.61 per cent per year. Critics of the 1998 census disagree with these results. They argue that the urban population is closer to 50 per cent. They point out that large peri-urban informal settlements are not included in the municipal areas of many cities and are considered rural. In addition, over 1,400 'towns', which have populations of more than 5,000, are also considered rural since they do not have urban local government structures. Critics also point out that ribbon development along major highways is considerable. These ribbons have urban characteristics and in almost all cases are informal in nature (Ali 2000). These issues are being increasingly debated in the press and through lectures and seminars.

Informal settlements and the legacy of Partition

Informal settlements are tolerated in Pakistan more than in the rest of South Asia. This is due to the legacy of Partition. According to the results of the 1951 census, 6.8 million out of a total population of 33.816 million were refugees from India. Over 45 per cent of the total urban population also consisted of refugees. As a result of this large migration, the population of many cities in Sindh and Pakistani Punjab increased by more than 100 per cent[1] and in a few cities in the North-West Frontier Provinces (NWFP), population registered a fall because of the outmigration of Hindus and Sikhs which was not replaced by migration from India.[2]

Migrants squatted on government land within cities. Alternatively, agricultural land near the city and urban properties belonging to the fleeing Hindus and Sikhs were subdivided and occupied. Over time, governments accepted the situation and the creation and acceptance of informal settlements became a part of Pakistani urban culture. The refugee activists and their local collaborators, who organised these settlements initially, became the middlemen and entrepreneurs who in latter years negotiated with corrupt government officials to create new informal settlements on government land. Today this activity is carried out by their apprentices and in many cases by a second and third generation of apprentices.

The reason for informal settlements

Estimates vary, but it is generally agreed that Pakistan today requires about 450,000 housing units per year for its increasing urban population and for dealing with a housing backlog. The formal sector can only cater for less than 125,000 units (Guerl *et al.* 1991). This demand–supply gap is met by the creation of squatter settlements on government land (known as *katchi abadis*) or through the Informal Subdivisions of Agricultural Land (ISALS) on the city fringe. There is a major difference between these two systems of development. *Katchi abadis* do not have security of tenure until they are notified for regularisation under the Katchi Abadi Improvement and Regularisation Programme (KAIRP) of the government. ISALs, however, are owned by the residents although they are unserviced and informal. Government documents describe ISALs as 'slums' and report that 35 per cent of Pakistan's total urban population lives in *katchi abadis* and ISALs.

However, this figure is disputed and according to conservative estimates about 58 per cent of Pakistan's urban population lives in informal settlements, 22 per cent in *katchi abadis* and 36 per cent in ISALs (Hasan 1995). Detailed studies are available for two very different cities, Karachi and Faisalabad.

Karachi is Pakistan's largest city and has a population of just over 10 million. Its annual housing demand is 80,000 units. Over the last five years, the Karachi Building Control Authority (KBCA) has issued an average of 26,700 building permits per year. The demand–supply gap is met by the creation of new *katchi abadis* which provide about 28,000 lots of land per year and encroach on an average of 400 ha.[3] The rest of the demand is met by densification in existing settlements or not met at all. Even where schemes are developed for the poor, they remain unoccupied for anything between 10 and 20 years and are subsequently occupied by the middle classes. It is estimated that 200,000 plots developed by the Karachi Development Authority (KDA) are lying empty in Karachi (Siddiqui 1992), most of them owned by speculators. Meanwhile, over 50 per cent of Karachi's population lives in squatter settlements.

Faisalabad is Pakistan's third largest city and a major industrial area. Between 1947 and 1998, its population increased by about 1.9 million. A minimum of 200,000 housing units would be required for this population increase. However, between 1947 and 1998 the government was able to provide only 38,785 plots and houses, whereas Faisalabad requires 12,000 housing units per year. As a result of this demand–supply gap, *katchi abadis* have developed all over Faisalabad and now, unlike Karachi, no more government land is available for their expansion. Therefore, housing is being developed through ISALs on the city fringe. According to official estimates, 2,000–3,000 plots are developed every year through this process. Informal developers, however, claim that the figure is closer to 6,000 (Alimuddin *et al.* 1999).

The reason for this state of affairs is simple: government planning is not compatible with the sociology and economics of low-income groups which constitute about 68 per cent of the urban population of Pakistan. Over the years, the government has initiated a number of innovative housing policies and projects for the larger Pakistani cities. However, the end product has never reached the poor, whose need constitutes over 60 per cent of the housing demand. The reasons why this demand has not been met are that the supply was far too small as compared to the demand; the end product was unaffordable for the poor and there were no credit schemes to purchase land but only to build a house; and the procedures for acquiring land and credit for house building were long and cumbersome and involved being viewed by officialdom with suspicion and hostility. In addition, development of a scheme can take up to 10–15 years after payment has been made, whereas most poor families want a home immediately. Even after all this, if a family does acquire a plot, they are required to get building plans approved and to follow complex by-laws and construction details that make no sense to them and are in almost all cases unaffordable. These complex procedures have to be bypassed and this has created a culture of patronage and corruption in the house-building process.

Katchi abadis: *process and actors*

Katchi abadis (meaning non-permanent settlements or more precisely unbaked settlements) are really squatter colonies on government land. After partition they were created through unorganised invasions. These settlements had neighbourhoods created around clan, tribe or extended families. In most cases they were ethnically homogeneous and organic in form. However, during the 1960s; migration into the cities due to the introduction of Green Revolution technologies in the rural areas and industrialisation in the urban, increased considerably. At the same time, the government also started to bulldoze *katchi abadis* in a big way and attempts at protecting state land were also made. As a result, the creation of *katchi abadis* became a commercial venture and unorganised land invasions were replaced by informal subdivisions (ISDs). The process of establishing these subdivisions has been studied in detail for Karachi. Observations and interviews suggest that similar processes, though not so well organised, have been followed for other cities as well. However, except for certain Sindh cities, where considerable public land is available, the creation of *katchi abadis* is difficult if no longer possible.

ISDs are the major source of housing in Karachi. Surveys suggest that plots for over 28,000 housing units are developed in Karachi every year through this process[4] which is facilitated by the fact that about 94 per cent of Karachi's land is in some form of public ownership. In this process an informal developer occupies state land with the support of corrupt government officials who become his informal partners in development. Low-income families are moved on to these unserviced plots and the developer arranges water for them through tankers. In addition, a contractor moves into the area to supply building materials on credit, cash credit and technical advice for house building. The local police takes money from the plot owners when they move on to the plot, build their compound walls, lay a proper roof to the house or get a service connection of any sort. In return, they offer protection from eviction (see Box 11.1).

The developer forms an organisation of the residents and gets it registered. Through the organisation he lobbies for acquiring water supply, electricity, roads and other facilities from the state. He and his unofficial partners are interested in the development of the area since they hold plots in the settlement for speculative purposes. Schools are established through community efforts or by entrepreneurs and the private sector moves in to provide curative health services. In a period of 15–20 years, these settlements acquire most facilities, with the exception of sewerage, as shown in Table 11.1.

Through a legal battle, residents of *katchi abadis* (even if not notified for regularisation) and other informal settlements have a right to acquire electricity, gas, telephone, post office and water connections. Politicians patronise this process since it helps them in getting votes and residents now know how to negotiate with the politicians before an election for these services.

The residents of most of these settlements are either migrants from the rural areas or families who have moved from the inner-city slums to the ISDs. The developers of the *katchi abadis* who are known as *dallals* or middlemen, have now started producing apprentices who are carrying on

Box 11.1   The informal subdivisions of Yakoobabad

Yakoobabad is an informal subdivision settlement of about 2,000 houses in Orangi Township. Before 1977, it was vacant land belonging to the Central Board of Revenue (CBR). The CBR had given it on a renewable one-year lease as pasture land to an elder of the Rind tribe (henceforth referred to as X).

Mr Y is one of a number of informal developers who have illegally developed more than 200,000 plots on government land in West Karachi alone, over the last 30 years. Like other developers he has close links with officials in the CBR, Karachi Metropolitan Corporation (KMC), police and other departments relevant to his work.

In February 1977, Y moved on to X's land with 100 'destitute' families. These families were transported in trucks along with bamboo posts and mats (supplied by Y) for the construction of shacks. Y had identified these families through his contacts in the settlements he had created earlier. As soon as the families started putting up their shacks, members of the Rind tribe arrived in jeeps carrying guns and tried to eject them. A scuffle followed and a number of Y's people were injured. It was decided between the two parties that no houses would be put up but the 'destitutes' could stay on the land until matters were settled.

The next day X hired a lawyer and made a case in a court of law against the occupation of his land. The case was admitted. Y on the other hand, filed a complaint with the local police saying that the Rind tribe had caused 'bodily harm' to his clients. After this the local *thana* (police station) arranged negotiations between the two parties. As a result, it was decided that the Rind tribe would receive Rs500 for every plot that was developed by Y. The plots being given to the 100 'destitutes' were exempt from this payment and Y also did not receive any payments for them. It was further agreed that Y would pay Rs200 per plot to the KMC officials from the sale proceeds and that the police would recover Rs200 or more directly from the owners when they converted their shacks into concrete wall constructions. After the negotiations were completed, the Rind tribe withdrew its case against Y.

Y then laid out Yakoobabad on a gridiron plan. His apprentices (he was also an apprentice once) helped him in this work. The roads were levelled by informally hiring tractors and a bulldozer at a nominal cost from KMC staff in West Karachi. Space for a mosque and a school were set aside and plots on the main road were allocated for shops and businesses. At this stage negotiations were entered into with representatives and touts of government officials who could be of help in the future development of the settlements. Thirty per cent of all plots were set aside for these officials for speculation purposes. Whoever purchased a plot in the settlement (except for the ones reserved for officials) had to construct a house in a month's time and move in, failing which he would lose his plot and the money he had paid for it. Thus Y prevented speculation and saw to it that the settlement would expand fast.

X appointed a *chowkidar* (caretaker) to keep track of the number of plots that were developed so that Y may not cheat him. In the same manner, the KMC officials also had their informal representatives visiting the site regularly. Accounts were settled between the parties every week.

Y engaged donkey cart owners to supply water to the settlement. These suppliers acquired water illegally from the KMC water mains in Orangi. The payment of the first supply of water was made by Y, after which the people dealt with the water suppliers directly. A few weeks after the first shacks were built, a contractor, Nawab Ali, established a building component manufacturing yard, or *thalla*, in the settlement. He started supplying concrete blocks and tin roof sheets for the construction of houses along with technical advice and small credit. As such, he became the architect and the House Building Finance Corporation (HBFC) to the residents of Yakoobabad. He also constructed a water tank for curing purposes and this tank became a source of water supply for which the residents paid. In the initial 10-year period, 92 per cent of families had built their homes with support from Nawab Ali and 62 per cent had made use of the credit offered by him. At the same time as Nawab Ali, another entrepreneur, Faiz Mohammad Baloch, moved into the area. He set up a generator and started supplying electricity to the residents at the rate of Rs30 per tube light or a 40 W bulb. This sum was to be paid in advance and an advance non-payment for the next month would lead to a disconnection. Later Faiz Mohammad Baloch opened a video hall where three films per day were advertised and illegally exhibited to an audience of about 20–50 people per show. The local *thana* permitted this and received *bhatta* (illegal gratification) for their support.

Y has formed a welfare association of all the households who have ever purchased a plot from him. This association is legally registered under the Societies Act. The Yakoobabad families became members of this association and through it Y and the Yakoobabad leadership have lobbied for infrastructure and improvements in the settlement. In this they have been helped by officials and politicians, who hold plots in Yakoobabad, since all improvements increase the value of property.

Most of the early residents of Yakoobabad were people who owned no homes or who could not afford to pay rent. Later residents came to Yakoobabad to escape from degraded physical, social and environmental conditions in the inner city. By 1989, Yakoobabad had become a 'proper' settlement and even lower-middle-income families started to shift here and as a result, the physical and social environment of Yakoobabad underwent a major improvement.

Today, 60 per cent of Yakoobabad has electric connections (acquired through the bribe market) and the rest of the 40 per cent either buy electricity from their neighbours or illegally tap the Karachi Electric Supply Corporation (KESC) mains by paying the KESC staff. There is still no piped water although water mains and trunk sewers under an Asian Development Bank (ADB) financed project have been laid.

Box 11.1 (cont'd)

However, water now comes through tankers and not by donkey carts. The area has 10 primary schools, 2 secondary schools, 6 clinics and many roads have been paved by the councillors. The education and health facilities are all in the private sector. Transport through minibuses and Suzuki pickups is available. In addition, 401 micro enterprise units provide employment to over 2,600 persons in the settlement. Most of the units are engaged in garment stitching and *zari* (golden work on cloth) work and employ women. Nawab Ali has shifted his *thalla* to a newer settlement and Faiz Mohammad Baloch has become a video shop owner. A plot that was sold for Rs900 in 1978 now fetches a price of over Rs30,000.

The people of Yakoobabad have paid far more through bribes and extortion for their land and its development than they would have for a government-developed housing scheme. But they have paid for this incrementally over time and in sums that were affordable to them. In addition, this struggle to improve their conditions has transformed them into a community.

There are over 700 settlements like Yakoobabad in Karachi, housing more than 50 per cent of the city's population. They grow at a rate of 9 per cent per year against an annual urban growth rate of less than 4 per cent. It is they, and not the state agencies, that are determining the future physical, social and political structure of the city.

Source: Research by the author.

---

their work. In addition, most of the work that they are doing tries to follow the rules and regulations of the KDA so that there may be no problems in regularising these settlements. The population of *katchi abadis* in Karachi has increased in the last decade by 9.2 per cent per year as against a total urban growth of 3.52 per cent per year.[5] Table 11.2 illustrates the growth pattern of *katchi abadis*.

Table 11.1 Relationship between age of house and housing quality and level of services – *katchi abadis*

	Age of house				
	0–5	*6–10*	*11–15*	*16–20*	*Over 20*
Reinforced concrete	22.7	16.4	14.4	21.1	19.4
Plastered concrete walls	56.0	55.6	58.7	63.5	55.4
With piped water	48.1	48.7	57.4	50.9	48.9
With electricity	63.0	71.5	81.2	87.5	83.5
With gas	27.7	28.2	39.3	43.1	41.1

Source: Tabulation of AERC Survey (1989).

Table 11.2 *Katchi abadi* population

	1970s (1978)	1980s (1985)	Most recent (1988)	2000 (projection)
Katchi abadi population	2,000,000	2,600,000	3,400,000	7,070,000
No. of *Katchi abadi* households	227,000	356,000	465,000	960,000

Sources: Karachi Land and Housing Study: Dr. D. Dowall/KDA-MPD, 1989; *Shelter for Low Income Communities: Inception Report on Sindh*, World Bank, October 1990.

Informal Subdivisions of Agricultural Land: process and actors

In the Punjab (where most of Pakistan's urban population lives), North-West Frontier Provinces and Balochistan, *katchi abadis* are now seldom possible as little or no state land is available except through the railways. In these provinces the demand–supply gap in housing is met through the informal subdivisions of agricultural land (ISALs). Detailed studies of this process are available for Faisalabad and the same process, with minor modifications, is operative in almost all other cities of Pakistan. These subdivisions usually consist of small schemes of between 50 and 200 plots. The size of the plots are of $2\frac{1}{2}$, 5, 7 and 10 *marlas* (one *marla* is 20.9 m^2) with the 5 *marla* plots being in the majority.

In Faisalabad, almost all the schemes are located on the roads connecting the city to the other cities of the Punjab. The price of a plot in these schemes depends on the road off which it is located; how far it is from the road; and how far it is from the city. Schemes that are more than 2–2.5 km from the inter-city roads, do not sell easily. This is because the intercity roads have transport on them and it is not easy to walk more than 2–2.5 km. As a result, ribbon development is taking place along the intercity roads (Alimuddin *et al.* 1999).

The most important player in the development of these settlements is the middleman or developer (see Box 11.2). He purchases land from a farmer, subdivides it into plots and sells it to prospective house builders. Alternatively, he enters into an agreement with the farmer whereby he plans the scheme, develops a bit of infrastructure, finds the buyers, who then make direct payment to the farmer and the middleman gets a commission. However, the most successful schemes are those where the farmer and the middleman enter into a 'joint venture'. These schemes are more successful because payment from the buyers can be recovered in monthly instalments over a two to four-year period. Because of this arrangement, the price of the plots can be increased and at the same time they sell easily since they suit the paying capacity of the buyers. The plots are transferred to the names of the buyers in the records of the revenue department since there is no law preventing this transaction. The strategy of recovering the cost of land in instalments was adopted by the developers only six to seven years ago as

Box 11.2 The informal developers of Faisalabad

Chaudhary Ghulam Rasool Cheema is a Faisalabad informal developer. His family came from Gurdaspur in India and lived near a village on Jaranwala Road. His first job was as a Water and Power Development Authority (WAPDA) storekeeper. He began this business because his salary was not enough to support his big family. He has been a member of the Pakistan Peoples' Party and later of the Pakistan Muslim League. To begin his business he sold a piece of land that he had in his village which is about 20 km from Faisalabad. He chose to work along the Jaranwala Road because the people of the area knew him through his political activities. He planned his first housing scheme in 1990, but work on it started in 1994. Until now he has completed five small schemes each having 70–150 plots. The size of the plots is usually 5 *marla* (104.5 m²) and measures 9.1 m by 13.7 m. The streets are 6–8.5 m wide. He raises the streets 0.6 m above the road level. If the streets are not raised then people do not buy the plots because they are afraid that the settlement will get flooded. The earthwork for the streets is done by the Afghanis who have trolleys and jack machines for this job. Local people do not do this work since they have no experience in it and no machinery. To set up his business he employs two persons as office staff. However, he hires a number of 'fieldworkers'. These fieldworkers contact prospective clients, prepare layout on site and supervise earth filling. When a project begins he usually has about 20 fieldworkers who provide forms to the clients at Rs10. If they sell 10 forms in a day they earn Rs100. For the advertisement of a scheme a pamphlet is prepared and is inserted in newspapers inviting young middle or matriculate educated boys to come and work as field staff. These boys go to the areas which are congested or where people do not have their own houses. They brief them about the scheme and try to convince them that they should buy a plot. Most of the boys who respond to Mr Cheema's advert already have experience in this field. They are given a further incentive of a commission for each plot that they sell. The planning of the scheme is done by Mr Cheema himself, after which the sketches are provided to a draftsman for further development. The draftsmen who work for him are Faisalabad Development Authority (FDA) employees and are hired by him on a per-job basis. The most important criteria for the purchase of land for the scheme are availability of transport, which means access to the main intercity road, and electricity. If the land is more than 2 km from the intercity road, the scheme does not sell. There is no attempt to develop corner plots or commercial plots. It is simply a 5 *marla* subdivision. In the smaller schemes Mr Cheema provides no services such as water, sewerage or electricity. People acquire water by hand pumps, which they later convert to piston pumps, sewerage through self-help (it invariably disposes into a canal) and electricity through lobbying with the WAPDA. The developer does not keep any plot for speculation, but 30 per cent of the plots normally remain unsold for a period of three to four years. There is a written agreement with the person who purchases the plot and proper records of receipts of instalments paid are maintained. People invariably pay regularly by coming personally to Mr Cheema's office. For the transfer of land from the landowner to Mr Cheema, both the parties visit the divisional headquarters where land records are kept. Here they pay the legal as well as the 'other' charges. In the revenue department ledger, land remains as agricultural and streets and roads are recorded as amenities. The cost of transfer of land to the developer is borne by the purchaser.

When Mr Cheema started his business in 1990, he had to look out for people who wanted to sell their agricultural land. Now that people know that he is in business and has an office where plans are displayed, landowners come to him themselves. Also, wherever he develops a scheme, he puts up a board on which the name, plan and details of the scheme are displayed. Mr Cheema says that the success of these schemes lies in the fact that the developers have understood what a poor man can afford to pay and they act accordingly. He also says that if the government could support this activity and provide the developers with some loans, then in two to three years' time there would be no one left in Faisalabad who was homeless.

Source: Observations and interviews by the author.

a consequence of the declining capacity of buyers to pay for the land in one go or to pay a large sum of money as down payment.

The developer has the plan of the settlement drawn up by a draftsman. He is usually an FDA employee. He is paid a lump sum for this work. Once the plan has been prepared, the marketing of the plots begins. The developer employs local educated young men to find prospective buyers of the plots (see Box 11.2 for details). Thus, the developer has no regular overheads for marketing his scheme and no regular employees either. According to the developers, the vast majority (70–80 per cent) of plot purchasers are from Faisalabad city who want space for their young married children or wish to escape from paying high rents. Many of them move to the new schemes because environmental conditions here are better than in the Faisalabad inner-city low-income settlements. However, in spite of the pull factor of the settlements, it takes 10–15 years before they are fully occupied. This is because people also buy property for their children and for speculation. To promote their work, the developers also print and distribute sensational leaflets.

The planning of the scheme is on a gridiron. The width of the street is kept between 6 and 7.6 m. The developers do not provide open spaces or plots for amenities. However, they do raise the level of the roads to 0.6–0.76 m above the level of the land so that flooding does not take place. This work of earth filling for the roads is let out on contract to the Afghanis who have the necessary machinery and expertise for it. The Afghanis acquire this earth from the fields of those farmers whose land is higher than the irrigation channels and who wish to lower the level of their land so as to avoid pumping water for their fields. In some cases, developers do provide a sewerage system, complete with underground pipes and man-holes. However, disposal is seldom provided and the sewage either flows into a cesspool, a natural drainage channel or into an irrigation canal. Electricity and water are not provided. Residents collectively struggle to acquire electricity after they have started living in the scheme. This they do by negotiating votes with their councillors, members of national and provincial assemblies during or before an election process. They tap the subsoil aquifer for water, and if it is saline, they make a bore adjacent to the nearest canal. Near the canal, the subsoil aquifer is usually potable.

Physical and social infrastructure

As mentioned earlier, *katchi abadis* which have de facto or *de jure* security of tenure and ISALs acquire physical and social infrastructure over a period of time. Much of the physical infrastructure is developed through self-help and through municipal councillor funded programmes. Infrastructure developed by self-help is usually substandard in nature since communities do not have access to technical advice and managerial guidance. Also, they do not have plans of their areas or of government schemes to which they can connect. Municipal councillors also identify small schemes in these settlements for which they receive grant-in-aid. Most of the development carried out through councillor funds is haphazard, non-coordinated and substandard in quality. This is because it is done piecemeal over the years since the funds allocated to the councillors are too small to develop the area at one go. As a result, drains built under this programme normally do not function, water does not reach the extremities of the distribution system and road paving is often substandard. Thus, large sums which could be better utilised are wasted.

The problems with the work of the councillors is that it is not according to a master plan for the settlement, as none is available. Consequently, the work that is carried out is done on an ad hoc basis with no relationship to larger planning considerations. Thus, paved streets are often torn up for laying gas, water and sewerage lines and drains empty their effluent on to neighbouring unpaved lanes. Then, in many settlements there is more than one councillor, and a lack of co-ordination between them makes it impossible to tackle issues that are of a common nature such as an access road or a secondary drain.

The councillor's decision to develop a certain street in a neighbourhood is politically motivated rather than need or planning related. This development is carried out for those who have helped him in the elections, his friends and relatives or those who can be of assistance to him in the future. That this development may be detrimental to a few other lanes (especially if people who belong to the 'rival' group live there), is of no concern to him. In addition, no detailed design or supervision for the works is carried out.

Often the contractors employed do not even have simple survey instruments and use unskilled labour for skilled jobs. Communities complain that the contractors' profit margins are exceptionally high and that the contract is given not on merit but on the relationship that the councillor has with the contractor. Where the community is organised and has building skills, it manages to get the councillor and the contractor to do a better job and gets lanes repaved after gas or sewerage laying has damaged them. The municipal executives are not concerned about the nature and manner of work carried out in the poorer areas, because these areas are not politically powerful.

A survey of 136 Karachi *katchi abadis* has been carried out by the Orangi Pilot Project-Research and Training Institute (OPP-RTI) (see Box 11.3). These *katchi abadis* have a total of 79,426 lanes in them, 81.6 per cent of these lanes have built sewer lines at their own cost and over 90 per cent of the homes have linked themselves illegally to government water supply systems. The people and their councillors have invested over Rs203 million ($3.4 million) in this work. However, this work is never integrated into official sewerage and water supply systems being planned and implemented under various programmes, some of which are supported by international loans. If they were, the projects would be a fraction of their present costs; they would be completed in a fraction of the time it takes to complete them now; and the poor, instead of the contractors and consultants, would be their beneficiaries (Orangi Pilot Project 1998).

Education facilities in the informal settlements in most cases are developed by the private sector, neighbourhood committees or by public-spirited individuals. In Orangi, the largest *katchi abadi* in Karachi, there are 76 government schools and 509 private schools. The vast majority of the private

---

Box 11.3   The Orangi Pilot Project, Karachi

Orangi township
Orangi is Karachi's largest *katchi abadi* and has a population of 1.2 million. The Orangi Pilot Project (OPP) was established here in 1980. In 1988 the project was upgraded into four autonomous institutions: the OPP Research and Training Institute (RTI); the Orangi Charitable Trust (OCT); Karachi Health and Social Development Association (KHASDA) and the OPP Society which channels funds in these institutions.

The OPP considers itself a research institution whose objective is to analyse outstanding problems of Orangi, and then through action research and extension education, discover viable solutions that overcome the constraints that government programmes face in upgrading informal settlements. The OPP does not fund development but by providing social and technical guidance it encourages the mobilisation of local resources and the practice of co-operative action. Based on these principles, the OPP has evolved a number of programmes, some of which are described below.

Box 11.3 (cont'd)

The Low-Cost Sanitation Programme

This programme is managed by the OPP-RTI. It enables low-income families to construct and maintain an underground sewerage system with their own funds and under their own management. For this programme, the OPP provides social and technical guidance (based on action research), tools and supervision of implementation. The OPP's work has shown that people can finance and build underground sanitation in their homes, their lanes and neighbourhoods. This development is called 'internal' development by the OPP. However, people cannot build 'external' development consisting of trunk sewers, treatment plants and long secondary sewers. This only the state can provide. In Orangi, people have invested Rs78.79 million on internal development (including 405 secondary sewers) in 5,987 lanes consisting of 90,596 houses (there are 104,917 houses in Orangi). The state would have spent over six times as much to do this work. The programme is being replicated in seven cities of Pakistan by NGOs and CBOs and in 49 settlements in Karachi by the Sindh Katchi Abadi Authority (SKAA). The OPP concept has been accepted by the Karachi Municipal Corporation (KMC) and SKAA and is being applied to their development plans.

The Family Enterprise Economic Programme

This programme is run by the OCT which was formed in 1987. The OCT borrows from commercial banks and then lends on to small family businesses but without red tape and collateral. These loans vary between Rs1,000 and Rs75,000. The aim of these loans is to increase production and generate jobs, which they have done. Loans are usually given to people who have expertise in what they plan to do or are already operating businesses. Interest is charged on the loans at the current bank rate of 18 per cent. Presently, there are 6,555 units being supported by OCT loans of Rs123,738,610. Out of these Rs97,327,482 have been paid back with a mark up of Rs22,999,610. The recovery rate is 97 per cent. The World Bank has also given a grant as a revolving fund for the programme.

The OPP's Low-Cost Housing Programme

This programme is operated by the OPP-RTI and provides loans and technical assistance (based on research) to *thallas* (building component manufacturing yards) in Orangi so that they can mechanise their production, improve their products, train their staff and increase their production. In addition, the programme also trains masons in using the new technologies and components that are being developed at the manufacturing yards. Also, house builders are given advice on how to relate to the manufacturing yards and masons and also advice on design, light, ventilation and other hygiene-related design aspects. To provide such advice, the OPP is in the process of training para-professionals who are mostly young unemployed youth from the Orangi communities who are paid by house builders or those who want improvement to their homes. The OPP housing programme thus tries to create a more equitable relationship between the actors in housing drama, as a result of which housing has improved in Orangi.

So far, 57 *thallas* have been mechanised due to which employment has been generated and machine-made blocks and roofing elements are being fabricated, not only for Orangi, but for the rest of Karachi as well. In addition, 33 masons have been trained and 2 para-architects after a training of 2 years at the OPP-RTI, have started working independently designing homes and community building and being paid for it.

Health Programme

The OPP's Health Programme is operated by KHASDA and originally consisted of developing women's organisations at the lane level where the sanitation system has been built. A mobile team of experts gave advice to such organisations, through discussions and meetings, on common diseases in Orangi, their causes and ways of preventing them. It also gave advice on hygiene, immunisation and family planning. As a result, 90 per cent of households that were part of this programme immunised their children and over 45 per cent families adopted birth control. However, the OPP could not reach more than 3,000 families through this method and the project was revised.

In the revised model the health programme is imparted through training to local lady teachers, managers of family enterprise units and doctors in private clinics, thus anchoring the programme institutionally in schools, private clinics and family enterprise units. A health centre is operated at the OPP office, which provides vaccines and family planning supplies to the activists in these centres. As a result of the sanitation and health programmes, infant mortality in those parts of Orangi that built their sanitation system in 1982 has fallen from 130 to 37 in 1991.

OPP's Education Programme

OPP's Education Programme is run by the OPP-RTI and through social and technical guidance it improves and upgrades the physical conditions and academic standards of private schools in Orangi. These private schools cater to the needs of the vast majority of Orangi school children. Physical improvements are made with loans from OCT and advice from OPP-RTI. Academic improvements are made by arranging teacher training through existing relevant organisations; provisions and use of libraries and audio-visual aids; and publication of manuals and guide books.

Financial support is extended during three stages of establishment of these schools. First, a small start-up grant of Rs3,000 to Rs6,000 is required to set up the schools. Second, within a year the school is institutionalised and then the need for physical expansion arises. This requires a further Rs20,000 to Rs30,000, support that is very important for the survival of the school. Third, a loan for upgrading is needed as the school is by now a formal education institution and can take loans which can be repaid through its income.

Box 11.3 (cont'd)

OPP has provided 364 loans to such schools. Teacher training through Allama Iqbal Open University is also being co-ordinated. The education entrepreneurs also hold their monthly meetings at the OPP office, where they share information on registration and teaching methods.

Significance of OPP programmes and their new directions
The OPP research, programmes and their documentation have provided NGOs, CBOs and government agencies with successful models for overcoming the physical, social and economic problems faced by low-income settlements and communities. These have been successfully tested through government–OPP–community participation projects but have still to become official policy. The infrastructure development models, in particular, reduce capital costs, ensure good-quality work (since communities acquire skills for building internal infrastructure, maintaining it and supervising government work for external infrastructure) and create a more equitable relationship between government agencies and poor communities.

Increasingly, the OPP is becoming involved in policy issues and promoting macro-level solutions, based on its models, to sanitation, health, housing and economic issues and seeking to integrate informal settlements into larger city plans.

---

schools were set up as informal institutions, but the majority of them have now become formal registered schools. They are affordable to the people because neighbourhood educated girls teach in them for very low wages. Almost all the private schools are co-educational, unlike government schools, and the vast majority of teachers are women (Khan 1994). A similar situation as in Orangi exists in the older *katchi abadis* in Karachi and Faisalabad.

Health facilities, too, are provided by the private sector. In Orangi, there are more than 400 private clinics and only 18 government and formal sector health facilities (Orangi Pilot Project Health Survey 1998). No attempts have been made by the state to support the private clinics although the clinic operators are interested in being involved in such a programme and in being supported. There is no preventive health programme either except the government's immunisation programme which has a major outreach in the low-income settlements of Pakistan's major cities.

Employment and social change

Very few detailed socio-economic surveys are available for informal settlements in Pakistan. However, a detailed survey of Orangi in Karachi is available, along with a number of sample surveys and observations by professionals. The Orangi *katchi abadi* have over 42,000 small businesses and workshops which employ about 150,000 people.[6] Because of this, almost 68 per cent of the Orangi labour force work within their homes or cycle to their workplaces.[7] In Faisalabad 52.68 per cent of the population is self-employed (Faisalabad Development Authority 1985). The vast majority of this employment is located in the informal settlements. These micro-enterprises often service the needs of formal sector industry and trade. Through middlemen, the garment, textile and carpet weaving industry operates in these settlements. Other enterprises consist of making cardboard boxes for packaging industrial goods; rope making; piece work for the formal shoe making industry; manufacturing components for light engineering industry; recycling of solid waste; manufacturing cheap copies of brand name soft drinks, soaps and cosmetics; retail outlets; and vendors and hawkers serving various transport and cargo terminals. In most of these enterprises, men, women and children of the family work together as these activities are invariably carried out at home.

These micro-enterprises have severe limitations. They cannot get credit from banks for their activities. Nor do they have access to technical and managerial advice. They can only borrow from the open market at interest rates of 10 per cent per month or more. Where NGO programmes have supported them with credit and advice, they have expanded their business and generated employment. In many cases (such as tanning and soap making), they create environmental degradation, not only for their settlements but for the city as a whole, and are resented by their neighbours.

In settlements where the second and third generation has now come of age, major sociological changes have also taken place. The younger generation is educated, is not entirely working class, and is increasingly finding employment in the formal sector. These settlements house young engineers, doctors, school and college teachers, receptionists, bank clerks and managers, and other white-collar workers. They are becoming the new leadership of these settlements as a result of which a change in the relationship of these settlements with local government and politicians is taking place for the better. In many cases, such as Karachi, this has also led to the development of extremist ethnic politics and organised confrontation between the state establishment and the political opposition.

The Katchi Abadi *Improvement and Regularisation Programme (KAIRP)*

The KAIRP is a government programme. It aims at regularising 2,320 *katchi abadis* in Pakistan which have a population of over 7.6 million. In addition, it aims at improving the settlements by providing water, sanitation, electricity, road paving and social facilities. The beneficiaries are supposed to pay for this improvement through land and development charges known collectively as 'lease charges'. At present, all settlements on government land that have more than 40 houses in them and which were established before 23 March 1985 are to be regularised provided they are not in ecologically dangerous zones. The programme in its present form has been in operation since 1978. However, it has been facing a number of problems which are discussed below.

According to a 1989 evaluation (there has been no national level evaluation since), the pace of work under the KAIRP had been slow. Only 13.85 per cent of the households had been regularised since 1978 and 22.41 per cent of *abadis* had been or were in the process of being developed. The reasons for the slow pace of work were given as a lack of funds; heavy government subsidies in development along with a lack of recovery of development charges which prevented funds from revolving; and a lack of community involvement in the programme. To keep up with the increasing backlog of *katchi abadis*, 100,000 households would have to benefit annually from the programme. However, only 30,000 houses were upgraded annually during the 1977–89 period (Asian Development Bank 1989). Since 1989, performance on an all Pakistan level has not improved.

Although funds are given as one of the reasons for the slow pace of work, according to the 1989 evaluation only Rs200 million of the Rs400 million ($13 million) allotted every year for the programme were actually utilised. This points to a lack of capacity on the part of the programme-implementing agencies to carry out the work. Conditions have not improved in the last 10 years.[8] Another important negative factor is that much of the funding for the programme has come from external sources as loans and this has added to the debt burden on Pakistan. This aspect makes the programme unsustainable in the long run.

The development programme frequently manages to develop on-site infrastructure, but it fails to provide secondary infrastructure to link the *abadis* with the town's or city's trunk and communication network. Because of this the infrastructure provided often does not function. In addition, there is a lack of trust between the government agencies and the people and no community participation in the programme. The development work undertaken is substandard and as such the residents do not wish to pay for it. Also, acquiring a lease is a long and complicated procedure involving many steps and open to corruption. These factors mean that residents do not come forward to acquire leases although they do desire a *de jure* secure tenancy. Another major negative factor is that once a settlement has been notified as a *katchi abadi* or is large enough to seek security in numbers, its residents are not under pressure to legalise their homes.

The planning standards developed for the KAIRP are also unnecessarily high and their application means the uprooting of up to 35 per cent of the *katchi abadi* population. The resettlement of this population poses major political, social and financial problems. As a result, the implementation of the standards does not take place and regularisation is being carried out by the government agencies in violation of their own standards. In addition, incorrect maps of the settlements are made by the government agencies and as a result a number of families and businesses are victimised and land grabbers benefited. To rectify these injustices and retain their homes, those affected informally pay the staff of the agencies concerned.

The situation in Karachi is similar to that which has been described above. Karachi has about 700 *katchi abadis* in which over 50 per cent of the city's population live; 539 *abadis* having 386,000 housing units have been identified as regularisable. Of these, 260 are under jurisdiction of SKAA and the other 279 under the KMC. The KMC has worked in 72 *katchi abadis* with its own resources and also in other *katchi abadis* of which there are no records (SKAA 1999). There are settlements that have not been notified but where leases have been given. There are also people who hold lease documents that are not recorded in the KMC registers. Then, there are settlements where development work has been done only on paper.

Proper records, however, have been kept for the Asian Development Bank (ADB) and World Bank-funded upgrading programme for which a total of Rs427.137 million were provided as a loan in 1984. This loan is meant for 101 *katchi abadis* in Karachi. So far, upgrading work through these funds has been completed in 33 *abadis* by the KMC. The total number of leases issued so far by the KMC is 98,245 and SKAA has issued another 10,000 (SKAA 1999). Meanwhile, new *katchi abadis* have been born. Given the constraints mentioned earlier, it is unlikely that even this small achievement could have been possible without international loans. However, a new strategy adopted by SKAA has resulted in making the programme effective and sustainable. The methodology of this programme and its achievements are described below.

Conclusions: informal settlements and urban sustainability

For the foreseeable future, informal settlements will continue to cater for the housing needs of low-income communities. This is because they are affordable for them and do not involve complex procedures for acquiring land and building a house. However, these settlements create a number of ecological and environmental problems for the cities of Pakistan. They develop as ribbons along major corridors out of the city, congesting them and making vehicular movement difficult. Alternatively, they develop in ecologically dangerous areas (such as marshes, areas prone to landslide or flooding, or quarries) which the formal sector does not wish to develop, or along natural water bodies into which they discharge their waste water and sewage. They are often located outside municipal limits and their transport links with the city are problematic. Industrial activity within them creates difficult living conditions, and parks and open spaces seldom exist within or alongside them. Such experience is not limited to Pakistan. Numerous case studies elsewhere in South Asia demonstrate that this is a region-wide problem, while other case studies have shown that urban environmental degradation is endemic in many Third World cities. Agarwal *et al.*'s recent study of environmental degradation in Ludhiana, India, demonstrates how growth of population and industry has polluted river and canal waters, while industrial, auto and domestic emissions have created alarming levels of atmospheric pollution (Agarwal *et al.* 1999: 211–18). Van Naerssen and Barten (1999: 231) have observed that 'In most countries of the South, the urban population, urban poverty and the use of energy continue to grow rapidly and thus the quality of the physical environment in the cities is deteriorating steadily.'

247

However, in Pakistan as in other parts of the South, the residents of these settlements are fiercely upwardly mobile and willing and capable of making investments incrementally to improve their living conditions. The government, too, has programmes for supporting the upgrading of *katchi abadis* and in the case of ISALs considerable council funds are available. But there is a mismatch between community efforts and government programmes since the latter do not support the former. In addition, these programmes are limited to the informal settlements and do not seek to integrate them in a larger city infrastructure and ecological plan. In many cases, these settlements lie outside municipal limits, although in physical and economic terms they are very much a part of the urban centres they are located around.

What is required to deal with this situation has been a subject of much debate and discussion between professionals, NGOs and government agencies. Although no consensus has emerged so far, a number of ideas and NGO programmes are accepted as possible solutions. It is felt that city structure plans should integrate the informal settlements into the larger city plan. For this, the physical mapping of these settlements and the infrastructure that has been built in them needs to be documented. It is also felt that the work of the *dallals* and the informal developers should be supported by the state. Locations for the creation of these informal settlements should be directed by providing major infrastructure and leaving the building of neighbourhood infrastructure to technically supported communities and informal developers.

The government at various times has also expressed its desire to redesign KAIRP on the model of the work done by the OPP institutions in Karachi. These institutions have developed models that have tried to overcome the constraints that governments face in upgrading and rehabilitating existing informal settlements (see Box 11.3). So far, this has not become policy except in the case of the SKAA and the Lodhran Municipal Committee. The KMC has also adopted this policy for certain projects. SKAA has over the last seven years evolved a new methodology of regularising and improving *katchi abadis*. The development of the infrastructure component in this methodology is based on the work of the OPP-RTI and leaves the development of neighbourhood infrastructure to the communities themselves and provides only major infrastructure. The nature, location, cost and design of the infrastructure are developed through a dialogue between neighbourhood organisations and SKAA. Costs of the projects are available to the community, and in many cases the work is managed by committees in which they are represented. In addition, getting a lease has become a simple one-window operation that takes place within the settlement itself where a lease camp is established, complete with a registrar, and the residents are informed well in advance. Community organisations are also made responsible for representing their community in this process.

In the initial stages, the OPP-RTI acted as consultant to SKAA, trained its engineers in OPP-RTI methodology and in negotiating with the community, designing appropriate infrastructure, documenting the work SKAA was doing and conditions in the settlements. However, this work is now done by the SKAA staff itself. The OPP-RTI meanwhile continues to document *katchi abadis* and their physical and social conditions, independently of SKAA.

The SKAA programme has created trust between communities and government agencies and introduced transparency and accountability in the development process. It has also simplified leasing procedures. Communities are now willing to acquire leases and their organisations have been strengthened. As a result, SKAA's programme has become self-sustaining, requiring no foreign loans or government financial assistance. So far, SKAA has spent Rs47.188 million on development in *katchi abadis* and recovered over Rs147 million.

The OPP-RTI methodology and SKAA's programme requires a very different mindset from that of conventionally trained professionals and administrators. However, the OPP–SKAA model is increasingly being accepted and the two organisations regularly present their work before relevant university departments, the National Institute of Public Administration, and before various government task forces on the subject. It is hoped that this interaction and debate will lead to the development of a more rational and sustainable urban development model, complete with appropriately trained professionals and administrators, for Pakistan.

One of the most significant innovations for the development of informal housing areas in rapidly growing cities in the South represented by the Orangi Project is the attempt to bring together government programmes, including those making use of external government and agency funding, with NGO and local community-based projects into an integrated whole. Such an approach currently seems to offer the best way forward in managing the bewildering complex of factors which have together restricted both the pace of development and the access to improved living conditions of a high proportion of the urban poor.

Notes

1. For example, Kaunpur, Rahim Yar Khan, Kamalia, Khanewal, Attock, Sargodha (in the Punjab) and Karachi, Mirpurkhas (in Sindh) all increased by more than 100 per cent.
2. Abbottabad, Haripur, Charsadda, Peshawar, Kohat, Bannu, Dera Ismail Khan, Tank (in the NWFP) registered a fall.
3. Urban Resource Centre, Karachi, estimates 1999.
4. Urban Resource Centre, Karachi, estimates 1999.
5. Figure quoted by government officials in seminars and workshops.
6. Calculated by the author from OCT Survey of Micro-enterprises (1998).
7. Worked out by the author from AERC's *Socio-economic Survey of Planned Areas and Katchi Abadis*: Karachi Development Plan (2000).
8. Author's observations and reports.

References

AERC (1989) *Socio-economic Survey of Planned Areas and Katchi Abadis*. Oxford University Press, Karachi.

Agarwal, A., Narain, S. and **Sen, S.** (eds) (1999) *The Citizens Fifth Report: Part 1 national overview*. Centre for Science and Environment, New Delhi.

Ali, Reza (2000) 'Urbanization in Pakistan'. Unpublished paper.

Alimuddin, Salim *et al.* (1999) *The Work of the Anjuman Samaji Behbood and the Larger Faisalabad Context.* Publisher unknown, Karachi.

Asian Development Bank (1989) *Pakistan Low Cost Housing Project report*, Manila.

Bastías Gonzales, M.V. (1999) 'Popular organisation, local power and development', in D. Simon and A. Närman (eds), *Development as Theory and Practice: Current perspectives on development and development co-operation.* Longman, Harlow.

Bradnock, R.W. (1989) *Urbanisation in India*, 2nd edn. John Murray, London.

Chapman, G.P., Dutt, A.K. and **Bradnock, R.W.** (eds) (1999) *Urban Growth and Development in Asia*: Vol. I *Making the Cities.* Ashgate, Aldershot.

Chapman, G.P., Dutt, A.K. and **Bradnock, R.W.** (eds) (1999) *Urban Growth and Development in Asia*: Vol. II *Living in Cities.* Ashgate, Aldershot.

Faisalabad Development Authority (1985) *Faisalabad Master Plan, 1985.*

Guerl, S. *et al.* (1991) *Housing Parameters.* Dawood College – Aga Khan Program, Karachi.

Gugler, J. (ed.) (1997) *Cities in the Developing World: Issues, Theory and Policy.* Oxford University Press, London.

Hardoy, J.E. and **Satterthwaite, D.** (1997) 'Building the future city', in J. Gugler (ed.), *Cities in the Developing World: Issues, theory and policy.* Oxford University Press, London, Chapter 17.

Hasan, A. (1995) 'Urban basic services programme in Sukkur'. Unpublished report.

Jacquemin, A.R.A. (1999) *Urban Development and New Towns in the Third World: Lessons from the new Bombay experience.* Ashgate, Aldershot.

Khan, Akhtar Hameed (1994) *OPP Programs*: OPP-RTI. *OCT Survey of Micro-enterprises, 1998.*

Orangi Pilot Project (1998) *Proposal for a Sewage Disposal System for Karachi.* City Press, Karachi.

Orangi Pilot Project Health Survey (1998) *Health Survey.* City Press, Karachi.

Siddiqui, Tasnim (1992) *Incremental Housing Scheme.* Hyderabad Development Authority, Urban Resource Centre, Karachi, estimates 1999.

SKAA (1999) *21st Quarterly Progress Report, March 1999.* Karachi Development Plan 2000.

Turner, R. (1962) *India's Urban Future?* University of California Press, Berkeley, CA.

Van Naerssen, T. and **Barten, F.** (1999) 'Healthy cities in developing countries: a programme of multilateral assistance', in D. Simon and A. Närman (eds), *Development as Theory and Practice: Current perspectives on development and development co-operation.* Longman/Developing Areas Research Group, The Royal Geographical Society (with the Institute of British Geographers), London, Chapter 10.

CHAPTER 12

Urbanisation, migration and development in Bangladesh
A gender perspective

Shahnaz Huq-Hussain

Introduction

Belying its image as not only one of the world's poorest countries but as one of the countries of the global periphery most enmeshed in the cycle of stagnation, low opportunity and gender inequality, since 1971 Bangladesh has shown a remarkable degree of social and economic change. Against a background of population growth exceeding 2.5 per cent per annum in what was already one of the world's most densely populated countries (ninth overall, according to UN statistics, 2000), Bangladesh has experienced radical social changes, notably in the role of women. In common with many countries in the South, these changes are most pronounced in the urban sector, which through the same period has grown from about 5 per cent of the total population to over 20 per cent of the current population of over 130 million. The rapid adoption of family planning, an initiative led by women, has contributed to a decline in the rate of population growth to 1.6 per cent in 2001, and promises to continue to contribute to a rapid decline in net total fertility and rates of population growth. Although industry as a whole still contributes only 11 per cent of the GDP, compared with the 32 per cent derived from agriculture, the ready-made garment industry, in which women play a crucial role, dominates the growing industrial sector.

These changes, which have seen a steady rise in income per capita and a growth in levels of education, have been associated with demographic changes of rural urban migration which has further accentuated the changing roles of women in Bangladeshi society. This chapter explores the factors contributing to urbanisation in Bangladesh, and the gendered significance of evolving patterns of migration and their impact on development for women.

The urbanisation process and urban growth

Urbanisation in Bangladesh is more poverty-driven than fostered by industrialisation. Although the rate of population growth has shown considerable decline since 1974, the date of the first census taken after independence

Table 12.1 Urbanisation trends and processes in Bangladesh

Average annual population growth rate	*1970–5*	*1975–80*	*1980–5*	*1990–5*
Total	2.77	2.83	2.73	2.69
Urban	6.74	6.76	6.57	6.14
Components of urban growth (as % of total)	*1980–5*	*1990–5*	*2000–5*	
Natural increase	37.6	39.9	41.9	
Migration and redefinition	62.4	60.1	58.1	

Source: Barkat (1997: 161).

from Pakistan in 1971, the cities are growing faster than ever. The urban population has expanded threefold over the past two decades. In 1974 only 8 per cent of the population were living in urban areas and by 2001 a quarter of the country's population, over 30 million, were urban. The rate of urban population growth since 1974 was 6 per cent per annum, and by the year 2020 it is projected that some 45–48 per cent of the population (60–80 million) of the country will be living in the towns and cities (Barkat 1997).

As has been the case across South Asia, urbanisation in Bangladesh is largely caused by the combination of high natural growth of urban population, migration of people from rural areas and the redefinition of urban areas accompanying urban areal extension. Internal migration and redefinition generated nearly 59 and 66 per cent growth of urban population in 1960 and 1970 respectively (Barkat 1997). However, the share due to natural increase will rise slowly and internal migration will continue to be the prime contributor to urban growth. It has been projected that migration will contribute up to 60 per cent of the total urban growth up to the year 2005 (Table 12.1). Urban 'pull' factors in the form of economic development played an important role in attracting migrants to the city, especially to Dhaka, the capital city.

According to the census of 1991 (the most recent for which data have been published), there were 522 urban centres in the country and 220 municipalities, including four city corporations, namely Dhaka, Chittagong, Khulna and Rajshahi. Nearly half of the national urban population is agglomerated in these four metropolitan centres.

Migration, urbanisation and development

Since 1947 migration has been an important feature of sociocultural change in Bangladesh and rural–urban migration has contributed significantly to urban growth. Rural to urban migration accounted for nearly 52 per cent of the growth in urban population in the 1990s (BBS 1997), and both rural push and urban pull appear to have been factors in the migration. Microstudies have shown that in recent decades rural push factors have been more important than urban attraction in certain sectors of urban economy. Nearly 67 per cent of the rural population is landless and the bottom 40 per cent of the poor group own a little over 2 per cent of the cultivable land (UNDP 1996).

According to some measures, women fare particularly badly. One estimate (UNDP 1999) suggests that women constitute a significant majority of the absolute poor. Female-headed households are among the poorest, with 45 per cent of them living below the poverty line. As such, landlessness, persistent unemployment and the lack of any other source of income in the rural areas are the major push factors in driving rural dwellers out of their rural homes. The routine occurrence of natural disasters adds to this. Severe floods, such as those of 1987 and 1988, and the extensive bank erosion which is an annual feature of the flood season in some parts of Bangladesh, destroy rural homes every year. Rural people with no other resources than tiny pockets of land or work on the land of others then migrate to the cities in search of a livelihood.

The urban centres, most notably the capital Dhaka, witnessed a major growth in manufacturing which offered some job opportunities, especially for women, unavailable in the rural areas. The opening up of global markets and the movement of international investment towards low labour cost economies produced a boom in the urban manufacturing sector. The rate of growth of large-scale manufacturing during the first half of the 1990s was 9.6 per cent year compared to only 3.7 per cent of the small-scale manufacturing sector, although a relative decline in the growth rate of the large-scale manufacturing sector has been observed since 1995–6. The share of the urban sector in the GDP of the country showed notable increase since 1972–3 when it grew from 25 per cent to little over 36 per cent in 1991–2 period and to over 40 per cent in 1999 (Islam 1999a). Despite the growth of large-scale industries, it is growth from an exceptionally small base, and offers limited scope for entry into the formal sector of employment. Hence there is also great demand for work in the informal sector of the urban economy. The Labour Force Survey (LFS) of 1995–6 suggests that the informal sector furnishes 70 per cent of the total urban employment. Employment in this sector grew at a much faster rate than manufacturing between 1990/1 and 1995/6. In contrast, the share of employment in manufacturing declined at a rate of 1.4 per cent per annum over the same period (Sen 2000).

Poverty, migration and urbanisation

Conventional perceptions of poverty include multidimensional problems of income, consumption, nutrition, health, education, housing, insecurity and lack of control or access to resources. By all these criteria Bangladesh has witnessed acute poverty, though since 1971 it has made some progress in overall poverty reduction. Lakshman Yapa has argued in Chapter 9 of the present volume that attempts to measure poverty are themselves a part of the problem, and he takes issue with the belief that there is 'a distinct, bounded poverty sector in the economy whose problems can be eradicated through economic development'. As he points out, such conceptualisations of poverty are widespread across the South. According to official figures in Bangladesh the level of poverty declined at the rate of 1.55 per cent annually and the proportion of the population below the official poverty line decreased from nearly 71 per cent in 1973–4 to about 47 per cent in 1995–6. Official figures suggest that the proportion of people in rural areas living below the poverty line decreased from 71 to 51 per cent during the same period, but urban poverty declined at a much faster rate, from 63 to 26 per cent (Sen 2000). But even with this declining incidence, the absolute number of urban poor remains at about 12 million. Despite the reduction in the levels of poverty, the per capita nutritional intake has gone down over the last 30 years from 2,118 calories in 1964 to 1,868 calories in 1995–6, figures which suggest a degree of self-contradiction in the different measures, and which add weight to Yapa's critique of the poverty–development conceptualisation. In Bangladesh the general figures of poverty also mask a significant gender disparity in the poverty level. Estimates from national surveys (BBS 1997) indicate that the proportion of absolute poor among female-headed rural households is greater (45 per cent) than among male-headed households (39 per cent).

Irrespective of Yapa's conceptual critique, such static figures fail to indicate some important dynamics of changes in poverty. Poor people show both social and geographical mobility, and there is growing evidence that through time, for those who have moved from rural areas to towns, poverty tends to decline with the length of stay in the city (Sen 2000; Huq-Hussain 1996, 2000). Most of the urban poor live in the poor clusters of slums and squatter settlements. According to Sen (2000):

> Poverty declines with the duration of stay in city. The length of migration is higher for the non-poor compared to the poor (16 vs. 13 years). The incidence of poverty in case of 'recent migrants' (with length of migration up to three years) residing in the urban slums is about 80 per cent compared with 68 per cent observed for 'old migrants' (with length of migration exceeding three years).

Dhaka: the city of migrants

From its role as a modest provincial capital during the latter years of the British colonial period, Dhaka megacity has become the primate city of the country. With over 36 per cent (over 9 million) of the total urban population in 1991 it is expected to contain nearly 16 million people by the year 2015 (Islam 1999b). While such estimates are subject to a significant margin of error, these rates of growth are comparable with those currently being experienced by major cities in India and Pakistan. India's 2001 census, for example, has shown that New Delhi has grown between 1991 and 2001 from 8.38 million to 13.78 million. Population density in Dhaka is about 20,000 persons per km^2 compared to 3,000 and 15,000 in medium and small towns of the country. However, population size is not only the main contributing factor to Dhaka's primacy. Many other functions are concentrated largely or exclusively in the city, including the larger administrative, financial and industrial institutions, and it has also benefited from a wide range of public sector investments.

A number of factors have contributed to Dhaka's pre-eminence. It was made the national capital in 1971 and its image changed rapidly. The effects of the liberation war were apparent in the fragile rural economy that showed no immediate scope for expanding employment. As such, many of the rural destitute headed towards Dhaka in search of a livelihood, and some of the estimated 10 million refugees who returned from India found their way into the city. Dhaka grew up as a city of migrants that contributed about 3 per cent per annum to the growth along with its natural growth of about 1.6 per cent per year.

However, Dhaka also proved to have geographic and strategic advantages for rapid growth. Central to the country and uniquely accessible to every region, it proved the most attractive location for newly emerging businesses, with most of the new industrial activity taking place around Dhaka. Its accessibility also increased its attraction to migrants. The already crowded capital city, now ranked 22nd among the megacities of the world, has been unable to accommodate and to provide jobs and services for this huge rural exodus. Rapid population growth and urban poverty have had a serious bearing on the physical condition of the city. For the majority of the city's population infrastructure, services and shelter have been deteriorating fast and the quality of life has seriously gone down. About 40 per cent of the population live in extremely overcrowded conditions in the slums and squatter settlements which only occupy about 11 per cent of the city's residential area. Here the density ranges between 2,500 and 6,200 per hectare.

It has been projected that if this pace of urbanisation continues, at least half of the city population will be living in the poor clusters of the city by 2015 (Barkat 1997: 169). Thus the general level of poverty remains very high, but gender disparities make the position of women in the urban population even more serious. Female-headed households are the worst sufferers in terms of services and housing, and the remainder of this chapter explores the ways in which women have been adapting to the socio-economic pressures and opportunities of Dhaka.

Women, socio-economic change and migration of females

Bangladesh has been undergoing a process of very rapid social and economic transformation. The economic fabric of the country is changing fast through continuing investment in agriculture, the textile industry, shrimp production and so on. The rural scenario is changing rapidly through infrastructure development which has resulted in a rural economy that is more cash oriented and service based. Despite such changes, rural employment remained stagnant. As such, a greater number of women have been coming out of their homes, whether being pushed by poverty or pulled by opportunity, and their greater independence has been reflected in a widespread social transformation. Further evidence of social change may be noted in the increasing number of female-headed households. The percentage of female-headed households increased from about 11 per cent in 1991 to 13 per cent in 1998. Women are thus playing an increasingly explicit and important economic role and are contributing significantly in the overall development process.

Rural women comprise an important component of the rural to urban migration stream (Huq-Hussain 1996, 1997) and female migration has considerably lowered the sex ratio in the towns and cities. In urban areas, the male–female ratio declined steadily from 150.7 in 1951 to 107.2 in 1998. Rural poverty and the introduction of various labour-minimising technologies have pushed women of very poor to poor households to work outside their homes in order to support their families. The rural economy was unable to absorb all these people. As such, more and more women are taking part in the family as well as independent migration particularly to Dhaka. The poor rural in-migrants, particularly the disadvantaged, widowed, divorced or deserted women, found it comparatively easy to get employment in informal and service sector work in the city.

Although earning cash for the family is vital, the lack of education combined with the responsibilities of household work and childcare restricts many women to choose between family responsibilities and competing for wage-earning activities. As a result many of these women are limited to part-time or self-employed home-based work, for which they may not earn as much as those who work full-time outside their homes. An increasing number of migrant women in the city are being absorbed in construction activities. As building activity accelerated it expanded the scope for employment of rural migrants, steadily accommodating the bulk of poor female migrants. One specific example is the high concentration of women in brick chipping. It may be noted that the greatest concentrations of female migrants, however, are in formal-sector ready-made garment plants. Some 46 per cent of the female industrial workforce are employed in the textile industries and garment plants alone. Female labour in this sector contributes about 60 per cent to the total foreign exchange earnings of the country. Thus in addition to bringing desperately needed income into poor households, female migration has been having a positive impact in the national economy. In this connection one study (Huq-Hussain 1996) concluded that:

> the rural migrants in the city are not as much the victims of the culture of poverty of Oscar Lewis. They are rather hard working and contribute significantly in the daily maintenance of the city life as well as themselves. Their involvement with services, informal and limited formal activities in the city have been found to improve their economic conditions and life chances. It may also be said that the longer their duration of stay in the city the better the chance of economic and life conditions.

Women and development

According to the UNDP's Gender Development Index (GDI), Bangladesh ranks 140th out of 174 countries (UNDP 2000). This low index value may be explained by the fact that women have low literacy rates and a very low share of earned income. However, with all these limitations women have contributed significantly to achieving the country's demographic objective of reducing the rate of population increase. Through the adoption of various family planning measures the annual population growth rate has been reduced from 2.5 per cent in the 1970s to about 1.8 per cent in the late 1990s. This is further reflected in the contraceptive prevalence rate that shows an increase from 4 to 49 per cent among the currently married women between 1972 and 1996. Thus the fertility rate also came down from 6.3 to 3.3 within the same period. The mean age at marriage for women was 19.2 in 1996, and there were about 10 million adolescents of age 15–19 years in 1991 with 50 per cent of the females of this category married. The proportion of never married women between 15 and 19 has continued to rise at a fast rate in urban as well as in rural areas and the proportion of the never married women between 20 and 24 is also increasing in the urban areas.

Women also made a remarkable contribution to the changing literacy situation of the country. Women's share in increasing the literacy rate is quite significant, showing a rise of 8 per cent during the 1981–91 period (from 17.5 to 25.5 per cent) compared to 5 per cent for men (34 to 39 per cent). Recent economic trends of the economy of Bangladesh show that an increasing number of women are becoming involved with the market economy. Women are also making some progress in the area of governance by participating in parliamentary elections, government organisations and services and in local government bodies. Thus women are important contributors to the social, cultural and economic development of the country.

Economic participation of women: national and urban perspective

Although not all women work for wages, their role in the subsistence economy has a great impact on the total economy of the country. As such there is a presence of visible and invisible labour in the total labour force of the country. Official statistics suggest that the share of female labour force in the country is 21.3 million compared to 34.7 million males. In the urban areas nearly 3 million women are in the civilian labour force against 7.4 million men (BBS 1998). However, these statistics fail to take account of the non-formal contribution of women to production. Women have always been at work, but the national definition of 'work' has not been realistic enough to incorporate in full the female contribution to the economy and society. The traditional norms of society emphasise the domestic and productive role of women. As such they remain a neglected group in the areas of wealth creation, education, health, nutrition and work. As in many developing nations, the economic participation of women is difficult to evaluate because the national definition of the labour force ignores the contribution of the unpaid labour of women. The Bangladesh census not only routinely excluded women's agricultural activities conducted from home but also neglects their informal activities. Only recently has the government developed an extended definition of work that includes household and home-based activities. The new data show that

Table 12.2 Civilian labour force (10+ years) by residence, sex and surveys (in millions)

Residence and sex	1991 census	1989 LFS*	1990–1 LFS*	1995–6 LFS*
Bangladesh				
Both sex	30.7	50.7	51.2	56.0
Male	28.4	29.7	31.1	34.7
Female	2.3	21.0	20.1	21.3
Urban				
Both sex	7.0	5.7	8.7	10.2
Male	6.3	4.2	6.6	7.4
Female	0.7	1.5	2.1	2.8

Source: BBS (1998: 172).
* BBS (1996b).

the female labour force participation rate was little less than 51 million in 1995–6 as against 18 million using the usual definition. The disparity in education levels between the male and female population is one of the main obstacles for women entering the formal job market to improve their social status.

The characteristics of female labour force indicate the predominance of the younger age group; about 35 per cent are under 20 years with another 48 per cent under 25 years of age. The unemployment rate as per the usual definition of labour force for females was a little over 6 per cent in 1995–6 compared to only 2.7 per cent for males. However, the extended definition of the labour force suggests that 2.3 per cent females fell under the unemployed category. The extended definition of work indicated that 29 per cent of urban women were economically active and the unemployment rate for females in urban areas was only 3.9 per cent.

Female employment and occupational patterns

The occupational pattern of the female labour force has registered a substantial change since the late 1980s. While agricultural occupations have remained relatively unchanged, the non-agricultural sector has witnessed both growth and diversification. The average growth rate of total employment from 1961 to 1974 was about 2 per cent, and rose to 2.7 per cent from 1974 to 1983–4. However, the average annual growth rate of labour force between 1990–1 and 1995–6 was 2.4 per cent. The towns presented a different picture, with an increase in the average annual growth rate of the female urban labour force of nearly 22 per cent between 1989 and 1995–6 compared to about 17 per cent of the males.

As Table 12.3 suggests, women's work is still concentrated in a restricted number of sectors. The largest number of women work in service occupations (46.3 per cent) both in urban and rural areas of the country. When

Table 12.3 Distribution of employed persons aged 15+ years by sex and broad occupation (1995–6)

Major occupation	Employed persons (millions)		
	Both sex	Male	Female
Total	34.8	29.8	5.0
Professional and technical	1.5	1.1	0.5
Administration and managerial	0.2	0.2	0.0
Clerical workers	1.2	1.1	0.1
Sales workers	5.8	5.3	0.4
Service workers	1.4	0.7	0.8
Agriculture, forest and fisheries	17.0	15.6	1.4
Production, transport, labour and others	2.5	1.3	1.2
Not adequately defined	5.3	4.6	0.7

Source: BBS (1996b).

other occupational categories are taken into account women's involvement, however, is found to be extremely low. Of the total occupations females constitute only 9.5 per cent. Even in the late 1980s, despite the already burgeoning growth of the female labour force in the textile industry, the employment situation of women remained marginalised, as highlighted in a World Bank report which stated that:

> Employment of women in the cotton textile, wearing apparel and leather industry has expanded rapidly, reaching 34 per cent of the total labour force employed in this sector in 1985–86. Women comprised 51 per cent of the labour in food, beverage, and tobacco, 45 per cent in wood and wood products, 22 per cent in mineral and metal products, and less than 20 per cent in other sector industries in 1985–86. In the export-oriented garment industries, about 90 per cent of the labour force employed are women. In public sector corporations, women comprise less than 3 per cent of employees, although they constitute about 59 per cent in the industrial estate of Bangladesh Small and Cottage Industries Corporation (World Bank 1990).

Thus the increased participation of the last 10 years started from a very low base.

Income and wage earning

Female incomes are very low all over the developing world. Excepting domestic service, wages have been less regular in the informal than in the formal sector. In Bangladesh the situation of poor women is serious. Wage rates for women are very low, approximately half of those of men. Along with receiving low and irregular wages, the problems of women in the country are compounded by men obtaining access to and control of their earnings. The type of informal work with which women are usually involved is either a shared activity between other family members, including males of the household who are mainly involved in marketing the products, or through a contract, in which case men are the intermediaries of such contracts. Monetary transactions related to contracts take place through men, which in turn results in male control over female income, and hence women remain totally dependent on the men. Men in South Asian countries feel that if the females are engaged in direct cash income activities their role in the family will be undermined and their control over women will be loosened. Men are also influenced by the cultural and ideological norms of the society and fear that women through their independent income misuse their independence and that it will lead them even to misconduct (Huq-Hussain 1996).

Women and work in the bastees of Dhaka

A study conducted among 399 women in the bastees located in different parts of Dhaka city indicates that the desperately poor migrants cannot afford any choice with respect to the type of work they do (Huq-Hussain 1996). Hence, they are ready to be involved with any type of work for the survival of their families. Nearly half of the female migrants surveyed intended to take any work available to them. The waiting time to get work may be an important factor in the adaptation of migrants to urban life. The rural poor, particularly women, generally lack proper skills and training and this may restrict their search for work immediately after arrival in the city. In Dhaka a varying pattern has been observed with regard to the length of time female migrants wait to get work. About 38 per cent of all the females had to wait less than six months and nearly a quarter of them had to wait between six months and one year. Those who waited for more than a year needed more time to establish friendships with their neighbours who could help them to find a job. Personal introduction plays an important role in procuring work for migrant women since 16 per cent of the respondents obtained their first job with the help of an introduction and assistance from their relatives and friends. However, a large number of women who are involved in home-based activity did not take any assistance from their friends or more distant relatives since their husbands or parents found the work for them. When procurement of work by their own efforts is considered, nearly 15 per cent of the females could find work on their own within six months of their arrival.

Occupational pattern

Details of women's occupational pattern are shown in Table 12.4. This chapter has divided female occupation into three broad categories. These are the service sector, the informal sector and the formal sector that involves mainly the ready-made garment industries.

Service occupations

Service occupations offer women by far their most important access to work, which can often be found relatively quickly and easily. One half of

Table 12.4 Occupational distribution of female workers (in millions), 1995–6

Occupation	Female	Total	Percentage
Sales and service	1.6	8.1	19.7
Construction	0.1	1.0	10.0
Community, personal and household sector	2.6	6.3	41.0
Day labourers	1.3	9.8	13.0
Garment sector	1.3	1.4	90.0

Sources: BBS (1996b: xv–xvi).

the urban workforce is engaged in such occupations. The service sector is an important contributor to urban adaptation and to modernising female migrants from the countryside. Some of the important female service occupations in which poor bastee women are engaged are housemaids, cleaners, cooks and related jobs, hospital *ayas*, low-status (class III and IV) office workers, and paramedical workers. Females constitute 79 per cent of the total employment in the household sector (World Bank 1990: 24). Nearly 40 per cent of the respondents in this study were engaged only in the domestic sector. As already mentioned, Huq-Hussain's study focused only on female migrants living in the bastees, therefore it was not possible to include the full-time housemaids who reside with their employers. Hence, the domestic workers in this study are those who work for other houses throughout the day and then come back to their families in the bastees after their day's work is over. These people provide many part-time household services such as cooking, dishwashing, grinding spices, and other work in the kitchen, day-to-day shopping of fresh items for the kitchen, washing clothes, cleaning, washing and wiping floors, looking after children, ironing, etc. They go to the employer's house or to several different houses a day depending upon the contract. These jobs involve only a verbal contract and so there is no job security as such. Their employer can fire women any time, and at the same time they also have the option of entering or leaving domestic service at different stages of their life.

Informal sector work

The concept of the informal sector, which was developed within the social sciences *inter alia*, to explore the scale, role and significance of economic activity which takes place outside those areas of structured activity recognised both by the state and by conventional economic measurements, plays an important role in the occupational pattern of the bastee women in Dhaka. The concept of the informal sector has been widely applied to the study of urban employment in the South. Schneider, working in Thailand and the Philippines for example, took it to refer to work in units which employ five or fewer employees, or self-employment, or work in unlicensed units (Schneider 1999: 238). Despite the importance of the informal sector very little research has been conducted to evaluate the role of this sector in the overall urban economy in Bangladesh. The scale and nature of women's activities in the informal sector in Bangladesh, however, remain unclear and their role is yet to be evaluated. A high concentration of female involvement with informal work is influenced by the location of work, i.e. the need to be near to the home so that women can perform their domestic chores, or involve themselves in such work where children can accompany them or to go for part-time work. Many women who cannot work outside their homes involve themselves in home-based activities. A considerable proportion (38 per cent) of the bastee women of this study are engaged in informal sector activities.

The pattern of informal sector work is varied. Women have been found to generate their own activities, depending on the local material in the informal sector. At least 12 different types of work were identified, most of which are home based (Huq-Hussain 1996). Examples of such activities include making bags out of recycled waste paper, wrapping sweets for small home-based factories for consumption within the informal sector, stitching plastic and jute bags, making rope and coir from coconut shells, preparing and packing dry snacks, cooking rice for selling on the city pavements, pickle-making, grinding spices for restaurants, collection and sale of ashes for cleaning cooking pots and pans and collecting and selling drinking water. All these items are consumed within the informal sector. It has been observed that many informal female activities are influenced by their familiarity with traditional rural industries. In such activities women have the flexibility of working hours and elderly women can also take part. Other important informal work undertaken outside their residence includes door and market sales, restaurant work, laundry work (washing and ironing clothes) and even shopkeeping. After 1971 construction activities accelerated quickly which also attracted rural migrants including females. The following section presents the role of women in this sector based on Huq-Hussain's study (2000: 37–43) conducted among 200 construction sector workers.

The role of migrant women in the informal sector construction work

Since the independence of the country the construction of multi-storey buildings expanded very rapidly. However, in a flat alluvial country, there is severe shortage of gravel and rocks that could be used to make strong building foundations, floors and walls. As such, burnt earthen bricks are broken into small pieces and brick crushing is one type of work dominated by women workers. In addition, women also work in such jobs as load-carrying, mason assistants, floor tile fitters and polishers. The migration pattern of female construction workers (Table 12.5) shows that nearly one-third of them moved independently, whereas over two-thirds moved with their husbands and families. Over half of them moved from their native villages as a result of loss of land due to river erosion. The rest were forced to move because of poverty and lack of work at their place of origin (Huq-Hussain 2000).

Thus, despite the efforts by the government and non-government agencies to promote rural development and poverty alleviation programmes, the hard core poor still find it difficult to survive in the rural areas.

Table 12.5 Pattern of migration by type of worker (%)

Pattern of migration	Construction worker	Factory worker
Independently	31	8
With family	26	26
With husband	41	24
With relatives	2	42
Total	100	100

Source: Huq-Hussain (1998, 2000).

The formal sector: ready-made garment industries

As mentioned earlier, the expansion of ready-made garment industries opened up new avenues for Bangladeshi women, particularly the poor rural migrants, into formal sector work. In the wage employment sector this industry showed a notable growth and 90 per cent of the labourers are women. A survey carried out on 120 female garment knitwear factory workers (Huq-Hussain 1998) shows that females are involved in jobs including machine operation, ironing, packing, mending, supervision and general help. The motive and migration pattern of these workers indicate that over three-quarters of them migrated to obtain a job in the garment industry and made their moves with their husbands, family or relatives.

Change of work and job satisfaction

It is striking that, even among the very poor, women exercise a degree of choice over their working practice. The results of the study demonstrate that not all the females stick to their first job and 70 per cent of the female migrants changed their job at least once. Family reasons topped the list for changing work, followed by change of work due to low wages, long work hours and shift of migrant's residence within the city due to increasing house rent and better salaries elsewhere in the city. This helps to explain the finding, reported above, that longer-term residents of the city have higher incomes than newly arrived migrants, as they actively seek out better alternatives to those they were prepared to accept on first arrival.

Another important finding is that a large number of all the female migrants were satisfied with their work while a quarter of them were dissatisfied. Factory workers are better off in terms of job prospects and can move up in the job ladder, since about two-thirds of those who initially joined the factory as helpers often learned other skills and in the process became machine operators. Upward mobility was less possible for construction workers as only 3 per cent had been absorbed as mason assistants or floor tile fitters/polishers. Construction workers were aware that they did not have much chance for upward mobility or to acquire proper skill and therefore were satisfied with their present work.

Earnings and expenditure

Informal sector work has commonly been found to be less well paid than formal sector employment. In one recent study, Schneider has observed with respect to two cities in Thailand and the Philippines, Chiang Mai and Baguio, that 'as a rule incomes in the informal sector are lower than in the formal sector' (Schneider 1999: 238), though detailed study showed that there were exceptions. He also observed the widespread pattern that new migrants are more likely to find work in the informal sector than the formal. Studies in Bangladesh indicate that there is great variation in the income pattern of the migrant women. This variation may be noticed in terms of their length of stay, type of work and skill acquired. Huq-Hussain (1996, 1997) found that formal sector workers have more regular and higher earnings compared to informal sector workers. In many cases, particularly in informal sector home-based work, women receive no cash income as such. Their male counterparts provide them with some items in 'kind' or a 'little pocket money' specially for 'betel leaves' or 'when needed or if they ask for it'. Interestingly, many of these women do not consider such labour as 'economic'. They consider it as part of their household responsibility. As for formal sector employment a report (UNDP 1996) based on different surveys in manufacturing industries reveals that women's rates of pay are typically less than half those of men, even when they are doing identical or comparable jobs.

Contribution to family income

Women contribute significantly to family income. Even in male-headed households the female contribution to family earnings may be between 25 and 50 per cent. According to a World Bank report (1990: 3), female earnings contribute one-quarter to one-half of family incomes. Also, female earnings alone are responsible for food security in 25 per cent of landless households. Evidence (Huq-Hussain 1996) suggests that up to a quarter of the total poor family income is being contributed by women in the bastees of Dhaka. With respect to the expenditure pattern Huq-Hussain's (2000: 41) study revealed that:

> some 30 per cent–40 per cent of the earnings of the factory workers went to food, accommodation had a share of 25 per cent–30 per cent, remittances had a share of 10 per cent– 20 per cent and only negligible on medical expenses...informal construction workers spent on two major items, viz. house rent and food accounting to almost equal share of 45 per cent–50 per cent.

In most other cases, however, the earnings of the poor women go into the family income pool. Despite the importance of women to the family income stream, the handling and control of female income reflects the generally subservient position of women in the family. The studies reveal that the husband, father or any other male household head is generally responsible for handling family income. Only in the female-headed households do women control the distribution and use of income.

It is clear that an increasing proportion of women are contributing significantly to the growth and modernisation of Bangladeshi society through their active participation in public life. According to government sources, female participation especially in low hierarchy jobs, i.e. class III and IV employees, has increased by two and a half times and twice respectively between 1994 and 1997. It should also be noted that a greater number of women, particularly poor women, are taking part in voting at national elections. All these developments have a significant and positive impact in the national development process.

Occupation, urban adaptation and modernisation

Women's involvement with urban economic activities has an important effect of their urban adaptation. In this connection Shah and Smith (1984: 297) argued that:

> A migrant's entry into the urban labour market has important implications for her adjustment. The type of economic activity that she enters into is crucial not only for her initial adjustment to the city but also for her future social mobility within the urban environment. Certain occupations may expose migrants more to the urban facilities and amenities, which in turn influence migrant's quick adaptation to the city life.

Huq-Hussain (1996) has noted that

> domestic service exposes women immediately to modern technologies in the cities. They become more aware about various aspects of life by working in the middle and upper income households. This helps them to expand their knowledge at a faster rate than those who are engaged in home based activities. Similarly, those female migrants who are working in the ready-made garment industries are also equally more aware and adapt at a faster rate to urban life. In many cases it has been seen that the women involved in the home-based informal sector cottage industries have broadened their knowledge from hearing from their fellow domestic and garment workers, vendors, construction workers etc. They also can find work and generate their own income from various knowledge thus gathered and make contacts to get materials for such work. In one case it has been found that a woman heard about scrap garment materials (scrap clothes, sponge etc) from a garment worker and she developed the idea to use these scraps to earn some money. She went with her contact to the factory to collect them. Later, she used such scrap pieces of cloth and sponge to make and fill pillows and sold them in the streets. Her husband however, did the marketing.

Conclusion

Across South Asia women are playing an increasingly important and explicit role in economic activity, and are claiming a greater degree of independence in their range of social and economic choices. A number of contributions to

Chapman *et al.*'s volumes *Urban Growth and Development in Asia* (1999) have highlighted both the significance of female labour in Asian cities and the dynamic of socio-economic change which is accompanying such development. While Anita Weiss has pointed to the continuing significance of cultural markers in the gendered division of space in Lahore, Pakistan, arguing that 'gendered space has not undergone as substantial a reconfiguration as has class-based space' (Weiss 1999: 191), Sharma has shown that despite a continuing significant differential between male and female participation in the urban labour force there is some evidence that in India the gap is narrowing, again largely as the result of female participation in the informal sector (Sharma 1999: 181).

This chapter has explored the ability of women in Bangladesh to adapt to changes in their socio-economic environment. The experience of Bangladesh reinforces the view that the poor, and poor women in particular, are not simply passive subjects of economic and environmental changes that go on around them, to their major disadvantage, but are active agents of their own development and wider change. Bangladesh has experienced profoundly important endogenous changes, notably in its continuing population growth and rapid rate of urbanisation. By the year 2020 nearly 50 per cent of the population is expected to be urban, and Dhaka alone will contain up to 20 million. It has also suffered major external shocks, including exceptionally severe floods such as those of 1987, 1988 and 1998, and cyclones such as the one in 1970 which was estimated to have killed over 250,000 people (Houghton 1997: 113). Even the normal annual flooding leads to widespread erosion and the loss of villages and agricultural land, enforcing migration. This chapter has shown how far women have responded to such challenges and taken an active role in seeking to increase their opportunities for security and socio-economic development.

The range of adaptation has been remarkable. In terms of employment, the urban informal sector will grow at a rate of 9–13 per cent against 5–7 per cent of the formal sector in the next period up to 2025. Migration will contribute up to 60 per cent of the total urban growth up to the year 2005. Participation of women in the migration will further increase with their greater involvement in the development of economic activities. As already seen, female participation rates have doubled during the 1980s, and by the year 2001 stood at over 30 per cent. Poverty has been a major factor in stimulating an enlarging trend in female participation rates which is inevitable in the immediate future (Huq-Hussain 1997).

As in many parts of the developing world, women in Bangladesh play a vital role in the production system. In addition to their household work they are an important contributor to the total household income, as well as to the wider society. Their household work is vital in terms of 'use' value. In a sex-segregated society women are confined to certain types of work. Poor women are restricted in their choice of economic activities because they do not have proper skills and training. They are still tied to the age-old profession of home-making and related work. Technical training still caters for men.

Even with such limitations women contribute significantly through their supplementary work. A bastee woman works hard and for long hours. Apart from those who work in the garment industry, much of this work is never included in estimates of the GNP of the country. Their contributions to the family and national economy are largely overlooked and are thus under-reported in government documents. Yet the contribution of poor women to the family has clear implications for alleviating poverty in households. Women's economic activities in the bastees of Dhaka are highly varied and they generate income using traditional methods and indigenous materials. As such their role in the overall programme for poverty alleviation in the country cannot be overlooked.

Even though the construction industry in Bangladesh is heavily dependent on female labour, the role of women is commonly underestimated. The positive involvement of women in productive activities both at home and outside can clearly promote the development of human resources and will contribute to the general development of the country. Women's exposure and participation in both the urban and rural labour force will increase their access to different opportunities and enable them to get the benefits of development.

The government of Bangladesh and various local and international agencies have launched a wide range of programmes to improve the overall condition of Bangladeshi women. Despite the serious gaps both in data and in the analysis of women's changing role in Bangladesh, it is clear that women are acting as crucially important catalysts of socio-economic change. Where the conceptualisation of poverty is increasingly coming to include social and political empowerment, this chapter has shown the variety of ways in which some of the most excluded groups in Bangladeshi society have opened up new opportunities for themselves and their families. While the particularities of Bangladesh's socio-economic and geographical environments have conditioned some aspects of female adaptation to urban challenges and opportunities, it is striking that there is also much common ground with the nature of socio-economic change in many urbanising areas in the poor South. In what some would regard as a society hostile to change and to enhancing the role of women, women in Bangladesh have shown that they are already acting as a force for development. Their potential is still widely under-recognised, and in achieving greater recognition it may be concluded that it is necessary to undertake a much wider range of systematic research to analyse their changing role effectively.

References

Ahsan, R.M., Huq-Hussain, S. *et al.* (1987) *The Invisible Resource: Women and work in rural Bangladesh.* Westview, Boulder, CO.

Barkat, A. (1997) 'Population distribution, urbanisation and internal migration', in *Population and Development Issues in Bangladesh: National plan of action based on ICPD '94 recommendations.* Ministry of Health and Family Welfare, GOB, Dhaka.

BBS (Bangladesh Bureau of Statistics) (1996a) *Report of the Poverty Monitoring Survey 1995.* Government of Bangladesh, Dhaka.

BBS (1996b) *Report on Labour Force Survey in Bangladesh, 1995–96.* Government of Bangladesh, Dhaka.

BBS (1997) *Summary Report of the Household Expenditure Survey, 1995–96.* Government of Bangladesh, Dhaka.

BBS (1998) *Bangladesh Statistical Pocket Book.* Government of Bangladesh, Dhaka.

Chapman, G.P., Dutta, A.K. and **Bradnock, R.W.** (eds) (1999a) *Urban Growth and Development in Asia:* Vol. I *Making the Cities.* SOAS Studies in Development Geography, Ashgate, Aldershot.

Chapman, G.P., Dutta, A.K. and **Bradnock, R.W.** (eds) (1999b) *Urban Growth and Development in Asia:* Vol. II *Living in the Cities.* SOAS Studies in Development Geography, Ashgate, Aldershot.

Houghton, J.T. (1997) *Global Warming: The complete briefing,* 2nd edn. Cambridge University Press, Cambridge.

Huq-Hussain, S. (1996) *Female Migrants' Adaptation in Dhaka: A case of the processes of urban socio-economic change.* Urban Studies Programme, Department of Geography, Dhaka University, Dhaka.

Huq-Hussain, S. (1997) 'Pattern of internal migration in Bangladesh: some new dimensions in female migration', *Oriental Geographer,* **39,**(1),(2) 1995, Printed in 1997: 86–105.

Huq-Hussain, S. (1998) 'Health condition of factory workers: a case study of a sweater factory in Dhaka', in Nazrul Islam (ed.), *Recent Urban Studies, Bangladesh.* Urban Studies Programme, Department of Geography, Dhaka University: 163–76.

Huq-Hussain, S. (2000) 'From villages to Dhaka: female migration as a strategy for poverty alleviation', *Asian Migrant,* **13**(2), April–June, Philippines: 37–43.

Islam, N. (1999a) *Urbanization, Migration and Development: Recent trends and emerging issues.* Paper 1, Centre for Policy Dialogue (CDP), Dhaka.

Islam, N. (1999b) 'Dhaka city: some general concerns', in Naved Hamid and Saleemul Huq (eds), *Asian Cities in the 21st Century: Contemporary approaches to municipal management,* Vol. 3, Asian Development Bank, Dhaka: 71–82.

Schneider, H. (1999) 'Livelihood strategies of urban households in secondary cities in Thailand and the Philippines – a comparison', in Chapman *et al.* (eds), *Urban Growth and Development in Asia:* Vol. II *Living in the Cities.* SOAS Studies in Development Geography, Ashgate, Aldershot, Chapter 15.

Sen, B. (2000) *Bangladesh Poverty Analysis: Trends, policies and institutions.* Asian Development Bank, Dhaka.

Shah, N.M. and **Smith, P.C.** (1984) 'Migrant Women at Work in Asia', in Fawcett *et al.* (eds), *Women in the Cities of Asia: Migration and Urban Adaptation.* Westerview, Boulder, CO.

Sharma, R.K. (1999) 'Female labour participation in urban India: some socio-economic correlates', in Chapman *et al.* (eds), *Urban Growth and Development in Asia:* Vol. II *Living in the Cities.* SOAS Studies in Development Geography, Ashgate, Aldershot, Chapter 12.

UNDP (United Nations Development Programme) (1996) *A Fork in the Path.* UNDP, Dhaka.

UNDP (1999) *National Gender Profile.* Dhaka.

UNDP (2000) *Human Development Report, 2000.* Oxford University Press, New York.

Weiss, A.M. (1999) 'The gendered division of space and access in working class areas of Lahore', in Chapman *et al.* (eds), *Urban Growth and Development in Asia: Vol. II Living in the Cities.* SOAS Studies in Development Geography, Ashgate, Aldershot, Chapter 13.

World Bank (1990) *Bangladesh Strategy Paper on Women and Development.* Report No. 7899BD, Population.

Conclusion: Development in South Asia
Ground-truthing constructions of regional change

Robert W. Bradnock and Glyn Williams

There has not been space in this volume to discuss all areas of importance for contemporary development in South Asia. Aid, for example, analysed by Anders Närman under the title 'Getting towards the beginning of the end for traditional development aid', in the first volume of this series has not been a focus of analysis within this book, even though for Bangladesh, Nepal and some other areas across South Asia foreign aid remains an extremely important part of the development budget. Clearly aid itself has undergone major changes since its inception 50 years ago, just as the conceptualisations of modernisation and development have themselves been subject to critical scrutiny. The increasingly sharp focus of national and international aid programmes on identifying and targeting poverty and achieving sustainable development reflects the extent to which critical analysis of aid programmes in the 1970s has influenced the agendas of international aid-giving bodies. The focus of World Bank thinking on poverty itself reflects a profound change from its policies of the 1970s. From within Bangladesh the example of the much discussed and finally abandoned Flood Action Plan illustrates the extent to which, far from being a straightforwardly top-down or paternalist programme, national aid programmes have become more reflexive and subject to international and national critiques and pressures. Yet aid itself is dwarfed in significance by the move towards neo-liberalism which became the hallmark of the 1990s and which has left its imprint on all the countries of South Asia, as it has across the South.

In the wake of this neo-liberal agenda, as we argued in our introduction to this volume, there is an increasingly widespread sense, expressed by Agnew in 1999, that old geographical divisions no longer matter, and that in an era of global economics and politics the internally differentiated, socially diverse and politically complex realities of individual localities and regions have lost their significance. We believe that the evidence of the widely differing contributions to this volume demonstrates the reverse to be true. Their critiques have played a part in re-evaluating the wider development agendas explored in this volume. While, as Corbridge and Harriss showed, global realities impinge increasingly strongly on national and regional political options, as they also demonstrated, and as has been powerfully illustrated by Craig Jeffery and Emma Mawdsley, global forces are often mediated by domestic political forces, themselves far from static and often undergoing major change. In India, for example, caste has not been replaced by class as either a social or an economic discriminator, and its political significance, dynamic as it is, remains profound. National politics in the countries of the South Asian region discussed in this book have followed widely differing paths in the 50 years since the colonial era ended. Between them, they have experienced every form of regime, from military dictatorship to democratic government, as well as undergoing traumas from natural disasters to war and civil war which have been experienced across the South. Each has its own national and subnational political structures, reflecting and shaping diverse social and economic realities. Even within the countries of South Asia it is possible to find areas which fall within any of the positions of the development spectrum, from rapidly industrialising regions like the Delhi region, Mumbai, parts of Tamil Nadu, Karnataka and Andhra Pradesh, whose development of high-tech industries is matched by rapid social change, including rapid growth in literacy rates and falling population increase, to parts of eastern India like Bihar, or the North-West Frontier of Pakistan where social and political forces have combined to produce negligible economic growth and continuing illiteracy, gender inequality and rapid population growth.

An important theme in Simon and Närman's introductory book to the series on rethinking development (1999) has been the concern that empirical research and the conceptualisation of development need constantly to be brought together. They recognised an unease which has run through development studies and development geography in the last decade, that all too often the theorising has appeared to have little relevance to realities on the ground, or to understanding the problems of development as practised. Michael Parnwell picked up this concern when he wrote of his own research interest in South-East Asia that: 'development theories... always seemed to lack the power and ability to explain what I was constantly seeing and experiencing on the ground...'. He goes on later to argue that 'much greater attention is now being paid by what Corbridge (1993: 450) calls "development activists" to the actual workings of development rather than an abstract representation of development processes' (1999: 84).

It was a fundamental premise of this book that any attempt to understand the 'actual workings of development' demands a commitment to field-based research. As Lakshman Yapa has shown, such research does not obviate the need for theoretical or conceptual understanding, and as has been shown throughout this book, such attempts to ground-truth existing hypotheses often expose them to radical reappraisal. Thus, while the relationship between the environment and development is a theme of undoubted global significance, as Bradnock and Saunders have shown, the implications of global environmental change cannot simply be assumed to have universal significance. If it is essential to understand both the physical geography of the Bengal delta and the subcontinental politics of water development, as Bradnock and Saunders argue, so as Chapman has demonstrated, the potential, costs and benefits of large-scale dams cannot simply be calculated from engineering formulae but are embedded in the social and political structures of specific places.

In attempting to reconstruct a regional geography for the current decade it is evident that the need for a development geography rooted in an understanding of place is greater than ever. This is not to emphasise uniqueness at the expense of the underlying process, and as the thrust of Lakshman Yapa's critical review of the concepts of development and poverty suggests, there is ample scope within a reconstructed regional geography for intense theoretical debate. However, as Michael Parnwell has argued, 'It would appear that an important criterion in areal specialisation is what Benny Farmer (1993: 9–10) has called "dedication to area"....Most academics who work in development will have undertaken a period of field-based research in a particular developing area.' The Eurocentrism which may be inferred from that quote can be put today in the context of an extraordinarily flourishing debate throughout modern South Asia of every conceptual and empirical issue in the development field. You need only to look at the publications such as the Citizens Reports on the Environments produced by the Centre for Science and the Environment (e.g. Agarwal *et al.* 1999) to recognise the scale and intensity of the interest in environmental issues alone. Universities and research institutes, NGOs and government agencies are engaged in extremely active debate, and while the internet is indeed transforming global communication and the spread of ideas, for academics researching development as outsiders South Asia offers the certainty of vigorous debate in every development field.

Yet it is not just to outsiders that Parnwell stresses the 'importance of contextuality in the "new development", or emphasises the contribution that the area specialist can make to weaving this into a broader understanding of the development process' (1999: 92). The importance of 'ground truthing theory' applies wherever you live, and has been amply carried through in this volume. The concerns with some of the most pressing contemporary issues in global development underlay the thematic structure of the book and its identification of three core areas of focus: environmental challenges to development, the politics of development, and issues of social inequality and social change. The contexts within which these issues are relevant range from the urbanising populations of Bangladesh, India and Pakistan through to the tribal populations of the still forested regions of central India. Where Shahnaz Huq-Hussain and Arif Hasan analyse patterns of social and economic change among some of the poorest urban communities of Bangladesh and Pakistan respectively, they are engaging with a debate that as the contributors both to Simon and Närman's *Development as Theory and Practice* and to Gugler's edited volume on cities in the developing world (1997) show, have resonances throughout the South.

In highlighting the dynamism of South Asia's contemporary patterns of socio-economic change and development, their interleaving with global development issues as well as their rootedness in local societies, we hope that this book will be of relevance beyond the South Asian region with which it is primarily concerned.

References

Agarwal, A., Narain, S. and **Sen, S.** (1999) *The Fifth Citizens Report: Part 1 National Overview.* Centre for Science and the Environment, New Delhi.

Agnew, J. (1999) 'Regions on the mind does not equal regions in the mind', *Progress in Political Geography,* **23**(1), 91–6.

Corbridge, S. (1993) 'Marxisms, modernities, and moralities: development praxis and the claims of distant strangers', *Environment and Development D: Society and Space,* **11,** 449–72.

Farmer, B. (1993) 'Geography, area studies and the study of area', *Transactions of the Institute of British Geographers,* **60,** 1–16.

Gugler, J. (ed.) (1997) *Cities in the Developing World.* Oxford University Press, Oxford.

Parnwell, M.J.G. (1999) 'Between theory and reality: the area specialist and the study of development', in D. Simon and A. Närman (eds), *Development as Theory and Practice.* Longman/DARG, London, Chapter 4.

Simon, D. and **Närman, A.** (eds) (1999) *Development as Theory and Practice.* Longman/DARG, London

Glossary

There is a vast social science literature on South Asia, much of it written for and by academics from the region. As a result, there are a number of South Asian terms that are not often translated. Some of the most common are given below:

adivasi	'tribal': used to describe the peoples and cultures of South Asia's indigenous groups. Often ethnically and linguistically very different from 'mainstream' South Asians, they have been socially and economically marginalised. Hence their labelling as 'Scheduled Tribes' for special government support
andolan	movement, uprising, political protest
Bharat	the official Indian name for India (in Hindi)
bhatta	bribe, illegal gratification
bastee	shanty town
chowkidar	caretaker
crore	ten million. In South Asia, large numbers are often counted in *crore* and *lakh* (100,000), and are written using these units, i.e. one *crore* is written '1,00,00,000' rather than '10,000,000'
crorepati	a millionaire
dacoit	thief, brigand
dallals	middlemen
des/desh	country, 'homeland'
desi/deshi	'local', provincial, of the countryside
dharma	socio-religious moral order, law and duty
dharna	a traditional form of protest in which the aggrieved demonstrate their cause through self-inflicted punishment, usually by fasting
gherao	a form of protest: holding officials 'hostage' in their office
goonda	violent criminal, often associated with organised crime
Gram Panchayat	lowest tier of local government (literal meaning = 'village council')
Hindu	a believer in Hinduism, the largest religion in India
jati	birth-group, 'caste'
jotedar	large landlord
katchi abadis	squatter settlements

khadi	hand-spun and woven fabric: still of symbolic importance because Gandhi promoted its production to challenge imports of British manufactured cloth
kisan/krishak	peasant
lathi	cudgel, police baton, e.g. 'police *lathi*-charged protesters'
lakh	100,000 (see *crore*)
lok	people
mahajan	moneylender
mahindar	farm servant/'attached' labourer
mandir	a Hindu temple
marla	25 square yards (20.9 sq.m)
masjid	mosque
maund	a unit of weight – regionally variable, but within Jharkhand (eastern India), equivalent to about 40 kg
mullah	a Muslim cleric
Naxalites	left-wing revolutionary activists, often inspired by Maoist teachings. Named after Naxalbari, the location of an important peasant uprising in the 1960s
Panchayat	traditional council (usually rules a village)
Panchayat Raj	'rule by village councils': a system of rural self-government and development
patta	redistributed land, the document of ownership of such land
pradhan	chief, chairperson of a *panchayat* (*sarpanch* is also used)
raiyat	wealthy peasant farmer
raj	regime, kingdom, state, rule, 'governance'. 'The Raj' usually refers to the former British government of India
sabha	council/committee. The Indian (federal) Parliament is divided into two sections – the *Lok Sabha* (House of the People – directly elected) and the *Rajya Sabha* (Council of the States – indirectly elected)
Sanskrit	ancient South Asian language – the regional equivalent to Greek in Europe
Sanskritisation	a sociological term describing a process of 'social climbing' for castes by religious/cultural change
sarkar	governance/rule
Swaraj	Self-rule, independence (now often refers to subnational groups)
thalla	building component manufacturing yard
thana	police station, the area that falls under the police station's authority
Untouchables	castes seen as being of such low status that they are beyond the pale of ordinary social contact. Untouchability is illegal under the Indian constitution, but still practised to some degree. Also known as *Scheduled Castes* (used by government), *harijans* ('people of god' – Gandhi's term) and *dalits* ('the oppressed' – increasingly their own chosen description)

varna   lit. 'colour'; the four major divisions in the caste system, namely *brahman* (priest), *kshatriya* (warrior/ruler), *vaishiya* (producer of wealth, trader) and *shudra* (worker). Each contains many *jatis*

zamindar   owner (originally rent collector) of a large landed estate

Index